ENERGY AND THE MAKING OF MODERN CALIFORNIA

TECHNOLOGY AND THE ENVIRONMENT

JEFFREY K. STINE AND WILLIAM MCGUCKEN

SERIES EDITORS

JAMES C. WILLIAMS

ENERGY AND THE MAKING OF MODERN CALIFORNIA

THE UNIVERSITY OF AKRON PRESS
AKRON, OHIO

Manufactured in the United States of America
First Edition
01 00 99 98 97 5 4 3 2 1

LIBRARY OF CONGRESS CATALOGING-IN-PUBLICATION DATA
Williams, James C., 1942–
Energy and the making of modern California / James C. Williams. — 1st ed.
p. cm. — (Technology and the environment)
Includes bibliographical references and index.
ISBN 1-884836-15-1 (cloth). — ISBN 1-884836-16-X (paperback)
1. Energy development—California—History. 2. California—Economic conditions—History. 3. California—Social conditions—History. I. Title. II. Series: Technology and the environment (Akron, Ohio)
TJ163.25.U6W536 1997
333.79'15'09794—dc21 96-29944
CIP

The paper used in this publication meets the minimum requirements of American National Standard for Information Sciences—Permanence of Paper for Printed Library Materials, ANSI Z39.48–1984. ∞

For Betsy and Clinton

Contents

Illustrations

List of Figures

List of Maps

List of Photographs

Series Preface

This book series springs from public awareness of and concern about the effects of technology on the environment. Its purpose is to publish the most informative and provocative work emerging from research and reflection, work that will place these issues in an historical context, define the current nature of the debates, and anticipate the direction of future arguments about the complex relationships between technology and the environment.

The scope of the series is broad, as befits its subject. No single academic discipline embraces all of the knowledge needed to explore the manifold ways in which technology and the environment work with and against each other. Volumes in the series will examine the subject from multiple perspectives based in the natural sciences, the social sciences, and the humanities.

These studies are meant to stimulate, clarify, and influence the debates taking place in the classroom, on the floors of legislatures, and at international conferences. Addressed not only to scholars and policymakers, but also to a wider audience, the books in this series speak to a public that seeks to understand how its world will be changed, for ill and for good, by the impact of technology on the environment.

Acknowledgments

This book has metamorphosed more than once, as colleagues and friends have stimulated my thinking about the subject. I began it around 1980, when energy was a more immediate topic than it is today. At the time, publication of a handful of compelling studies and some important energy history conferences appeared to mark the subject as one of sustained importance. While these things helped cement my decision to move forward with this book, I would not be writing this acknowledgment if it had not been for the guidance of two colleagues, in particular. Recognizing that scholars working in energy history seemed to ignore the American West, my mentor and dear friend Carroll Pursell encouraged me to fix on California's experience. He shepherded my early efforts and, over the years, prodded me to complete this work. More recently, my friend Jeffrey Stine read a draft of this entire manuscript more carefully than any writer should expect of a colleague, then worked with me to sharpen my focus on the relationship between technology and the environment. I owe Carroll and Jeff a debt of gratitude that I shall never be able to repay properly.

I could not have completed this project without support over the years from various organizations. In 1982, when simply beginning seemed the most difficult task in the world, a grant from the University of California Appropriate Technology Program allowed me to take needed time from teaching to focus on research. Gavilan College granted me a year's sabbatical and followed it in Spring 1983 with an additional quarter's leave, while the University-wide Energy Research Group of the University of California Energy Studies Program provided me with additional financial support. Between 1985 and 1992, my work at the California History Center and Foundation at De Anza College left me little time to concentrate on writing, but another sabbatical leave during the first half of 1992 at last permitted me to finish a first draft. I thank these institutions for fostering my work.

A number of librarians and archivists guided me to resources and made valuable suggestions. I am grateful to staff members at the libraries of the University of California at Berkeley and Santa Barbara, Stanford University, Gavilan College, the California History Center Foundation, and De Anza College. I am also grateful to staff at the Library of Congress, the California State Archives, and the National Archives' Pacific Sierra and Suitland branches. Some of the material in this book stems from my work as a public historian in the field of cultural resources management. My thanks for these opportunities go to Biosystems Analysis, Pacific Gas and Electric Company, Public Anthropological Research, Professional Archaeological Services, Region Five of the U.S. Forest Service, Southern California Edison, and Theodoratus Cultural Research.

I also want to acknowledge that parts of this work have been presented as conference papers and/or appeared in print elsewhere. Portions of chapters three and seven appeared in "California Energy Supply Systems: Two Cases In Regional Development," *Energie in der Geschichte (Energy in History): The Topicality of the History of Technology,* Proceedings of the 11th Symposium of the International Committee for the History of Technology (ICOHTEC), September 2–7, 1984, 2 vols. (Dusseldorf, Federal Republic of Germany, 1984), 2:527–534. Parts of chapter nine appeared as "Regional Development in the Technical Sciences in California's Electric Power Industry, 1890–1920," *Technik und Technikwissenshaften in der Geschichte,* Proceedings of the 12th Symposium of the International Committee for the History of Technology (ICOHTEC), Dresden, August 25–

29, 1986 (Berlin: VEB Deutscher Verlag der Wissenschaften, 1987), 313–320, and as "Hydroelectricity and the Modernization of California," a paper presented at the III Congreso Latinoamericano y III Congreso Mexicano de Historia de la Ciencia y la Tecnología, Mexico City, January 1992. Parts of chapters seven and twelve were included in "Energy, Conservation, and Modernity: The Failure to Electrify Railroads in the American West," a paper presented at the XIXth International Congress of the History of Science, Zaragoza, Spain, August 1993. An earlier version of chapter ten was delivered as "Purveyors of Technology: The California Electrical Cooperative Campaign, 1917–1930" at the annual meeting of the Society for the History of Technology, Raleigh, North Carolina, October 31, 1987. Chapter eleven appeared in similar form as "Otherwise a Mere Clod: California Rural Electrification," *IEEE Technology and Society Magazine*, 7 (December 1988), 13–19.

The final preparation of a book is an involved process, and, in this case, it stretched out over almost three years. I am grateful to The University of Akron Press and Jeff Stine and William McGucken, editors of the Technology and the Environment Series, for taking on this project. I thank them and an anonymous reviewer of the initial manuscript for their insights and suggestions. I also thank the press director, Elton Glaser, for his editing of the final manuscript, and Marybeth Mersky for guiding it through the various stages of production.

All who do it know that research and writing is largely a solitary activity. Nevertheless, one shares his or her work regularly with colleagues and, if one is receptive, finds renewed strength for more writing in every encounter. I am exceedingly grateful for the continued support of my many friends and colleagues, most of them members of the Society for the History of Technology, the International Committee for the History of Technology, the California Council for the Promotion of History, or the National Council on Public History. I trust they will forgive me for not mentioning each and every one of them by name—to try would only result in omitting someone—but they must know that they have my deepest and most sincere thanks for their willingness to listen to me, for their suggestions, and most of all for their sustained friendship.

Finally, there are a few people in one's life who deserve special thanks. I am grateful to Ted Hinckley, who nurtured me at the start of my career and at moments when I most needed it. I regret deeply that Robert Kelley, my friend and mentor in public history, did not live to see this book. If my writing has any style at all, it is because he taught me, and it is one of the greatest gifts I have ever received. As well, I am profoundly grateful to my colleague and dearest best friend Molly Berger, whose own intellectual accomplishments have inspired me and whose friendship provided the nourishment and encouragement I needed to complete this book. Finally, I thank my longtime friend Mary Sylvain for sharing with me so much of the journey that led to this moment. Without her, I probably would not even have embarked on the trip.

ENERGY AND THE MAKING OF MODERN CALIFORNIA

Introduction

In 1973, Americans were suddenly jolted by a vision of their vulnerability to foreign forces: Arab nations imposed an oil embargo on the United States in retaliation for American support of Israel in the Yom Kippur War. The embargo sparked an energy crisis that lasted into the 1980s. It occurred just at the time America's environmental movement was coming into its own and taking on issues such as nuclear power, offshore oil drilling, and air and water pollution. As people lined up at gasoline stations to refuel their automobiles, sometimes waiting for hours, they experienced how dependent they were on fossil fuels and on complex energy systems. The twenty-year-old recognition, by geologist M. King Hubbert, that energy use in the twentieth century was "the most abnormal and anomalous in the history of the world" finally received attention. People began considering seriously the proposition that society's basic energy resources—oil, gas, and coal—were finite. They pondered warnings that even the smallest growth rate is exponential, and that all exponential growth curves must reach an equilibrium, followed by a decline, possibly an accelerating decline. If this were true, people no longer could continue consuming fossil fuels at the current rate. Moreover, society's general energy consumption would not be able to increase perpetually.[1]

The energy crisis went to the core of America's existence. Two hundred years earlier, Americans had abandoned the mercantilist ideology of a finite amount of wealth in favor of capitalism. Fully embracing the idea of progress and confronted by a continent which appeared to contain infinite natural resources, they opted to achieve a just society by increasing individual economic opportunity. They set out to create new wealth rather than to redistribute what already existed, by expanding across the land, by taking up the emerging technologies of the new industrial age, and by redefining natural resources as commodities so as to bring them into the market place.[2] In the process, they came to believe, however erroneously, that energy consumption was fundamental to achieving the economic growth they desired: without energy, society would not progress forward.[3] Overall, indeed, growth seemed to serve Americans well, and they assumed that the growth rates they experienced would last forever. After all, what is more American than a steady 5 percent rate of growth? Yet, if fossil fuels ran out and ended growth, would not American civilization itself be in jeopardy of coming to a ruinous end? At the very least, would not Americans face significant, perhaps catastrophic, societal changes?

These sorts of questions give the history of energy a special currency in the modern world, particularly if society is to deal successfully with the modern energy dilemma. It is the thesis of this book that the nature of human society's energy experience is shaped by technology, the natural environment, and population growth, and that at the nexus of these three factors are people, their values, and their appetites.[4] Across the world, environmental complexity and versatility constantly have challenged people's technological inventiveness. In harnessing energy resources, a reciprocal interplay has developed between technology and the environment. People's energy choices always involve consideration of environmental endowments and available technologies, and these same energy choices, in turn, influence the environment. In the end, people make energy-related choices based on a variety of continually shifting factors.

On the Pacific Coast of North America, in the region called California, a salubrious climate and a rich diversity of natural resources provided an especially attractive environment for human habitation. After America acquired it from Mexico, in 1846, a steady stream of people flowed into the territory. Energy resources were tapped, the region's population soared, and with this growth came changing attitudes toward the environment.

Technological progress in energy development became essential to the good life, the same good life that made the environment important to people in different ways, through travel, outdoor recreation, and similar leisure activities.[5] Thus, the interplay between technology and the environment was further modified as people's values changed. Because of California's diversity, and because of its relative historical autonomy within the nation, its energy history is a window through which the complex relationship of technology, the environment, and human values can be seen for the United States as a whole, as well as for other parts of the world. The California experience reveals the capacity of both technology and the environment to empower as well as to constrain, and it helps to clarify changing values concerning the environment and technology.

Energy Myths and Characteristics

Historians have suggested various characteristics about America's energy history. Historian of technology George Basalla, for example, observes that most Americans have believed in persistent energy myths. They have assumed that energy consumption is a measure of life's quality—the more the better. Historically, they have believed each new energy resource to be without fault, to be infinitely abundant, and, therefore, to have the potential to effect utopian societal change. They perceived coal in this way during the nineteenth century and, more recently, invested nuclear and solar energy with these attributes. Moreover, people's faith in an energy resource seems to persist until it has developed to the point that its shortcomings are obvious, and only then do they see it cannot bring utopia. In other words, each energy resource has been in its turn a panacea, and Americans often have held a sort of "single-source mentality" about energy. Yet the failure of a resource to live up to their expectations has not seemed to dampen their enthusiasm, as they simply transfer the myth from a fallen energy resource to the next resource appropriated for use.[6]

Another characteristic of America's energy history relates to the long-standing tension between two types of technology: authoritarian and democratic. "The first," wrote Lewis Mumford in 1964, "is system-centered, immensely powerful, but inherently unstable, the other man-centered, relatively weak, but resourceful and durable."[7] The nation's energy history reflects these two modes of technological development. People started with locally controlled and easily understood energy resources: water, wind, wood, and animal power. The technologies harnessing these were democratic in nature, being relatively simple, localized, and appropriate for the tasks to which they were applied. Over time, however, interdependent energy systems developed, evolving from simple to complex, mirroring the evolution of society itself. Technological and organizational changes in energy production were undertaken to take advantage of economies of scale. Fossil fuels—coal followed by oil and natural gas—displaced other energy resources in the nation's energy budget, and sophisticated electric power networks evolved. A certain momentum developed. Not necessarily a deterministic force for shaping social and economic structures, the evolution of regional as well as nationwide energy systems in America certainly reflected the conscious or unconscious visions people held about how society should be structured. The social construction of various energy systems grew out of competing views. In the United States, efficiency, the provision of inexpensive energy, and a preference for free enterprise as opposed to socialism took on enormous importance.[8]

A third national feature concerning energy is that all Americans seem to share in an obsession with the value of time, perhaps making it more important than energy itself. Today, saving time by driving an automobile a short, easily walked distance is a common occurrence in almost every American's life. As people discovered that they could save time by expending energy, they increasingly sought energy technologies and systems that demanded little expenditure of their individual time and required little personal ingenuity. They also sought reliability in technology, a concept closely related to the idea of saving time, and this led them to embrace energy resources, technologies, and systems that appeared to be inherently more reliable and resilient than others. Therefore, with the intention of saving time and effort, as well as gaining in convenience, people gradually gave up localized energy production based on technology such as windmills in favor of centralized production and distribution systems that promised affordable and reliable energy and service, such as electric utilities. As people discovered, however, this choice had a threefold impact of contestable value, for centralized energy systems evolved into extremely complex and thereby vulnerable systems, undermined user control, and ultimately demanded that people pay dearly for sophisticated back-up systems to avoid energy intermittency.[9]

California has shared in all these national energy characteristics, but they tell only part of the region's story. In general, national studies of America's energy experience ignore developments in the West.[10] Perhaps this is not surprising, considering that California and other western states have been perceived as having grown up as economic and social colonies of the rest of the nation, but it is unfortunate. Regional differences that are in themselves interesting and instructive and that, in some cases, are quite important beyond the region, are obscured. For example, scholars studying the history of electric power generally have passed over developments on the Pacific Coast. Yet the evolution of California's electric power industry contributed distinctively to developments in hydroelectricity, long-distance power transmission, rural electrification, and marketing of electricity, and to the political structure in which the industry operates.[11]

National energy studies also obscure basic differences between regional energy patterns and those of the nation as a whole. Over time, in both production (figure 1.1) and consumption (figure 1.2), California's energy resource budget deviated markedly from the national model. On a fundamental level, its citizens' energy choices and consequent energy developments often varied from those made in regions with different resource availabilities, technical developments, comparative prices, output of goods and services, and consumer preferences. California's natural resources, geography, and economic and technical conditions differed in distinctive ways from those in the East, Midwest, and South. Absence of coal deposits, properties of oil, location and character of rivers and streams, geographic isolation, climate, the nature of agricultural development, availability of labor, and pattern and pace of population growth each influenced the energy choices made by Californians.[12] Among these, environmental conditions emerge as particularly important in energy history.

The environment shapes human societies and people alter the landscape to suit their needs. The ubiquity of wood in colonial America compared to England, for example, led European-Americans into a "Wooden Age" that uniquely defined their material life. Historian of technology Brooke Hindle observes that "Americans used wood prodigally, as a fuel

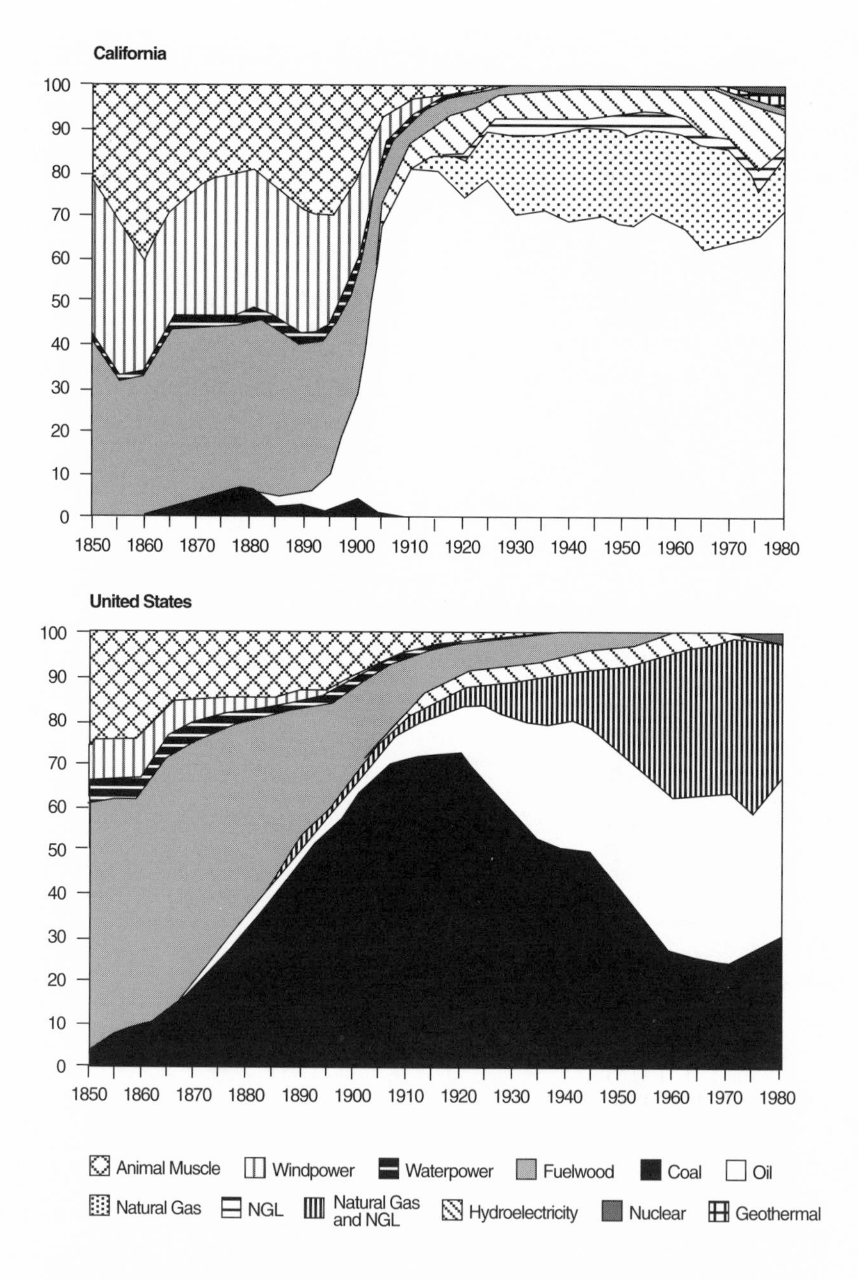

Figure I.1. California and United States Energy Production by Resource as a Percentage of Total Production, 1850–1980. Sources: *Supra,* Table A–3; U.S. data drawn from J. Frederic Dewhurst and Associates, *America's Needs and Resources, A New Survey* (New York: Twentieth Century Fund, 1955); Sam H. Schurr and Bruce C. Netschert, *Energy in the American Economy, 1850–1975* (Baltimore: Johns Hopkins Press for the Resources of the Future, 1960); and William Liscom, ed., *The Energy Decade, 1970–1980: A Statistical and Graphic Chronicle* (Cambridge, MA: Ballinger Publishing Co., 1982).

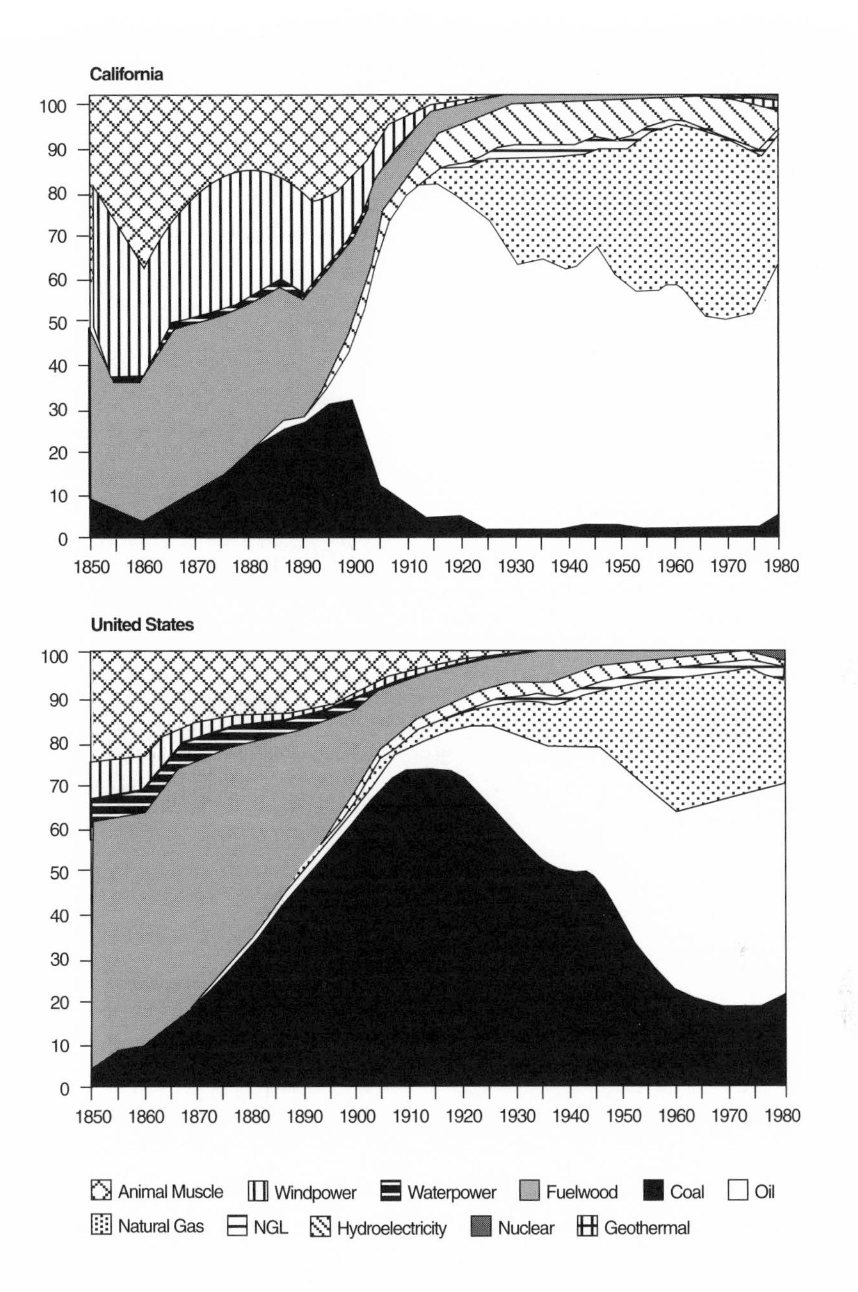

Figure I.2. California and United States Energy Consumption by Resource as a Percentage of Total Consumption, 1850–1980. Sources: *Supra,* Table B-3; U.S. data drawn from J. Frederic Dewhurst and Associates, *America's Needs and Resources, A New Survey* (New York: Twentieth Century Fund, 1955); Sam H. Schurr and Bruce C. Netschert, *Energy in the American Economy, 1850–1975* (Baltimore: Johns Hopkins Press for Resources of the Future, 1960); and William Liscom, ed., *The Energy Decade, 1970–1980: A Statistical and Graphic Chronicle* (Cambridge, MA: Ballinger Publishing Co., 1982).

and as the chief material from which they fabricated their buildings, their transportation systems, and most of their technology."[13] They exploited the environment, commodifying virtually every natural resource. In doing so, they reshaped nature. As environmental historian William Cronon points out in his study of Chicago, people saw "natural advantages" in the environment and sought to improve it according to their vision of what it should be. Each "improvement" added to the cluster of human things superimposed on the landscape, inscribing a kind of "second nature" over the original "first nature." Exploiting natural advantages, Chicago's railroad system, spiraling out in all directions from the city, appeared less a human artifact than "a force of nature. . . . To those whose lives it touched, it seemed at once so ordinary and so extraordinary—so second nature—that the landscape became unimaginable without it."[14] Like the railroad, energy development also occurred as part of the human interaction with nature that Cronon so marvelously depicts. Oil and gas pipelines, highways, waterpower delivery networks, electric power lines, and other energy systems became second nature, too.

Environmental historians focusing on the American West particularly have seen the relationship between technology and nature. Prominent among them is Donald Worster. In his collection of essays, *Under Western Skies,* he gives technology a crucial role in western history: "Far from being a child of nature, the West was actually given birth by modern technology and bears all the scars of that fierce gestation, like a baby born of an addict." Worster's own work, focusing on human domination of water resources, leads him to see technology as a tool to "instrumentalize nature," to see the West as "especially receptive to the vision of a technologically dominated environment."[15] His creation story of a leviathan-like hydraulic society in the West is linked inextricably to Lewis Mumford's conception of authoritarian technology.[16] In order to subjugate the vast and diverse territories of the North America Continent, Americans needed a type of technology suitable to the task. Authoritarian technology emerged and, along with modern corporate capitalism, provided Americans with the necessary tools to conquer the land—tools that took many forms, from railroads, mass-produced farm equipment, and irrigation systems to complex organizational bureaucracies and sophisticated energy networks. Yet, insofar as "technology is a product of human culture as conditioned by the nonhuman environment," Americans adapted to nature.[17] They did so unconsciously in terms of the megamachine of authoritarian technology, but the West's "tradition of appropriate technology . . . , from windmills and solar energy to ideas about ecologically sustainable agriculture," represents conscious and intentional adaptation to the environment. Thus, while environmental conquest is part of California's energy story, the perspective of environmental history demands that one see it also as a saga of "reciprocity and interaction rather than of culture replacing nature."[18]

The give-and-take between technology and the environment mingles over time, and in various ways, with the distinctively regional factors in California's energy experience. Together they encourage or retard exploitation of energy resources. This combination of factors, in turn, coalesces with the influence of energy myths, the tension between authoritarian and democratic technology, the value of time, and concerns for efficiency, reliability, and societal structure. This final mixture comprises the crude calculus of ever-changing advantages by which Californians make their energy-related choices. It provides, at any given time, the *weltanschauung* fundamental to understanding the historical evolution of energy technologies, systems, and use patterns.

Energy Models: Problems and Solutions

The chapters that follow examine California's evolving energy landscape, human exploitation and use of energy resources, and the interplay between technology and the environment, all in the context of a steadily growing population (figure 1.3). Focus falls first on the nineteenth century and on people's use of muscle power, wood and coal, hydraulic and wind power, and other natural energy resources. After 1850, coal became the primary fuel in America, outpacing wood in industrial and household use and undermining waterpowered manufacturing. But Californians discovered they had few coal reserves. Although many of them dreamt of replicating eastern manufacturing development on the Pacific Coast, inconvenient waterpower sites and expensive steam fuels stifled such efforts, and most people remained dependent on wood for fuel. As the national energy pattern shifted, only in agriculture, which relied on human and animal muscle power, did Californians adhere to the same energy model as the rest of the nation.[19]

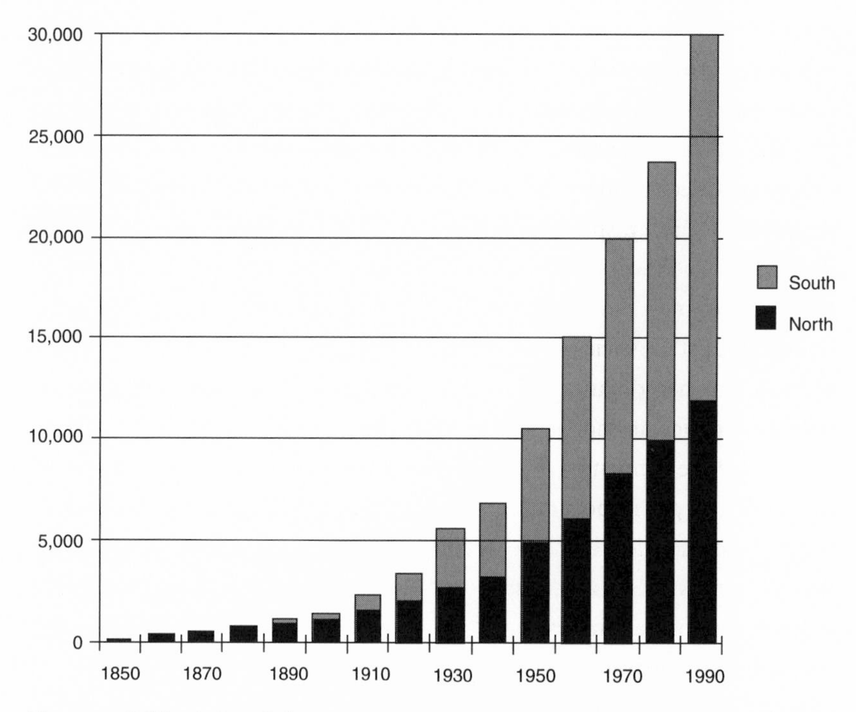

Figure I.3. California Population, 1850–1990. Source: *Supra,* Table C-24.

During the 1890s, entrepreneurs earnestly began developing California's abundant petroleum deposits, and they pioneered hydroelectric power development, drawing on the rich knowledge of hydraulic engineering which they had gleaned from four decades of gold mining. Petroleum, hydroelectricity, and, after 1910, natural gas formed an energy triad that virtually eliminated all other resources from the state's energy budget. Moreover, oil, gas, and hydroelectricity freed Californians from dependence on imported energy resources and opened enormous opportunities for economic and industrial development. Oil replaced coal and wood as steam fuel in industry and transportation, and a hydro-based electric power network delivered energy throughout the state for domestic, agricultural, and manufacturing use. Motor vehicles and electricity transformed both landscapes and lifestyles, and, as energy companies came to be among the largest and most influential in the emerging urban/industrial world, issues of power and control over energy resources surfaced in the region.

World War II wrought an economic transformation in California that played itself out during the cold war era.[20] The war stimulated enormous military and industrial investment in the state and attracted tens of thousands of new immigrants. The explosive population and economic growth, which it spawned, continued unabated into the 1980s. Southern California's decentralized urban environment, in part attributable to the petroleum industry, spread across the southland and to other parts of California. An electronics industry, partly grown from regional electric power research and development, provided a foundation on which Californians built an industrial society quite unlike the eastern model of which so many of them long had dreamed. Through it all, California's energy regime confidently supplied all the fuel and electric power people wanted.

Yet the state's enormous growth brought equally immense energy problems. Oil and gas imports exceeded domestic production, and hydroelectricity became secondary to giant steam-turbine generating plants. Metropolis-dominated and energy-hungry in the second half of the twentieth century, California joined the nation in facing an uncertain energy future. Air pollution, offshore oil drilling, nuclear power, and other energy-related issues increasingly fell afoul of the rising environmental movement that was itself prompted by the good life engendered in part by abundant, cheap energy. Amidst the 1973 energy crisis, the state's energy regime and environmentalists collided. As a result of their collision, Californians embarked, once again, on a different energy path. As the symbiotic relationship between energy, technology, and the environment became clear, they rediscovered solar energy, wind power, minihydro, biomass energy, and cogenerated electricity.

As California enters the twenty-first century, there should be no doubt that human use of energy will continue to play a crucial role in shaping both lifestyles and the natural world. For over 150 years, people's access to and application of energy, plus the interplay between technology and the environment, distinctively helped to sculpt the economic, social, and environmental life of California. The abundance or scarcity of energy served either to stimulate or to retard economic development, its use inextricably wove itself into the social fabric, and the tapping of it indelibly etched the natural world. In sum, energy played a fundamental role in the making of modern California.

Map 1.1. Physiographic Map of California. Reprinted, by permission, from David Hornbeck et al., *California Patterns: A Geographic and Historical Atlas* (Palo Alto, CA: Mayfield Publishing, 1983), 11.

CHAPTER I A Land Apart

"ON THE RIGHT HAND of the Indies, there is an island called California, very close to the side of the Terrestrial Paradise." So wrote Garcí Ordoñez de Montalvo in his 1508 romantic novel *Las Sergas de Esplandián.*[1] Ever since, for tens of millions of people, California has been another world, "a land apart," separated from the eastern half of North America by a vast expanse of desert wasteland and from the Far East by the Pacific Ocean. Its arbitrarily drawn political boundaries, geographer James J. Parsons observed, mark quite accurately "the only area of winter rain and summer drought in North America," and within this area is found a classical Mediterranean climate shared by only 1 percent of the earth's coastal land areas.[2]

California materialized over 400 million years as tectonic plates collided, compressed, and slipped by one another. For most of this time, the Pacific Ocean Plate slid under the North American, melting as it went and generating volcanic activity across a widespread subduction zone. Sediment, eroding from the continent, mixed with lava and ash and settled on the floors of two shallow undersea troughs which comprised most of the present-day region. During the Jurassic and Cretaceous epochs, 150 to 90 million years ago, the boundary between the plates collapsed, squeezing, folding, and tilting sediment and rock along the western edge of the continent to bring forth the Klamath Mountains in the northwest, the Sierra Nevada, and the Peninsular Ranges just east of today's southern California counties of Orange and San Diego.

About thirty million years ago, the westward-moving North American Plate overrode the long-established subduction zone and came into contact with the East Pacific Rise portion of the Pacific Plate. This created a boundary that allowed only for lateral slippage. The San Andreas fault system was formed, and subsequent intense compression and folding further thrust the Sierra Nevada upward and created the Coast Range, Great Central Valley, and the Transverse Ranges. Deep fractures in the crust brought extensive volcanism to the northeastern Modoc-Cascade region, the basin area east of the Sierra Nevada, and the Mojave Desert in the south. Finally, during the Pleistocene Epoch, two million to ten thousand years ago, advancing and retreating sheets of ice etched the final details of California's contemporary landscape.[3]

The final retreat of ice opened the region to human settlement. The first people to find their way into California came behind the melting glaciers of the late-Pleistocene Epoch. Entering from the southwest, small groups gradually spread through the territory, adapting to the warming climate and learning how to use the region's resources. Once arrived, they discovered that the forces of creation had established natural barriers around California that isolated them from outsiders. To the west, the Pacific Ocean cut them off. The Klamath Mountains, Cascade Range, and Modoc Plateau made entry into California from the north difficult. The arid Great Basin stretched east from the Sierra Nevada to the Rocky Mountains and, coupled with the formidable barrier of the Sierra Nevada itself, denied access to all but the boldest adventurers. Soon, even the land across which these people had wandered turned into a barren desert that ranged south from the Tehachapi Mountains and east from the Peninsular Ranges, across the Colorado River, and into Sonora, Mexico.

The geologic processes that created the island of California combined with climate to endow the region with an environment distinctively rich for human habitation.[4] The California Current emerges from the Japanese Current, near the latitude of the Canadian-United States border, to carry water from the north Pacific south along the coast, cool in summer and relatively warm in winter. Prevailing westerly winds sweep air masses ashore, moderating California's climate. Suspended a thousand miles off the coast, a high air-pressure zone called the Pacific High regulates storms that cycle from the ocean across the North American continent. From April to October, it drives storms northward across Washington and British Columbia and on to the Great Plains. From November to March it steers them into California, where the Klamath Mountains, Coastal Range, and Sierra Nevada capture the moisture, creating a rain shadow along the western side of much of the Central Valley and depriving the arid Great Basin and desert regions to the south of any significant rainfall.

For 1,264 miles, California's shoreline sweeps south to southeasterly. It falls from redwood rain forests along the rugged and windswept coast of Mendocino and Sonoma counties and skips blindly past the Golden Gate to curve through the broad, temperate sweep of Monterey Bay. It pauses at the lone cypress tree landmark that stands today along the famous Seventeen-Mile Drive, a point marking an east-west line above which almost 90 percent of the region's rainfall occurs. Then it tumbles along the Big Sur Coast to Point Concepción, where it turns sharply east and traces the Santa Barbara Channel. A coastal plain widens and warms as the shoreline runs south and east past Los Angeles and down to San Diego, and the distance between the ocean and the increasingly hot, almost waterless interior narrows to just a few miles. It is along these coastal low-lying areas and valleys that one finds the region's Mediterranean climate area.

Diverse landscape, variegated climates, and California's geographical position together result in a land of many resource-rich microenvironments. The region has over five thousand native plant species, three times as many as the northeastern United States. Although its broad-leafed, deciduous trees cannot compare in variety to those found in the eastern woodlands, it has three-quarters of the seventy-three species of conifers found in the American West and one-quarter of the world's pine species. Abundant fisheries along its coastline and in its bays, rivers, and deltas—until Europeans arrived—sustained vast populations of birds and sea mammals. The land also provided rich habitats for scores of animals. Pronghorned antelope ranged from San Diego northward, grazing in the region's valleys and along the eastern slope of the Sierra Nevada. Spanish explorers discovered tule elk in the Monterey area and around San Francisco Bay, but they were most conspicuous in the Central Valley. To the north, Roosevelt elk ranged the forests and valleys stretching into Oregon. Before systematic brush burning, tree cutting, and prairie overgrazing occurred, elk and antelope greatly outnumbered today's more commonly seen mule deer.

California's mountains are an especially important resource. A storehouse of precious minerals, the Sierra Nevada is perhaps known best for the gold rush and subsequent mining industry that sparked the region's first population boom. As importantly, however, the mountains serve as watersheds that collect winter precipitation, store it in the soil, and feed underground aquifers. Vast forests flourish there, slowing runoff and increasing water-storage capacity. Pine and fir engulf much of the Sierra Nevada's slopes, oak covers its foothills and the smaller coastal ranges, and redwood still thrives in the north coast region, although first-growth forests are all but gone. The Smith, Klamath, Trinity, Eel, and Russian rivers drain into the Pacific the enormous quantities of rain from the Klamath Mountains and northern Coast Ranges. In the Cascades and Sierra Nevada, almost two million acre feet of water a year are stored naturally in the form of snows that can be as much as fifty to seventy-five feet deep. As the snow melts through the spring and summer, water gradually is released into rivers tumbling from heights of eight to ten thousand feet. Flowing into the Sacramento and San Joaquin rivers, it eventually snakes through the Central Valley to the delta and San Francisco Bay. Along the coast south of San Francisco, however, streams flowing from the Coast, Transverse, and Peninsular ranges carry comparatively little water.

Use of Resources by Indigenous People

Native Americans drew liberally on California's resources. They relied on a wide variety of rock and organic materials to make tools. They chipped projectiles, scrapers, and knives from obsidian, fused shale, and chert. They carved cooking pots, bowls, platters, and frying griddles from soft sandstones and steatite (soapstone). They ground granite into mortars, pestles, and milling stones. On the northwest coast, they used stone tools and fire to carve dugout canoes, and, in the Santa Barbara Channel, they used wedges to split planks for thirty-foot-long canoes that they caulked and waterproofed with natural tar or asphaltum. Needles, awls, and harpoon points were shaped from bone, fishhooks made from shells, and wedges from elk antlers. Plant and animal species also provided them with abundant food resources. Shellfish, trout, manzanita berries, wild iris bulbs, buckeyes, rabbits, seals, sea lions, and scores of other species provided dietary diversification. Some species became important staples. Valley people depended on hard seeds, acorns, and venison. Coastal dwellers relied on bonito, jack mackerel, sardines, and other schooling fish, as well as harvested seeds and acorns. For mountain people, pine nuts and venison became staples, while others, living along California rivers, counted on the annual salmon run and gathered acorns.[5]

The native peoples learned to balance their use of various natural resources with the environment's capacity to yield them. The Pomo practiced birth control when food resources became scarce, Sierra Miwok carefully avoided harming oak trees when gathering acorns, and others practiced rituals that prevented overhunting and overfishing. As historian Arthur McAvoy observes, "fishing Indians carefully adjusted their use of resources so as to ensure the stability and longevity both of their stocks and of their economies." Some groups used fire to clear undergrowth and to promote the yield of grasses that would, in turn, stimulate the propagation of more plants and animals. In the Owens Valley, just west of the southern Sierra Nevada, the Paiute learned to irrigate parcels of land to stimulate the growth of native grasses, tubers, and bulblike corms. Yet, just as among Indians in New England, the social definition of need developed by California's native communities was inherently limited. Over time they acquired a comprehensive understanding of the life-giving uses of resources around them and harvested little more than they consumed or traded for other like items.[6]

Because California's Indians lived richly by desiring little, their energy needs remained limited. Food gained by hunting, fishing, and gathering supplied the caloric energy to fuel the human labor needed for carrying out these tasks, as well as most others. Fuelwood provided them with heat for cooking, warmth, and illumination. It was available readily, in moderate quantities even in relatively dry southern California. Indians obtained it in a passive manner, gathering only dead wood for fires. The warm climate meant that a small amount of wood easily provided them with sufficient fuel for heating and cooking. They prepared and cooked food on outdoor hearths in the dry season, and relying on an indoor fireplace for cooking, as well as warmth, during the rainy season. They also used fuelwood to smoke fish on racks. Finally, they actively drew on solar energy to dry salmon and other fish, whereas semisubterranean pit houses built in the Central Valley were passively insulated against summer heat and winter cold.

Use of Resources by Spanish Settlers

Spain colonized California in the 1760s, more than two hundred years after making contact with the region. Fears that other European powers might occupy it, dreams of new revenues for a faltering new-world empire, and the intense personal ambitions of the Visitor-General to New Spain, José de Gálvez, finally prompted the enterprise. In the end, serious threats from rival powers did not materialize, and deficits rather than revenues plagued the California venture. Nevertheless, a string of twenty-one missions was planted along coastal valleys from San Diego to Sonoma to subdue California's 300,000 indigenous people. With four lightly staffed presidios (or forts) supported by three pueblos (or towns), the missions served as New Spain's northern buffer zone against encroachment by England or Russia. Isolated and ignored, few people knew California.[7]

Spanish settlers altered both California's energy use pattern and its natural environment. They introduced active exploitation of wood, bringing metal tools with which they could and did cut quantities of timber. In sparsely wooded southern California, their removal of local riparian forests for fuel and building materials caused wood shortages and some flooding. More significantly, they added animal power to California's energy budget, using horses and donkeys for transportation and to herd cattle. Native annual bunch and sod grasses, green throughout the year, provided nutritious grazing for cattle and horses. The animals multiplied rapidly, soon trampling away native grasses, eroding hillsides, driving off elk and deer, and disrupting Indian subsistence patterns. In doing so, they opened the way for new grass species, the seeds of which accompanied the Spanish in grain imported for livestock, as well as in their baggage and the clothing they wore. Wild oats, mustard, foxtail fescue, nitgrass, red brome, cheatgrass—"hard plants, adapted already to the summer-dry, Mediterranean climate of California [and] . . . also to thousands of years of heavy pressure from livestock grazing"—replaced and choked out most native grasses.[8]

During the 1770s, mission padres initiated widespread crop irrigation. Supervising Indian laborers in the building of simple *zanjas* (or ditches), plus brush, dirt, and rock diversion dams, they diverted water from the coastal streams. Over time, improved *zanjas* and permanent stone and mortar water-system facilities were built: aqueducts, filtering plants, underground clay pipelines, and fountains "en beneficio de la humanidad" (for the benefit of humanity). Because streams in the southern coastal region could all but dry up during late summer and early fall, padres provided storage capacity with masonry dams and reservoirs. Although machine-aided manufacturing remained virtually nonexistent in Hispanic California, padres did marry simple, waterpowered, grist mills to some of the mission water systems. The earliest mill, at Santa Cruz, was constructed in 1794. Another was erected around 1816, a little above the crest of El Molino dam near Mission San Gabriel. The spring that filled the reservoir turned its wheel. Missions San Luis Obispo, San Antonio de Padua, San Diego, and Santa Barbara also had waterpowered grist mills. According to a description by Alexander Forbes, in 1839, all the Hispanic mills were equipped with small, primitive horizontal wheels, following designs that dated to medieval Europe.[9]

Albeit at the expense of the Native Americans, whose population and way of life were devastated by the Spanish invaders, the mission system flourished. Each year the Franciscan padres slaughtered thousands of surplus cattle for hides and tallow, which they exchanged for manufactured goods smuggled by British and American seafaring merchants. After news of Mexico's independence from Spain reached California in 1822, mercantilist trade restrictions were lifted. This and other changes ushered in a so-called halcyon period of rancho life. Under Spanish rule, the Franciscans possessed sufficient influence to stave off political pressure to secularize their missions. Leaders of the new Mexican government, however, saw the missions as relics of colonialism and made league with colonials in California, who coveted the mission system's vast landholdings. Under Spain, only twenty land grants had been made to private citizens, but in just over twenty years Mexican officials had secularized the missions and redistributed their lands to private rancheros by means of some five hundred grants averaging about 17,500 acres in size. Hides and tallow then became the staple industry of rancho life. Indians either found work with rancheros, drifted into towns, or ran away to the interior. Settlement remained along the coastal region, land grants in the Central Valley stirring the interest of only a few naturalized foreigners.

Meanwhile, American interest in California increased. It had first developed around 1800 through hunters of sea otters and fur seals. In a score of years, Yankee seafarers had decimated the plentiful otter and seal populations and turned to whaling; California became an occasional port of call. Development of the mission hide and tallow trade established regular contact between Americans and Hispanic California. A few visitors chose to settle there, and with secularization more arrived, some acquiring grants of land. Reports of the region's agreeable climate, coastal communities, and rich grazing land increasingly reached Americans through newspaper accounts and books, such as Richard Henry Dana's *Two Years Before the Mast,* and the first of several overland settlement parties arrived in 1841. California had become an object of the United States's territorial ambitions. In the view of Waddy Thompson, American Minister to Mexico during the 1840s, California was destined to be "the granary of the Pacific." Its climate and reportedly boundless resources prompted Thompson to describe California as "the richest, the most beautiful, the healthiest country in the world," although he had never been there.[10] The region's 160,000 square miles and its extended coastline, graced by San Francisco Bay, offered lustrous promise. During the summer of 1846, American military forces and opportunistic settlers wrested California from Mexico, as a part of the larger war with Mexico.

Hispanic and American Occupation

In less than a century of Hispanic occupation, California changed significantly. Its native population declined by two-thirds to about 100,000. Their subsistence lifestyle, one that had existed in relative harmony with the natural environment, disappeared wherever the Hispanic newcomers settled. A grazing and agricultural society replaced the Indian way of life and domesticated the environment along most of California's Mediterranean coastal area. Native animals declined in number or disappeared in the face of cattle, sheep, and horses. Grasses and other native plants were choked out by imported grasses and cultivated food crops. Trees were cut for fuelwood, waterpowered mills constructed, and animal power employed to herd cattle, pack goods, and pull carts, expanding utilization of energy resources.

In 1846, however, the land of California stood on the brink of even more prodigious change. Unlike the Hispanics, the Americans brought a much stronger determination to commodify natural resources. They saw gold and other minerals, timber, animals, fish, land, rivers, bays, and even climate in terms of products to be extracted from the overall ecosystem and transformed into manufactured products, traded, or otherwise subsumed into the market economy. They were a technologically aggressive people, bent on exploiting natural resources and subduing wildernesses. They demonstrated a voracious appetite for energy and a proclivity to believe that, wherever they settled, inexpensive energy resources would be generously abundant and easily subjugated. The destiny of Montalvo's island of California, so near the Terrestrial Paradise, fell into their hands.

CHAPTER II Sweat, Toil, Invention

ON THE EVE OF THE MEXICAN WAR, perhaps seven thousand people, other than full-blooded Native Americans, resided in California. They were scattered along the coast and around the Central Valley land-grant outposts belonging to John Marsh and Johann Augustus Sutter. Although some American settlers coming upon the Central Valley late in the dry season, or in the spring when the Sacramento River was in flood, reacted negatively to California's resources and possibilities, glowing perceptions far outweighed pessimistic ones. "The deep rich, alluvial soil of the Nile, in Egypt," wrote Lansford W. Hastings in his 1845 *Emigrant's Guide to Oregon and California,* "does not afford a parallel." Many observers saw the region's natural resources as unrivaled in commercial, agricultural, and industrial potential. California seemed to be an earthly Eden with a unparalleled climate, and potential settlers read effusive praises in descriptive and promotional literature.[1]

As American newcomers explored California, they uncovered its resources, sometimes by happenstance. The discovery of gold on the American River, in 1848, was one such incident, and seemed to seal the region's promise of abundance. Within a year, argonauts inundated the territory. The military occupation force found itself unable to deal with the population explosion. San Francisco, Sacramento, and Stockton materialized as

Photo 2.1. Harvesting grain near Stockton with a mule-drawn combined harvester and thresher. (Postcard courtesy of the San Joaquin County Historical Museum, Lodi, California.)

cities. Farms appeared by the thousands in counties surrounding the cities. Americans established local governments, often built on Hispanic tradition and form, created their own mining laws, and in 1849 wrote a state constitution. A diversified, urban and rural society took root almost overnight, residents undertaking formidable energy-consuming tasks as they remade the landscape. They cut timber and milled it to construct domestic, commercial, and industrial buildings. They extracted gold and other valuable minerals from the earth. They cleared and cultivated farm land, harvested crops, butchered animals, milled grains, supplied water, and cut, split, and stacked fuelwood. They fabricated and repaired tools and machinery for farms and industry. They built roads and bridges to facilitate transportation between countryside, towns, and cities. They constructed river harbors and seaports to smooth commerce with the rest of the world.

At first glance, it appeared that these and other energy-consuming tasks could be handled easily. When California became a state, in 1850, residents had little reason to believe that energy resources with which they were familiar in the East would not be plentiful and easily harnessed to the growing number of mechanical devices that characterized the evolving American society. The nation's major inanimate energy resource, wood, appeared ample. Waterpower, although located largely in the Sierra Nevada and its foothills, was touted as being "sufficient to run a hundred thousand large mills, and still have an equal quantity held in reserve."[2] Prospectors discovered signs of coal in various locations around the state, and oil was soon found in both the East and West. Moreover, California's wonderfully salubrious climate promised a life of relative comfort. "Considering how little fuel and extra warm clothing are required in California," reported the state mineralogist, Henry G. Hanks, ". . . the laboring man can afford to live better here, . . . than he could in any other country." A small family required only about half the annual fuel necessary in the East, argued one state booster, and at three-fourths the cost.[3]

But idyllic expectations were not easily met. Americans discovered that the energy resources they knew and understood were either scarce, difficult to employ, or expensive. Fuelwood was plentiful but costly to harvest and to transport to cities and towns. Coal, fast outpacing wood as fuel in the East, was found in only a few locations, and its quality was poor. Waterpower sites neither were the same in character nor as numerous as those in the East; the few good ones lay in the Sierra Nevada foothills, some distance from early population centers. Finally, local petroleum resources would be discovered during the 1860s, but would take over twenty years to develop. Although many settlers dreamed of establishing California as a leading manufacturing province, lack of waterpower and reliance on expensively imported coal mitigated against them.

Instead, through the nineteenth century, the region developed as an economic colony of the older, more established eastern half of the nation. Its residents extracted a variety of raw materials for trade, and farming, ranching, and fisheries became crucial industries. The caloric energy of their products fueled the human and animal labor on which Americans, like the Native Americans and Hispanics before them, depended to meet the bulk of their energy needs.

Human Labor

Although not easily quantifiable, human labor readily accounted for the largest energy expenditure in California, during the nineteenth centu-

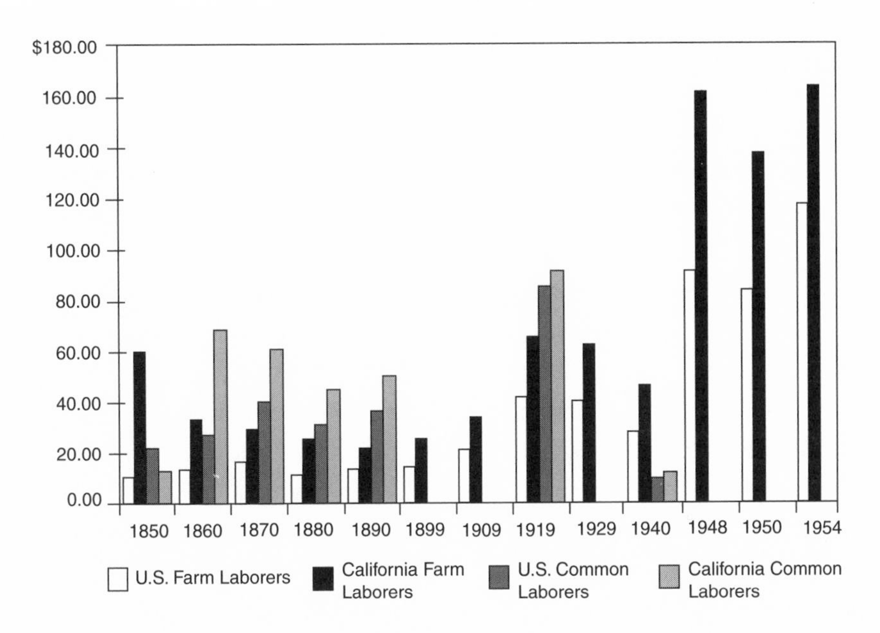

Figure 2.1. California and United States Farm and Common Laborers Comparative Monthly Wages, 1850–1954. Source: *Supra,* Table C-1.

ry. Prior to American acquisition of the region, Native Americans provided such labor for the Spanish, erecting missions, pueblos, and presidios. Under Spanish guidance, they constructed dams, aqueducts, and ditches to transport water for agricultural and domestic use. They planted and harvested sufficient crops by the 1820s to make the region a good deal more than self-sufficient, and they had performed the hard labor of the hide and tallow trade that enriched Hispanic Californians. When the Americans arrived, the Native Americans continued their work, often as indentured servants of the newcomers. As the century progressed, Americans turned to Chinese workers to perform other physical labor, constructing railroads and building levees in the Central Valley. In the twentieth century, the practice of using inexpensive labor persisted, and Japanese, Filipino, Mexican, and other immigrants were employed in hard agricultural work and, as California became more industrialized, in various urban sweatshops. But the European-Americans added their energies as well to the region. In 1850, Californians between the ages of fifteen and sixty-five performed perhaps 258 million hours of work, the equivalent of burning 1.25 million tons of coal. Twenty years later, human energy expended on work may have equaled 5.4 million tons of coal, with people working on average ten hours a day, six days a week.[4]

Unless one worked for his or herself, the cost of paid white labor during the gold rush era was especially high. Gold fever caused a scarcity of labor for hire, which pushed wages 450 percent above the national average for common laborers and farm workers. Even as more workers came on the market with the waning of independent gold mining, the gap narrowed only to 140 percent in 1860 and never closed entirely (figure 2.1). Following the Civil War, when new immigration from the East brought a surplus to the labor market and the depression of the 1870s spawned a serious unemployment problem, skilled and unskilled workers sought to retain their advantages. White workers pursued strikes, formed unions, and ultimately sought political change through the Workingman's Party and the 1879 state constitutional convention. As a result, common laborers' wages remained high until 1900, when they finally fell to about 25 percent above the national average. Farm laborers fared even better, earning between 75 and 120 percent above the national average through the 1880s and between 60 and 75 percent more by 1900.[5]

Animal Labor

Americans also brought thousands of work animals to California and relied much more heavily on the energy of work animals than had Hispanic Californians. Oxen that had pulled pioneer wagons across the overland trails were particularly prized, and, by 1860, there were as many as 26,000 in the state (figure 2.2.). Because oxen were steadier and stronger than mules and horses, midwesterners had become used to using them to clear and break virgin land. But the need to pull heavy supply wagons from Central Valley towns to miners in the Sierra Nevada harnessed most oxen, especially during the dry months from May to November. Depending on their condition, oxen could be very valuable. In 1852, Sampson Wright of Santa Rosa advised his brother in Missouri that a pair could range from $75 to $200, and that a friend of his recently had sold four yoke for $900. Oxen also became essential in the logging industry. From Humboldt County to the Santa Cruz Mountains and the Sierra, loggers used teams of

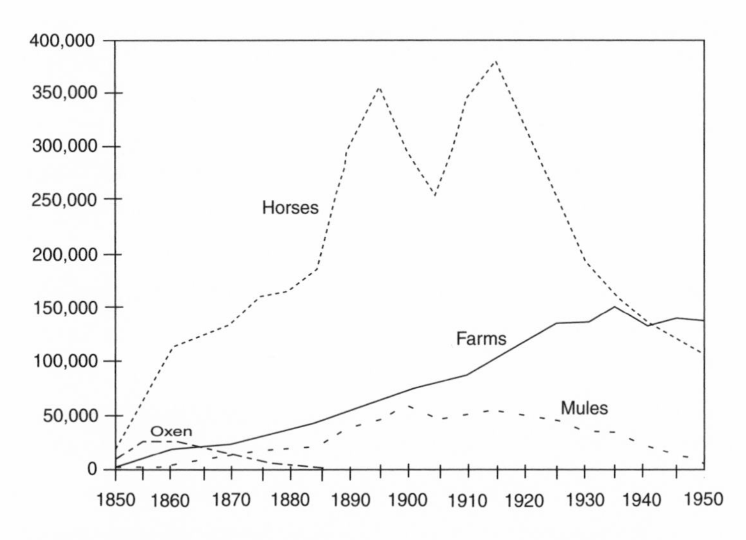

Figure 2.2. Work Animals in California, 1850–1950. Source: *Supra,* Table C-2.

eight to twenty oxen to skid strings of freshly fallen logs, weighing as much as one hundred tons, down to the sawmills.[6]

Heavy mules and horses also pulled wagons to California, as well as finding utility in the mining country alongside the smaller mules and burros bred and used for many years by Hispanic Californians. Like oxen, the largest mules were prized for pulling wagons laden with supplies, and a good pair brought an even dearer price. While smaller Hispanic-bred mules cost from $75 to $85 each, Sampson Wright noted that the "best quality of large wagon mules [went for] from 300 to 500 Dollars per pair." Mules' surefootedness made them extremely valuable as pack animals, carrying goods to mountain camps inaccessible by wagons. Pack trains of mules, as well as smaller burros, became a common sight on California trails, with seven hundred operating out of Sacramento in 1855. Meanwhile, at the mines, horses and mules were applied to a variety of other tasks, such as powering arrastras to crush gold-bearing quartz, turning windlasses to operate mine hoists and water pumps, and hauling ore carts and moving earth. Indeed, into the twentieth century, wherever construction went on—building roads, railroads, and aqueducts—they pulled scrapers, carts, and loaders.[7]

Not only animals were valued; scarcity of wagons, too, elicited exorbitant prices. The heavy prairie schooners brought by overland immigrants were converted into the first freight wagons and fetched from $1,500 to as much as $8,000 in mining country at the outset of the gold rush. As with other things, gold rush prices soon evaporated. By 1852, the cost of a good two-horse wagon had dropped to between $125 and $150, while large, heavy, two- to four-ton wagons could be purchased from $250 to $500, yet these prices remained higher than in the East.[8] Not surprisingly, entrepreneurs opened wagon shops and, in 1860, over 150 businesses, eighty-one of which were located in Sacramento and gold country counties, produced over two thousand wagons and some five hundred carriages. But, by 1881, production in San Francisco had dropped to half that of 1860—about five hundred spring wagons, five hundred buggies, and 125 various other conveyances. High wages, a decline in the mining industry, and the need to import good eastern white oak and hickory for wheels stifled the nascent industry. As with so many manufactured goods desired by Californians, wagons and carriages could be imported more inexpensively than they could be produced locally. Imports from the east coast climbed to some seven thousand per year. Yet, despite these obstacles, an expansive geography, far-flung population, and a high valuation of time led Californians to develop a good road system and "to own and use an exceptionally large number of wagons and buggies."[9]

Throughout California's towns and cities, horses pulled private carriages and omnibuses. The latter, the earliest form of urban public passenger transportation, first employed in New York in 1798, came to Sacramento in 1850. A decade later, entrepreneurs began installing the next most promising American public transportation system of the nineteenth century, the horse-drawn street railway. Horsecars gave a relatively smooth ride over rails, at speeds of six to eight miles an hour, and required but one or two horses, a great leap forward over omnibuses bouncing along dirt or poorly paved city streets. By 1868, progressive San Francisco had seven horsecar lines with almost thirty miles of track, and, in another decade, boasted ten lines using forty-five miles of track, 275 cars, and a stable of 1800 horses. Los Angeles, Stockton, Sacramento, and Oakland similarly had lines ranging from two and a half to thirteen miles long. As communities grew outward, horsecar lines became even more essential. By 1890,

California had 208 miles of horsecar railways, a mileage exceeded by only six other states, and its railways used 2,238 animals, which were always being replaced.[10]

Horse-drawn wagons and delivery vans also were essential for transporting goods in California cities, as in all other nineteenth-century cities. Wagons clogged downtown streets to deliver and to pick up goods at stores and manufacturing establishments, and scores of milk wagons, fruit vendors, smithies, and other tradespeople roamed residential areas. Additionally, urban residents had their own horses. At the century's turn, California's towns and cities stabled 94,171 horses in 37,306 enclosures. In San Francisco, 16,483 horses meant one for every twenty-one residents, and Oakland's 3,851 meant 29 percent more horses plodding its streets than in cities of similar size elsewhere in the nation. The 8,065 horses in Los Angeles amounted to one horse for every thirteen people, exceeding the national average by 87 percent for communities of its size and foreshadowing the city's future as a harbinger of individualized transportation. Horses were everywhere and, in an important environmental drawback, left their mark wherever they went. Each one excreted about two gallons of urine and some twenty pounds of fecal matter in an eight-hour working day. In Los Angeles, where rainfall was minimal and fuelwood extremely scarce, the potential health hazard of dried horse dung may have been ignored because of its value as a local fuel resource. Elsewhere the environmental impact of this animal-based urban transportation system could be summed up, at best, as a serious sanitation and health problem.[11]

Horsepower and Machines in Agriculture

In California, horses played perhaps their most important role in agriculture, the region's leading industry. "Nearly all the farm work in California, where draught animals are necessary, is done with horses," reported John Hittell, in 1879. "Mules are too dear and oxen are too slow." Although the best European draft horses, Percherons and Clydesdales, were not imported to the United States in large numbers until the 1880s, Californians acquired enough of them to cross-breed with their smaller horses. While a Clydesdale weighed approximately two thousand pounds and an average California horse around eight hundred pounds, a half-Clydesdale weighed thirteen hundred pounds and was quite suitable for farming. Thus, Californians slowly bred their own draft horses, and advertisements for them appeared in farm magazines through the early 1900s. But good draft horses were still quite expensive, and wherever they were unaffordable or unavailable, farmers simply employed multiple hitches with four or more smaller California horses to pull plows and other implements.[12]

The average farm workhorse worked between thirteen hundred and fifteen hundred hours per year and had a life of expectancy of about ten years. Rating, in fact, only about one-half horsepower hour for each hour worked, the average farm workhorse provided somewhere between 650 and 750 horsepower hours annually. While some farmers worked their animals as much as three thousand hours per year, this reduced the average working life of the horse by as much as two years. Estimates, not made until the 1920s, suggest most farmers paid about $170 per year to maintain a workhorse and reduced the largest single cost of horses, feed, by pasturing their animals for five months out of the year and using bought grain only as a supplement for hay the rest of the year. The average cost per horse-hour equaled 11.03 cents (10.06 cents for field work and 14.42 cents for orchard and vineyard work), which included feed, bedding, shoeing, veterinary service, time devoted to caring for the animal, taxes and insurance, interest and depreciation, and cost of shelter and equipment used by the animal.[13]

During the nineteenth century, farm workhorses performed a variety of chores in addition to pulling wagons, plows, and other field equipment. They powered treadmills and sweeps that ran machines for threshing grain, drilling wells, cutting feed, churning butter, separating cream, grinding, and pressing. They powered rural grist mills and orchard sprayers, and pulled stumps. After 1879, when San Francisco mechanic William H. Smyth introduced a horse treadmill, they operated crosscut saws to fell and to cut trees into logs.[14] Their service on irrigation pumps became an especially important task during the summer on dry California farms. Dr. E.S. Holden, President of the San Joaquin Agricultural Society in 1861, used one of his horses to operate a double-action Douglas pump that raised water for vines, vegetables, and shrubs on his farm near Stockton. W.T. Garratt and Company of San Francisco adapted a centrifugal pump to horsepower and claimed that one horse could lift four thousand gallons twenty feet per hour and fifteen thousand gallons five feet per hour. Another San Francisco company advertised a horse-powered pump with flywheel and walking beam for stock watering, irrigation, and household use. Actual practice may not have been quite so productive. A.J. Bowie pointed out that the average mule-powered sweep pumped only twenty to twenty-five gallons a minute with a thirty-foot lift, and "you have to keep after him all the time to keep him up to the twenty-five gallon rate. . . ."[15]

Because the ratio of farmers to farm laborers in California stood at one to two versus one to three in Ohio and Illinois and one to four in Indiana, and because extremely large farms and ranches became the norm in California, farm horsepower took on great significance in labor-intensive field work. Innovative mechanization based on animal power became a characteristic aspect of the state's agriculture. Unlike in other parts of the country, historian Reynold Wik has observed, mechanization "was among the very first needs, instead of being a late fruit of long discussion and experience." Local mechanics, working with owners of large farms set in fertile valleys, pioneered the use of newer and bigger machines rather than concerning themselves with refinements. And, though not all Californians were archetypical inventors, every town seemed to have a local genius turning out some particular implement.[16]

Stockton was just such a town. Because of the ease with which California soils could be broken, farmers quickly turned away from single share plows and developed gang plows with two or more shares side by side. Mechanics at Stockton's Shaw Plow Works became leaders in constructing them and produced some twenty thousand gangs during the period 1852 to 1886. The "Stockton gang plow" became known nationwide and contributed to the development of wheat "ranches" in California, some of which encompassed thousands of acres. One man could cultivate eight acres in a day with a five-gang plow drawn by eight or ten horses and cutting ten inches deep. He saved an estimated $.40 to one dollar per acre and dispensed with the work of more than one laborer. He also could attach a machine sower plus a spring-tooth harrow behind the plow, thereby fully preparing the land for a grain harvest in a single operation. Small farmers, of course, like Santa Rosa's W.S.M. Wright, could not afford the horses necessary to pull a gang plow, and they often hired plowmen to complete the task. In 1876, for example, Wright paid Jack Peterson $167.25 for plowing and harrowing with eight animals from November 13 to December 30, 1876. Peterson provided his own animals with grain.[17]

In California, where wheat became a staple export crop, farm horses also played an important role during reaping season. They pulled the reapers, and once the wheat was cut, they often powered the stationary threshers that separated the grain from the chaff. As with plowing, farmers could contract to have wheat harvested and threshed, which W.S.M. Wright did in 1872. He hired Wesley Brown to reap his 135-acre field, near Santa Rosa, and thresh 2,108 bushels of wheat for $286.82. Additionally, Wright paid a crew of seven men for ten days work, six more men for three of those days, and hired a cook for one week, an additional expense of $141. He also provided 2500 pounds of barley and 3100 pounds of hay at a cost of $60.75, presumably for Brown's horses, and he supplied 486 grain sacks at a cost of $70.19. Finally, he paid $77.50 to haul the wheat to a buyer, where he sold the crop for $996.78. Wright netted $360.52 for the year's crop, his overall expense totaling $636.26, or $.30 a bushel. While small wheat farmers, like Wright, had to rely on many workers, large wheat ranchers ultimately abandoned the reaper and stationary thresher. In place of these machines, they adopted the combine, a revolutionary new horse-powered machine that required much less human labor.[18]

The first successful combine was developed by Hiram Moore of Kalamazoo, Michigan, in 1836. Five years later, Andrew Y. Moore purchased one of the implements and used it for several years with mixed results. Michigan's damp weather tended to spoil the undried grain. In 1854, Hiram Moore sold a half-interest in the combine to George Leland and shipped it around the Horn to San Francisco.[19] Near Mission San José, John Horner, a farmer who had unsuccessfully sponsored an experiment with another combine in 1853, invited Leland and Moore to try out theirs. Powered by sixteen horses, the Moore combine harvested some six hundred acres on Horner's and a neighbor's farms, a successful agricultural operation because California's hot summer air had dried the wheat thoroughly, but a financial loss to Leland and Moore when the farmers failed to pay them. The next year their machine sat idle while Leland hunted gold, but they reactivated it in 1856. This time, failure to lubricate the bearings resulted in friction sparking a fire that destroyed both the wheat field and the combine.[20]

During the next several years, Horner continued to experiment with combines, investing as much as $10,000 each in three machines. From 1867 to 1869, he and his brother William oversaw the building of the combines by wheelwright A.O. Rix and blacksmith H. Crowell. For the 1868 harvest in Livermore and Mission San José, they advertised public demonstrations of their three "Monitors" in the *California Farmer*: "Three men and twelve horses have Cut, Threshed, Cleaned, and Sacked, in good work-man-like manner, fifteen (15) acres of grain per day—making five acres per man—a feat, we believe, never performed in America before! . . . FARMERS! Come and see if our claims are well founded."[21] The Horners' advertisements and demonstrations, along with the work of mechanics like William Marvin and H.H. Thurston of Stockton, spread the combine idea. In 1867, the *Stockton Independent* ventured that the combine "is considered one of the greatest labor-saving inventions ever produced."[22]

The combine greatly reduced the labor force required for harvesting and threshing wheat. Five or six laborers and a combine could do the work of a score of workers employing reapers, headers, and stationary threshers. Scenes like that at Richard Threlfall's Livermore wheat ranch in 1867, in which ninety thousand bushels were harvested off four thousand acres by sixty men, five reapers, three headers, two steam threshers, and 140 mules and horses, eventually disappeared. Between 1858 and 1888, at least twenty-one California individuals and companies built combines, and the "California Style" combine, pulled by scores of horses or mules, soon replaced vast armies of laborers and separate reaping and threshing machines. Prominent firms, such as the Houser-Haines Company and Shippee Harvester Works of Stockton, produced standard combine models, and wheat ranches reached phenomenal sizes. One San Joaquin County ranch planted thirty-six thousand acres, with one field seventeen miles long, and Dr. Hugh Glenn's 66,000-acre Colusa County ranch produced one million bushels in a season. Whereas farmers in New England, in 1800, invested seven minutes of labor per kilogram of wheat harvested, Californians using the combine plus gang plows and spring-tooth harrows spent less than thirty seconds per kilogram. By 1900, the combine began to find its way to other wheat states, yet it remained an amazement to visitors to California. "I shall ever cherish the trial I enjoyed in the San Joaquin valley," wrote New York resident A.S. Theath to the *Pacific Rural Press*, "in riding around a 5,000-acre wheat field on a forty-horse reaper, and seeing the reaping, threshing, winnowing and bagging performed with the regularity and exactness of clockwork."[23]

In a national trend, starting around 1900, the horse received strong competition on and off the farm from steam and internal-combustion traction engines. In 1914, the U.S. Department of Agriculture reported that the numbers of urban light driving horses, as well as those of farm horses, were dipping "owing to an unusually extensive and general use of autotrucks and traction engines." But Gordon True, of the University of California's Farm at Davis, felt the state's farmers were going against the trend, as opposed to urban dwellers. "The fact that good heavy horses were never harder to buy in California than at the present time," he wrote in 1916, "is sufficient evidence that there is a demand for the better classes of draft horses."[24] Wayne Dinsmore, Secretary of the Percheron Society of America, validated True's perception. Despite the World War I military need for draft animals, which had caused almost a million horses to be exported from America to Europe in 1915 and 1916 and resulted in "automobiles and light delivery trucks . . . displac[ing] nearly all driving and delivery horses," workhorses continued in wide and popular use on California farms. Even in the 1920s, between 17 and 18 percent of California's farm power was delivered by over 300,000 work animals, and some vendors in city neighborhoods, either unable or unwilling to adopt trucks or vans, continued using horse-drawn conveyances.[25]

Seeking Powerful Technologies

During the nineteenth century, Californians relied on human and animal labor for most of their work activities because muscle power, fueled by the caloric energy of food, was the most available and most easily employed energy resource. Factors governing the use of other energy resources were more complicated. Inexpensive good coal was not available and fuelwood was not always economical. Waterpower was not easily accessible. Petroleum still was being developed as fuel. Energy from wind was limited to powering ships and pumping water. Technologies that reliably and efficiently could harness noncaloric energy resources for many tasks awaited either inventors, engineers, entrepreneurs, investors, or a combination of all four.

Nevertheless, Californians sought technologies powerful enough to quickly subdue the land, to commodify many of its resources, and to increase per capita production. Wherever possible they employed animal power in place of their own toil, and they concocted machines which, using animal power, accomplished most efficiently the work at hand. Historical geographer Vaclav Smil suggests that millennia of improvements in agriculture, through substituting animal for human labor, peaked in California during the late-nineteenth century. "Heavy gang plows and combines took animal-drawn cultivation to its practical limit."[26]

Hard physical work led to invention. In addition to replacing their own caloric energy with that of animal power, Californians went on to replace it with other energy resources. They devised tools to use with engines that burned wood and fossil fuels. They experimented with ways to get the most heat out of poor quality coal. They discovered how best to use the Sierra Nevada waterpower. They learned how to utilize the territory's oil as fuel. They explored ways to harness energy from the sun and to better use wind power. Through the rest of the nineteenth and into the early twentieth century, they endeavored to subdue and to domesticate California. In doing so, they confronted a variety of factors that constituted a crude calculus of ever-changing advantages by which they made energy-related choices.

CHAPTER III Fuelwood and Coal

WOOD AND COAL PROVIDED two basic energy resources to nineteenth-century Californians, and absence of the latter particularly influenced the region's development. The number of inhabitants in the region exploded with the gold rush. In northern California, where most Americans settled during the nineteenth century, instant cities as well as mining centers were created. Residents relied on initially plentiful wood supplies for their fuel, but urban and mining energy demands soon surpassed the local environment's ability to sustain them. Thus, they also looked for and drew on local coal deposits, but scarcity and poor quality led them, in turn, to import large quantities of expensive coal. South of Santa Barbara, a significantly smaller population benefited from a warmer, drier climate which lessened fuel needs, but this advantage also limited natural wood supplies. Furthermore, coal deposits in the south were even poorer than those in the north, and trading conditions begot even higher prices for imported coal than in San Francisco. Consequently, southern California fuel prices were high from the start.

Fuelwood served domestic heating and cooking needs, as well as providing fuel for steam engines in manufacturing and transportation (use of steam power will be discussed in Chapter Six). Upwards of 90 percent of the fuelwood harvested was used in households. A superabundance of

Photo 3.1. Using oxen to haul fuelwood from the Santa Cruz Mountains. (Postcard courtesy of the California History Center Foundation, De Anza College, Cupertino, California.)

timber had shaped American life in the East, and immigrants from there to the Pacific Coast carried with them a wood-intensive disposition. Residents of new towns and cities procured wood from surrounding timber stands, and farms were planted in coastal valleys and along rivers and streams where trees stood in abundance. Redwood, pine, and fir forests covered coastal mountains and the Sierra. Oaks spread from foothills into the valleys, providing fuelwood for several decades. According to the California state mineralogist, oak trees often attained "large proportions, single trees when cut up making as much as thirty or forty cords of firewood." In 1882, for example, a mammoth oak yielded 65½ cords of stovewood.[1]

Annual per capita fuelwood consumption through the 1870s amounted to about three cords, 80 percent or less of national use estimates. A family of four to five persons consumed perhaps nine to ten cords of oak, pine, Douglas fir, madrone, or willow. A cord of oak, the most popular fuel, provided 97 percent of the heating value of a ton of coal, while other woods provided downward to 55 percent. Open fireplaces provided heating for some homes, as illustrated by plans published in *The Pacific Rural Press* for a cheap country house with six fireplaces, but fuel-saving wood stoves for heating and cooking were available. In the 1880s, California had two stove manufacturers who supplied about 25 percent of the market demand, while eastern manufacturers served other consumers.[2]

Meals were cooked on a wood-fired kitchen stove, which provided a well-heated room in the winter months and a stifling environment in the summer. Water for bathing and washing was heated in pots on the stove, making the Saturday night bath a major undertaking and clothes-washing a daylong affair. Urban residents and farmers who were able to pipe water directly into their homes installed waterbacks to their stoves. Water circulated through metal coils looped through the back of the stove firebox and into an adjacent holding tank. While a great improvement, the water heated in the stove could be stored only for a short time in the uninsulated tank, so one still had to fire up the stove every time hot water was needed. As late as 1916, one observer wrote: "The most economical [stove] for the average farm is a moderate-priced steel range with the hot water reservoir at the fire end ($50 to $75)."[3]

Wood fuel took a great deal of human energy to harvest and to prepare, requiring more caloric intake of food than was derived in comparable British thermal units from the wood fuel's end uses. A skilled axeman could produce a hardwood cord of 128 cubic feet in a day, whereas an unskilled worker might take three days to do the same. Although the California cord was a quarter to half the size of the traditional eastern measurement, the required labor still led most Californians, certainly urban dwellers, to purchase rather than harvest it. The cost of a sixty-four-cubic-foot cord varied greatly, selling for $3.00 in the early 1870s in rural Sonoma County on the north side of San Francisco Bay, and for $6.00 in the 1880s in Solano County. In urban Sacramento and San Francisco, a cord brought $10.00 to $12.00, and in Los Angeles it cost $15.00. The highest prices occurred in the Comstock Lode mining region just outside California in Nevada, where heavy timber demand for mining in the mid-1860s pushed the cost of a cord of stovewood to $30.00. Considering a common laborer earned around $56.00 per month, prices much above $10.00 were exorbitant for the average householder.[4]

Conversion of wood into charcoal, the essential heating fuel used by blacksmiths, tinsmiths, and in smelting operations, required even more human labor. Charcoal burners cut wood into four-foot lengths and piled it thirty feet high in a circular stack. They covered the stack with a layer of brush followed by a layer of wet soil and then ignited the pile's center from the bottom. After fifteen to twenty days, a well-managed pile would burn out, yielding as much as 1,000 sacks (about thirty tons) of charcoal that sold for about $18.00 per ton. The principal charcoal-producing area serving San Francisco was near Sebastopol in Sonoma County, where perhaps eighty Italian immigrants dominated the trade. Scattered through the Sierra Nevada foothills, charcoal burners provided smithies and smelters with fuel, and a few farmers in the San Joaquin valley made charcoal. In the 1880s, the California Iron Company tapped one of the state's few iron deposits at Clipper Gap and for a time produced pig iron using locally made charcoal. The Sisson and Wallace Company ran the largest operation in California, serving smelting works in Virginia City, Nevada, and Utah. Near Truckee, they employed 350 Chinese charcoal burners who met weekly orders for up to 58,000 bushels, over 90 percent going to the Central Pacific Railroad's Utah smelters.[5]

Although few farmers produced charcoal, many cut cordwood, hauling it for sale to nearby communities. But as distances increased between stands of timber and urban areas, professional wood cutters and dealers supplied most of the wood to towns and cities. In 1861, Sierra Nevada foothill oaks were stripped for tan-bark, then cut into cordwood and shipped by boat at a cost of $1.00 to $2.00 per ton from Stockton to San Francisco. Three wood dealers operated in the small Sierra Nevada mining town of Sheep Ranch, while Stockton supported six wood dealers. Cordwood also reached San Francisco from the mountains between Santa Cruz and San José, carried by wagon to Alviso on the southern tip of the bay and shipped by boat up to the city.[6]

Small and large logging operations, plus independent wood cutters, worked throughout California to meet the need for both fuelwood and construction lumber. Northwest coastal redwood provided the principal lumber supply, shipped to the San Francisco Bay Area by sea. In the Sierra Nevada, V-shaped flumes allowed cutters to reach deep into otherwise inaccessible areas and to deliver wood to valley communities. Developed by James W. Haines of Nevada in the late 1860s, the V-flume made easy the waterborne transfer of logs, lumber, and cordwood. Logs knocked out the sides of standard box-shaped flumes when negotiating curves, but they easily skipped off the slanted sides of the V-shaped flume. Haines's twelve-mile-long flume to Virginia City carried up to twenty thousand cords in 1869, and the Sierra Flume and Lumber Company used some 150 miles of flume in 1875 to deliver lumber and cordwood from ten separate mills to Red Bluff and Chico.[7]

The introduction and expansion of railroads to California, which were great fuelwood consumers, also opened new timber areas. The Napa Wood Company, for example, purchased fifteen thousand acres of oak-covered land along a new rail line. "In November 1867, there were 3,000 cords of wood piled up [at Oakville] for shipment to San Francisco."[8] A few years later in Merced County, the Southern Pacific Railroad picked up shipments from several locations, including a logging company that delivered wood down fifty-two miles of V-flume to Madera. Wherever the railroad reached a wood-rich area, wood suppliers used it to ship cordwood to towns and cities. In 1889, the Southern Pacific Railroad delivered 110,110 cords to Sacramento, 49,181 cords to Los Angeles, and 42,596 cords to Marysville. Shipping rates ranged from $1.00 to $2.00 per cord in the 1880s, reaching an extortionate $10.80 per cord in the state's Laguna Dam Reclamation District in 1911, but rail transportation became essential for fuelwood delivery.[9]

While many Californians may have agreed with the State Mineralogist's view in 1884 that California would never suffer "a timber dearth," scarcity of wood was recognized as early as 1850. The tremendous crush of immigrants stimulated by the gold rush brought unprecedented exploitation of California's timber stands. In San Francisco, 300-foot-wide lots were sold for as much as $500, with scrub oak firewood on them being the main inducement to some buyers. The urban demand for wood resulted in speedy exhaustion of local supplies. "The trees which formerly stood on the hills of Contra Costa and Alameda, visible from the bay, are to be seen no more," wrote the *California Culturist* in 1860. "The magnificent oak groves of Oakland and the Encinal have been so thinned and mutilated to furnish a supply of fuel," that residents had to cease cutting to preserve the remaining few trees as shelter from winds.[10]

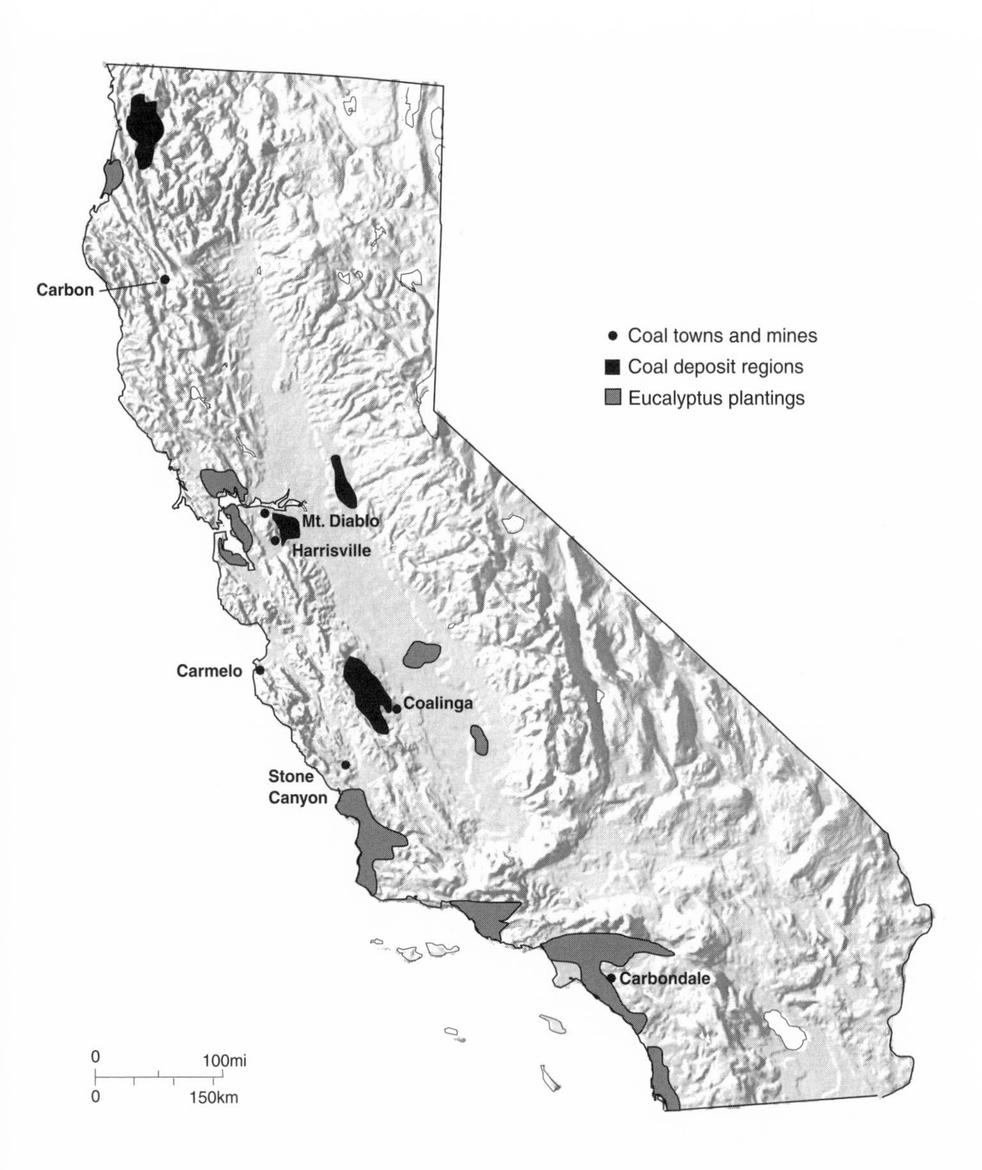

Map 3.1. California Coal Resources and Intensive Eucalyptus Plantings. Adapted from Dan L. Mosier, *California Coal Towns, Coaling Stations & Landings* (San Leandro, CA: Mines Road Books, 1979); U.S., Federal Power Commission, *National Power Survey* (Washington, D.C., 1935), 120; Jaqueline Anna Lanson, "Eucalyptus in California: Its Distribution, History, and Economic Value" (M.A. thesis, University of California, Berkeley, 1952), 167; C.H. Sellers, *Eucalyptus: Its History, Growth, and Utilization* (Sacramento: A.J. Johnson, 1910), 50; and David Hornbeck *et al.*, *California Patterns: A Geographical and Historical Atlas* (Palo Alto, CA: Mayfield Publishing, 1983), 115.

By 1870, the consumption of timber for fuelwood and construction reached alarming proportions. Charles F. Reed, president of the State Board of Agriculture, estimated that a third of the state's accessible timber had been cut. If annual consumption continued at the same rate, he argued, the state's timber would be exhausted in forty years. But, since timber cutting exceeded population increase by twofold between 1850 and 1860, "the whole available lumber of our State will be consumed and destroyed in twenty years instead of forty." A natural resource crisis, he suggested, faced California.[11]

Planting woodlots provided one solution. *The California Farmer* recommended this as early as 1858, the *California Culturist* suggested East Bay residents plant fast-growing locust trees and, heeding Reed's appeal, the State Board of Agriculture offered a $50 premium to the best forest plantation planted in 1870. A popular, fast-growing tree was found in the Blue Gum Eucalyptus, first imported to California from Australia in the mid-1850s. San Francisco's Golden Gate Nursery advertised the trees for sale in the *California Farmer* as early as June 26, 1857, and imported seed was sold during the 1860s. Hubert Bancroft reported that the first Blue Gum grove was set out in 1869 in Castro Valley. Thinning the over-ten-acre grove seven years later yielded more than $900 for fuel and telegraph poles, a higher return per acre than for wheat. Similarly, *Resources of California* reported that thinnings made between 1875 and 1880 at James Stratton's twenty-acre grove in Hayward yielded 1,361 cords of fuelwood at $6.00 each, plus over seven hundred telegraph poles.[12]

Although some Blue Gum consumers complained that it burned poorly, displaying "a perpetual detonation and fizzing going on, as though there were a display of fireworks imminent," the trees continued to be planted and used. Ranches around Denvertown, five miles south of Suisun in Solano County, raised them for their own use, and the *Pacific Rural Press* urged all farmers to plant them. Throughout tree-barren areas of southern California many groves were laid out, but a truly popular movement did not sweep the state until after 1900. Though commercial plantations were set out then in large numbers, by 1908 Blue Gum was in general use for fuel only in the southern portion of the state. Still, despite renewed statewide interest, evidenced by an increased number of articles about eu-

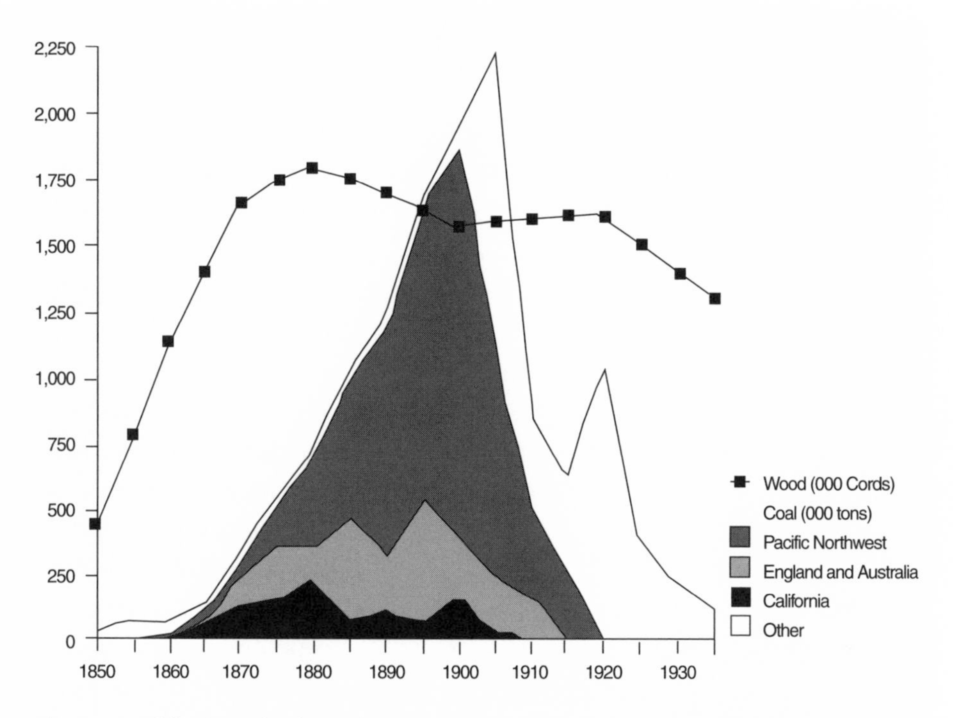

Figure 3.1. California Coal and Wood Consumption, 1850–1935. Source: *Supra,* Table C-3.

calyptus in the *Pacific Rural Press* from 1905 to 1910, in northern California the wood did not compete well with valley oak.[13]

Charles Reed's gloomy prediction that California would run out of wood in twenty years did not come to pass. While wood continued to be a domestic fuel into the twentieth century, overall cordwood consumption peaked in the 1880s and individual consumption declined steadily (figure 3.1). It dropped at a rate of approximately one-half cord per decade from 1870, when annual per capita consumption stood at three cords, to 1910, when consumption reached slightly over one-half cord per person. As elsewhere in the nation, wood as fuel gradually became a luxury to all but rural and small town Californians. In 1908, farmers burned 65 percent of the state's cordwood, people in towns with between one thousand and thirty thousand residents consumed 24 percent, and city dwellers used 5 percent. The remaining 6 percent was used in the mining industry. (On a national level, farmers consumed 81 percent, town residents used 15 percent, and city folks burned 2 percent.) Around San Francisco Bay, cities shifted away from fuelwood before the turn of the century. In 1889, when Los Angeles, Sacramento, and Marysville received a combined net total of 201,887 cords of wood via the Southern Pacific Railroad, San Francisco and Oakland received only 3,128 cords and wood-rich Santa Clara Valley shipped out a net 19,090 cords from San José.[14] Instead of wood, many bay area residents turned to domestic and imported coal as a principal heating and cooking fuel.

Workable Coal Deposits

The search for coal in California began as the gold rush delirium waned during the 1850s, many discoveries of the black substance made by disillusioned gold miners. Because coal for industry and transportation had to be shipped by sea from the East at great cost and averaged $24.00 per ton in San Francisco, local discoveries provoked real optimism that domestic "supplies of that indispensable necessity of all civilized communities—a good article of coal"—would push forward California's economy.[15] The opening of mines in the early 1860s at Mount Diablo in Contra Costa County signaled to some people a real possibility that Californians soon could cease importing coal.

Early reports suggested coal deposits of substantial quantity and quality, and Josiah D. Whitney, head of the State Geological Survey that began in early 1861, encouraged this view. In his March 1861 address to the state legislature, Whitney reported confidently that "workable deposits of coal of fair quality and great extent exist in Contra Costa County in a very favorable position for mining and shipping to all points in the interior of the State. . . . It is a matter of congratulation that the State will soon cease to be dependent upon other regions for her supplies of fuel."[16] Since Whitney's words were guided by his constant political struggle to get continued state funding for his scientific survey from practically oriented legislators, his views reflected what Californians wanted to hear. Few listened to Dr. John B. Trask, first State Geologist from 1853 to 1856, who accurately observed that coal deposits appeared relatively scant and tertiary in character.[17]

Prospecting at Mount Diablo began as early as 1855, and Francis Somers and James T. Cruikshank discovered the first important vein in December 1859. Followed by more prospectors and, soon, investors, communities such as Somersville and Nortonville developed, their Welsh residents working the local mines. In the mid-1860s, the Black Diamond and Pittsburg companies, the two most successful firms, constructed short-line railroads to carry ore from the mines to landings on the Sacramento River. Annual production increased steadily between 1861 and 1869, from 6,620 to 148,722 tons, and by 1900 the Diablo mines had produced 4.7 million tons, 80 percent of California's total coal production. Much of it went to San Francisco, where it sold for between $6.50 and $8.20 per ton and became a standard by which to measure the quality of other California coal.[18]

Excitement about coal spread as prospectors made new discoveries from Amador and Monterey counties in the north to Orange County in the south. Coal outcroppings at Corral Hollow on the line between Alameda and San Joaquin counties were developed in the 1860s, and new finds at Harrisville prompted an Alameda County newspaper to observe in the 1870s that "Livermore is destined to be the NEWCASTLE OF CALIFORNIA."[19] During this same period, mines also opened near Ione in Amador County and south of Point Lobos and at Stone Canyon in Monterey County, and in 1878, the discovery of coal in Orange County gave birth to the town of Carbondale. Noted California lawyer and eventual Federal District Judge, W.W. Morrow, speciously advised participants in the 1881 San Francisco Mechanics' Fair that California coal fields "are well nigh inexhaustible" while "the quality of our coal improves with each new discovery."[20]

Coal for domestic use became attractive where wood costs were high or a sense of progress imbued people. Some rural residents temporarily switched from fuelwood to local lignite coal, while city folks found a ton of coal to be as cheap or cheaper than the equivalent amount of wood. Although most Diablo coal was shipped to San Francisco, some Contra Costa County farmers purchased it for $3.00 per ton at the mine, and mine company agents retailed it locally for household use. In Alameda County, farmers coming "with their teams from as far away as the Mission [San José de Guadalupe] and San Ramon" also purchased coal at $3.00 per ton from mines near Livermore. People in and around Sacramento and Stockton purchased lignite from the Ione mines. Unfortunately, some promising California mines, such as those at Corral Hollow and Livermore, did not have the rail or water transport essential to serve urban markets, so they failed.[21]

Although California coal provided for over 30 percent of the state's coal consumption during the 1860s and 1870s and over 5.5 million tons were mined during the nineteenth century, it faced stiff competition from superior quality imported coal (figure 3.1). In the flush of the gold rush, the levy on imported coal reflected the exorbitant prices of the time. A ton of good East Coast anthracite, with an equivalent heating power of more than a full cord of oak firewood, cost $24.00 in San Francisco. By the late 1860s, as shipping and trade normalized and gold fever disappeared, its price fluctuated between $10.00 and $18.00. And bituminous coal from new mines in British Columbia and the territory of Washington cost from $6.50 to $11.00 per ton. As local timber stands disappeared and cordwood prices climbed, most bay area residents turned to coal for domestic fuel.[22]

Imports from mines in the Pacific Northwest increased during the 1860s and 1870s, with almost 300,000 tons delivered to San Francisco in 1880. The following decade brought 1.7 million tons. Mines near Nanaimo, British Columbia, and Seattle, Washington, were the first to produce good lignite and a fair quality bituminous coal for the San Francisco market, and, after 1874, Coos Bay, Oregon, mines contributed to the flow. By building steamships and facilities especially for the coal trade, some of the companies overcame the slow, tedious task of manually unloading one hundred tons per day from single hatches into waiting wagons. Fitted with four and five hatches each, the steam colliers were loaded mechanically and unloaded at specially built docking facilities in San Francisco that were equipped with machinery capable of unloading three thousand tons per day. Rather than using wagons to haul the coal to storage bunkers, coal cars were loaded at the dock and moved over elevated trestles to storage facilities. The Oregon Improvement Company's facilities, wrote one contemporary, were "probably the most complete and extensive on the coast."[23]

The imported coal trade had a profound impact on domestic coal. As technological improvements reduced freight rates for Pacific Northwest coal, California mines lost their competitive edge. With almost twice as much coal coming in from the northwest as California produced in 1877, Watson Goodyear saw that "it is unquestionably to the mines of Washington Territory and of British Columbia that this Pacific Coast must look." The Diablo mines' days were numbered. In 1883, the Pittsburg Mine outside Somersville closed, followed two years later by the Black Diamond Mine at Nortonville. Although a strike at the Nanaimo mines and a slowdown of other coal imports, in 1887, caused some local miners to take heart, domestic production declined steadily after 1882. Pacific Northwest coal supplied Californians with over half of their coal, close to one million tons annually, between 1887 and 1902. San Francisco and Oakland consumers would use Diablo and other California lignite only when scarcity drove up imported coal prices. On the other hand, the cost of transport inland from port of entry kept up the retail price of imported coal, so many residents beyond the bay area continued to use cordwood or what little California coal was produced.[24]

Ironically, the availability of coal from two other sources, Great Britain and Australia, owed much to Californians who lived outside the bay area. During the 1850s, demand for manufactured goods in California's gold fields and new cities resulted in the arrival of hundreds of ships with imported goods, many of them English. At first, most left California for England and other ports with rock dug for ballast from San Francisco's Telegraph Hill, but this was hardly a profitable cargo.[25] Farmers altered this situation.

California's climate, split between a wet and dry season, prompted farmers to raise wheat. They could plant in late fall and harvest at summer's end, taking advantage of natural winter and spring rainfall to provide all the water necessary for a crop. The grain turned out to be extremely high in quality, as it matured and dried fully on the stalk before harvest. But the California market was limited, and farmers were isolated from the wheat-deficit areas in America's domestic economy. As Morton Rothstein has observed, "West Coast wheat farmers and traders had no choice but to seek foreign markets." Opportunely, Great Britain had repealed its Corn Laws the year before the gold rush, opening its domestic market to foreign foodstuffs. Since the hard, dry, white California wheat kernels were ideal for shipping long distances over water and made superior flour, Great Britain became a major buyer. Wheat bagged in hundred-pound lots soon replaced rock as ballast on outgoing English ships.[26]

By 1868, San Francisco handled a little over one-third of the United States' wheat exports, and within two decades exports accounted for four-fifths of the value of California's wheat crop. Farmers with the gang plow and newly devised combine at their disposal optimized wheat farming. Grain peaked as the principal crop in counties around San Francisco Bay and along the Sacramento and San Joaquin Rivers in the 1870s, and during the next two decades, production peaked in Fresno, Tulare, and Kern counties, as railroad expansion opened up the lower San Joaquin Valley to the international market. Grain acreage reached 2.5 million by 1879, and within ten years California was the second largest wheat state in the nation. British trading firms established themselves in San Francisco, and several miles of warehouses and docks dotted the Carquinez Strait from Port Costa to Crockett. Some 500 ships loaded wheat in the peak seasons of the 1880s, and in 1884 almost half the ships clearing San Francisco for foreign ports carried grain.[27]

While the wheat production of California was in itself significant, more consequentially, noted the U.S. Treasury Department in 1885, "one of the most important incidents of the transportation of wheat from California to Europe is the fact that a large number of vessels engaged in that trade bring coal to San Francisco."[28] Starting with only 14,490 tons in 1860, annual coal deliveries from Wales, Scotland, and Australia reached 200,000 tons during the 1870s and 350,000 to 400,000 tons during the last two decades of the nineteenth century. In 1892, shipments reached 549,840 tons. An international energy supply system emerged based on wheat carried to England, where Welsh coal was loaded for California, or where manufactured products were shipped to Australia, and thence coal to the West Coast. Great Britain and Australia provided a third of the coal consumed in California through the 1890s.

Ultimately, this unique energy supply system, stimulated by California's climate and its geographic isolation from the rest of the United States, came to an end. United States coal import tariffs, though not large, combined with other factors to slow and, finally, to end British and Australian coal imports. Shipments of wheat from new fields in India, and later Argentina, caused grain prices in England to fall by more than half between 1882 and 1894. This meant lower profits for farmers, who faced the added dilemma of lower yields per acre as a result of two decades of cropping without fertilization. Large wheat farmers also found it attractive to convert their land holdings to smaller farms, a changeover made possible by California's growing population, improving transportation, increasing capital, and advancing irrigation technology. As wheat acreage and yield diminished, fewer ships carrying coal arrived. Finally, rising maritime insurance rates for coal cargoes and steady competition from British Columbian and Washington coal stifled the trade.[29]

By 1890, virtually all mining of California's lignite coal had ceased, the competition with imported coal simply being too great. But John Treadwell's purchase that year of the Eureka Coal Mining Company and other coal properties in eastern Alameda County's Corral Hollow district marked a final effort to turn around the state's lignite industry. By decade's end, his San Francisco and San Joaquin Coal Company had opened the Tesla Coal Mine, built a thirty-six-mile rail line to Stockton, and extracted almost 200,000 tons of coal. Treadwell counted on a new technique to make his venture successful—coal briquetting.[30]

Attempts to enhance lignite by molding it into fuel briquettes began in the 1850s in England and Belgium. Briquetting machines were displayed at the 1867 Paris Exhibition, and within three years Germany attained commercial briquetting success. A preponderance of lignite in Germany pushed that nation's efforts forward, and by 1906 she produced sixteen million tons, while France, Belgium, and England each produced over a million tons. Eastern U.S. interest in the technology remained sporadic through the 1890s, efforts to apply the technique discouraged by the availability and low price of good coal. But in California the presence of lignite and the high cost of imported anthracite and bituminous coal caused more serious interest in briquetting technology.[31]

In 1898, San Franciscan Robert Schorr designed a continuous-acting briquetting press. Three years later, John Treadwell opened a production plant in Stockton that employed Schorr's innovation and employed six workers to produce 125 tons of briquettes daily. Coal from bunkers supplied by rail from Treadwell's Tesla Mine moved into the plant on conveyors. Deposited into a hopper, the coal dropped into a disintegrator for crushing, after which it was mixed with screenings from imported coal

Photo 3.2. The Tesla Coal Mine, Alameda County, 1899. From *California Mines and Minerals* (San Francisco: L. Roesch Co., 1899), 389.

purchased from San Francisco. Finally, the mixture passed via chutes to five driers capable of evaporating nine thousand pounds of water per hour. The dried and heated coal then was mixed with a binder of asphaltum pitch (residue from petroleum distilled at the plant), cooled, and conveyed to two Schorr presses for compression into seven-ounce, round, convex "boulets."[32]

Treadwell's was the first commercially successful coal briquetting plant in the United States, and, although fire destroyed it in 1902, he rebuilt. Flushed with success, Robert Schorr took his press to Oakland, where he designed a plant for the Western Fuel Company. Opened in 1905, it used imported coal yard screenings mixed with pitch purchased from Richmond crude oil refiners to produce cubical rather than convex briquettes. Another Oakland plant opened with a Schorr press the same year but closed because of an accident, while a third plant in San Francisco employed the press until it was destroyed in the 1906 earthquake. Meanwhile, Charles R. Allen designed and built a new press and plant at Pittsburg, a Mount Diablo coal-mines-river-landing town located at the juncture of the Sacramento and San Joaquin rivers. Although the nearby mines had closed, Allen had faith in his system and went forward using coal screenings shipped upriver from San Francisco. He modified earlier processes by using a high-temperature retort to mix oil pitch with the screenings before pressing briquettes, and he claimed that "in his process the nature of the fuel is changed so that subbituminous coal partakes of the character of bituminous coal."[33]

These successes in coal briquetting led E. Woltman of San Francisco to try it with peat, which was abundant in the tule islands of the San Joaquin River delta. Drawing on similar efforts in Europe, he opened in 1905 an experimental peat briquetting plant in which he dried peat, mixed it with oil pitch, and pressed briquettes. A year later, the United States Briquetting Company built a plant at Stege in Contra Costa County to make peat briquettes: "They give promise," wrote Edward W. Parker of the U.S. Geological Survey, "of a method of using California oil as a domestic fuel, the peat on account of its spongy character acting as a carrying vehicle for the oil and at the same time performing duty as a fuel." Although the Stege plant failed, unable to expel excess water satisfactorily from the peat, experiments briquetting peat and agricultural waste continued. Ultimately, charcoal briquettes using apricot and peach pits, almond shells, and other waste products as binders were developed during the 1930s out of these earlier experiences.[34]

In the end, coal briquetting failed to become a successful industry within California. A couple of firms operated sporadically, but lignite briquettes gave way to oil-based briquettes after 1906. That year a San Francisco gas engineer, E.C. Jones, devised a press to make briquettes from lampblack, a soot by-product of oil gas. His employer, the San Francisco Gas and Electric company, built a manufactured gas plant to use the lampblack briquettes as fuel. Similarly, the Los Angeles Gas and Electric Corporation made lampblack briquettes, and in 1911 the firm produced and retailed for household use some ten thousand tons. Their briquettes overshadowing coal as a domestic fuel in Los Angeles for a short time. Although Californians never consumed more than 1 percent of the national briquette production, the effort to exploit the state's lignite coal gave birth to a product that did catch hold beyond its borders. The coal briquetting industry expanded to the Pacific Northwest and the eastern United States, and by the 1940s Americans consumed over a million tons of fuel briquettes per year, principally as domestic fuel.[35]

Summary

Wood and coal played an important part in California's energy budget during the nineteenth century. Fuelwood was plentiful enough for rural inhabitants, and residents of San Francisco and other urban areas could get it, although they sometimes had to reach far afield for it. "That indispensable necessity of all civilized communities," coal, made up between 21 and 27 percent of total state energy consumption during the 1890s. A unique energy supply system grew up based on imported coal exchanged for grain, which in turn was based on success in wheat farming made possible by California's distinctive two-season climate. Yet Californians still remained comparatively coal-starved. During the same decade, coal accounted for over 50 percent of national energy use. While the nation's per capita coal consumption climbed from .6 to 3.5 tons between 1860 and 1900, Californians individually used 35 percent less coal than their countrymen and paid some 40 percent more for it. Despite the fact that California coal imports comprised 57 percent of the nation's total coal imports during this period and actually accounted for 97 percent in 1891, Californians still burned less than 1 percent of all the coal used in the nation.[36]

The high cost of imported coal, plus disappointing domestic coal deposits and inadequate subregional fuelwood supplies, made manufacturing growth difficult. To some extent, the state's distinctive Mediterranean climate moderated discomfort or inconvenience caused by coal and fuelwood shortages. Californians never went cold at home in winter, nor did they lack fuel for small manufacturers. But to achieve the growth and prosperity of which boosters boasted and immigrants dreamed, they ultimately had to find other energy resources.

CHAPTER IV Candles, Kerosene, and Gas

ELECTRIC LIGHTING has become so pervasive in modern civilization that we take it for granted, but nineteenth century Californians had a much more difficult time illuminating their homes and workplaces. Evenings at home offered time for a variety tasks, provided there was light: sewing and knitting, repairing household tools, and, with the spread of literacy through public education, reading and writing. An illuminated home offered a sense of warmth and security for family and visitors. Artificial light in offices and commercial establishments extended business hours and lighted streets offered people safety. Wood and coal provided satisfactory heat but ineffective light, so people sought other means of illumination.

In the United States, a variety of lighting sources were utilized long before the conquest of California. Whale oil, for example, had been burned in lamps for several decades. New England whaling fleets arrived in the Pacific soon after 1800, and vessels visited San Francisco frequently between 1850 and 1870. In addition, shore whaling occurred at eleven stations along the California coast, from Monterey to San Diego. Because four factories refined whale oil in San Francisco during the 1850s, the product was available to Californians, but it was expensive. In both East and West, sperm oil never fell below $1.00 per gallon wholesale after 1847. Retail prices climbed as high as $2.50 per gallon in the East and $3.00 in California, thus limiting the market for whale oil to the well-to-do.[1] Those who could not afford whale oil turned to traditional tallow and wax candles or camphene, an illuminating fluid derived from turpentine.

In cattle-rich Hispanic California, locally made tallow candles were abundant, and tallow, along with cattle hides, was exported in great quantities. The gold rush ended this hide and tallow trade. Cattle became most valued as food in the mining towns, and local tallow became scarce. In the tradition of economic colonialism, argonauts willing exchanged their gold for candles and other familiar products shipped in from the East. During the 1850s and 1860s, imported candles dominated the California market. As commercial links became more firmly established, their price dropped from $.75 per pound in 1848 to between $.16 and $.46 per pound, depending on quality, during the 1850s. A twenty-pound box sold for $6.00 in 1865, and fourteen ounces (or about six Patent Sperm candles) averaged 32.5 cents. Californians imported 166,600 boxes annually between 1853 and 1856, but two San Francisco candle manufacturers, using local tallow as well as tallow imported from Australia, eventually gained over half the candle market. During the early 1880s, they produced 135,000 boxes, whereas imports declined to ninety-eight thousand. But even as California tallow production rebounded in the 1880s, California's state mineralogist observed that "we continue making heavy shipments of tallow to the East, where it is made up into soap and candles, brought back and sold to us with cost of freights, manufacture, interest, and insurance added."[2]

While most city residents purchased manufactured candles, many rural families made their own. J.A. Graves, who grew up on a farm near Marysville during the 1850s, remembered making them in molds, using a half-and-half mixture of mutton tallow and beeswax, the latter obtained from wild-bee hives:

> Our only lights were candles. We made them ourselves, because of the economy, and those we made were better than we bought. Fortunately, in a box of bed-clothing which was sent from Iowa, around the Horn, two candle-molds were packed. They were tin affairs, with a handle to hold them by while filling. Each one made four candles.

But miners were the biggest users of candles, for there was no effective substitute for candles as an illuminant in underground mining. Candles also served as a warning device for miners, since they flickered and then extinguished when oxygen became insufficient. During the 1870s, each mine in the Comstock Lode used between six and eight tons of candles per month, and five thousand tons were imported just for gold and silver mining in the decade's middle years.[3]

In California's cities, however, camphene became the leading illuminant at mid-century. It retailed at prices ranging from $1.00 to $2.00 per gallon, comparable to prices in the East, and was made by distilling turpentine over lime. While evil-smelling and highly flammable, it burned in lamps designed specifically for it, as well as in sperm oil lamps that were modified for a better draft. By 1859, there were seven camphene distilling firms in San Francisco, producing over fifty thousand gallons per month from turpentine shipped around Cape Horn from South Carolina (about 1.6 gallons annually for each Californian). When the Union blockade of the South during the Civil War interrupted shipments of turpentine in 1862, the state legislature offered a premium of $500 for the first one thousand gallons of camphene made in the state. Thus encouraged, Californians turned to their own sources of turpentine, principally the yellow or Ponderosa pine. A distiller in Marysville won the prize, while a Butte County company alone tapped over twenty thousand trees. These entrepreneurs satisfied the demand for camphene during the war years, but prices remained high because of freight costs from the interior to the cities. When imports resumed at the war's end, local manufacturers were unable to compete.[4]

Kerosene: the New Lighting Fluid

After the Civil War, stiff competition from a new lighting fluid, kerosene, all but eliminated even imported camphene from California's illumination market. Kerosene had entered the eastern market during the late 1850s. Distilled first from coal and then from petroleum, it offered greater safety from sudden explosion than camphene. Kerosene quickly swept the nation and was destined to become a basic illuminant for several decades in California. By 1860, eastern companies produced twenty-five to thirty thousand gallons daily, and users had purchased an estimated 1.8 million burners.[5]

Despite California's relative isolation from the East, it was not to be denied the new product. Kerosene, announced the *California Culturist,* in August 1859, is "for the first time being introduced into our state." With Hale's Patent Kerosene Burner, "genuine coal oil" could be purchased at Bragg and Company in San Francisco or Sacramento and the buyer treated to "unequaled light." Other dealers, such as the Stanford Brothers, received large shipments from the East via the Horn. But, just as the Civil War ended turpentine imports to California, it also disrupted Pennsylvania petroleum production and greatly impeded kerosene shipments. Stimulated by the new product, Californians began to seek out their own petroleum deposits, particularly looking near tarry bitumen seepages near the Santa Barbara Channel community of San Buenaventura.[6]

In 1859, George S. Gilbert, a transplanted Brooklyn whale-oil merchant who, in 1851, had established the Pacific Oil Works in San Francisco, visited tar beds near Los Angeles. Within a year he set up a four-hundred gallon retort to distill kerosene from the liquid asphaltum, but a quarrel with Major Henry Hancock, owner of the now famous La Brea Tar Pits, cut off his supply. Impressed with the possibilities of distilling kerosene from what seemed to be "heavy oil," Gilbert moved to Rancho Ojai in 1861. There he tapped a similar tarry oil after tunneling sixty feet into a hillside, built a retort, began producing kerosene and lubricants, and finally settled as a storekeeper in nearby San Buenaventura.

Gilbert's activities attracted others to the region, included among them mineralogist Edward Conway and chemist Captain James H. White, both from the coal county of Contra Costa. White acquired seventy-eight thousand acres of land around San Buenaventura, in early 1864, with hopes of finding oil. In April, Benjamin Silliman, Jr., the noted Yale University scientist and consultant who had completed the first scientific analysis of Pennsylvania oil in 1855, arrived in San Francisco on his way to investigate mining in Nevada and then Arizona. Conway and White urged Silliman to look at their holdings on his way to Arizona, and he landed in Santa Barbara in June. After visiting the Ojai area, seeing seeps, and hearing Gilbert praise the local oil, Silliman submitted a glowing report in September: "I am of the opinion that the promise of a remarkable development at Buenaventura is far better than it was in the Pennsylvania or Ohio Regions—since so famous."[7]

Silliman's promising opinion sparked a California oil boom. Speculators in and out of the state leaped into the fray, despite the *San Francisco Bulletin's* caution to be "a little skeptical about the assumption of the astute Professor Silliman that California will be found to have more oil in its soil than all the whales in the Pacific Ocean."[8] Historian Gerald White noted that seventy-five associations with capital roughly equaling $50 million were "actively engaged in exploring for oil" by the end of 1865. A repeat of the Pennsylvania success was widely anticipated, and all along the coastal mountains, from Humboldt through Santa Clara to Kern counties and Los Angeles, the search proceeded. Local journals reported oil indications "of the most flattering character" and printed articles on all aspects of petroleum, from discoveries to "how to sink wells."[9]

The California Petroleum Company organized to exploit the Ojai area's good fortune and sold some of their oil for distillation into kerosene to the Stanford Brothers' San Francisco refinery. The yield was half that of eastern petroleum and excessive carbon gave a smoky, dull flame, but throughout the state perhaps twelve thousand barrels of oil were gathered from seeps or from tunneling efforts in 1866. The next year, unfortunately, the price of eastern kerosene dropped to $.54 a gallon from an 1865 high of $1.70. Imports trebled from 55,321 to 132,252 cases of two five-gallon tins. Californians tapped the ready supply of kerosene from New York refiners. With the availability of lamps delivering from five to fourteen candlepower for less than a $1.00, replacement chimneys for $.25 each, and safe fuel at not much more than $10.00 per year, kerosene illumination was a bargain, although considerably more expensive than in the East. J.A. Bostwick and Company's Daylight and Gaslight kerosene plus Devoe's Brilliant quickly captured the California market. California oil and refining activities all but ceased, and other experiments to distill kerosene from shale proved a complete failure. Yet, even though California's product was at best marginal and readily marketable only when eastern kerosene was scarce, the promise of better lighting from local resources loomed large for a short time.[10]

It is hard to imagine today how great a leap forward illumination took with kerosene lamps, for they are primitive in comparison to electric lighting. Lamps had to be refilled every few days, wicks trimmed regularly and changed frequently, and soot cleaned daily from lamps and lamp chimneys. While remarkable compared to candles and even camphene, at ten feet or so the common lamp provided inadequate lighting for close work such as reading, writing, sewing, or knitting. Finally, the lighted lamp could be a great hazard, acting as an incendiary bomb if dropped. Many a barn, home, or business must have been destroyed as a result of accidental stumbles. But little kerosene lamps, recalled J.A. Graves, were "such an improvement over the candles that we felt very much puffed up when we got to burning them."[11]

Through the century, lamps improved. Early flat-wick lamps provided about sixty hours of fourteen-candlepower light on a gallon of kerosene. Argand burners or round-wick lamps had a central draft tube reaching up from the base of the lamp, which provided more even distribution of air around the wick. They gave about eighteen-candlepower but almost doubled fuel consumption. After the turn of the century, a mantle lamp was introduced. It had a cylindric mantle made of cotton soaked in nitrates and then collodian to give it a firm shape and was suspended just above a round wick. With careful attention to trimming and regulating the wick, the heated mantle provided an incandescent soft white light of fifty to sixty candlepower. Mantles lasted up to five months, cost only $.25, and fuel consumption improved considerably over Argand lamps. The popular Aladdin mantle lamp advertised in 1917 that the "New KEROSENE LIGHT Beats Electric or Gasoline."[12]

Toward 1900, kerosene also found another use where fuelwood and coal were in short supply, providing an alternative fuel for heating and cooking. Although wood and coal cooking ranges dominated the market, the availability of Iowa Standard Oil's "Red Crown Deodorized Stove Gasoline" led some people to purchase stoves that could burn kerosene or gasoline. An energetic Standard Oil advertising campaign pointed out the comforts of cooking with petroleum products: no splitting of wood or hauling wood and coal, elimination of soot and ashes in the kitchen, and a fire easily turned on and off. Wallace Hardison, a major figure in California's Union Oil Company, observed in 1900 that "when the means of burning the various products of petroleum shall have been perfected, and universally introduced, it will be the housekeeper's delight." By 1906, industry advertising and testaments gained credibility through a U.S. Department of Agriculture endorsement of recent oil stoves as economical, clean, and safe "if managed with care."[13]

Standard Oil claimed its kerosene and gasoline provided even heat, always kept the kitchen cool, eliminated long chimneys, and neither smoked nor smelled. The sales pitch worked particularly well in warm regions, and southern California housewives, joined by others in the San Joaquin and Sacramento valleys, converted to kerosene and stove gasoline. Standard Oil sold large quantities of its New Perfection Oil Cook-Stove and its Perfection Oil Heater, a space heater advertised as the "The Modern Fireside." A competing firm marketed Wickless Blueflame Kerosene stoves, and, by 1910, an Oakland company offered Daggett's Oil Burner for converting wood stoves into modern kerosene ranges.[14]

A shift from wood and coal to kerosene and stove gasoline was particularly swift in southern California. There, thousands of easterners in search of a healthy climate, who were attracted west by reduced railroad fares, outstripped local fuelwood supplies during the century's closing decades. Immigrant Julia Carpenter testified, in 1882, that at Los Angeles "coal is 14 dollars per ton and poor at that price." A little over a decade later, a southern California correspondent for the *Oil, Paint and Drug Reporter* observed: "Fuel is so scarce and high here that it is the practice of local manufacturers to utilize for their furnaces the dried manure of the streets, and even office sweepings, carted for the purpose to their doors."[15]

Until the 1920s, when gas and electricity became competitive through much of the interior areas of the state, kerosene sales for lighting, cooking, and heating remained steady. A number of protests to the State Railroad Commission during the 1890s over a proposed raise in the rail shipping rate for petroleum products revealed the dependency of some Californians on the product. Bay Area residents used perhaps one gallon a year, observed a committee of the San Francisco Paint, Oil, and Varnish Club, an organization representing the oil and varnish business. But this was not so on farms in the San Joaquin Valley. "Coal oil is one of [the farmer's] heaviest household expenses—not only does he use it for light, but as fuel for cooking purposes, and for heating his family sitting-room on cold, rainy days in the long, dreary months." The President of the Club, J.P. Jourden, and Sam Magner, the group's Secretary, amplified this view. "Many of the farmers and residents of interior towns (where there are no electric light or gasworks) use coal oil lamps for heating, thereby taking the place of coal as fuel. Naptha and gasoline are also the products of petroleum, and large quantities are used for heating and cooking purposes."[16]

Much of the kerosene coming to California from New York and Pennsylvania came by rail, and during the 1880s shipments fluctuated from 3.6 to 5.7 million gallons annually.[17] Additional heavy shipments arrived regularly in San Francisco by sea, but eventually the railroad had to carry it to inland markets. Although the freight rate from the East by sea was $.50 for a ten-gallon case and that by rail $.77, railroads remained in control. "As but a small portion of this product is consumed in this city," stated the Paint, Oil, and Varnish Club, "it naturally must find its way to the large fertile valleys of the interior." Railroads provided the principal transport, and to deliver kerosene third class from San Francisco to Bakersfield, they charged $.74 per case. This added substantially to the cost for kerosene shipped by sea, and the requested rate change to first class would increase an already high rate to $.81. Plainly the rural and small-town users suffered.[18]

The Southern Pacific, Southern California, and San Francisco and North Pacific railroads argued, on the other hand, that kerosene should be rated first class rather than third class because of the hazard in handling it. They emphasized, too, that "while it is an article of general consumption for illuminating purposes, it forms a very small share of the living expenses of an individual consumer or family." A raise in rates "would scarcely be noticed in a year's expenses of an individual family."[19] Their arguments failed. After a full hearing and investigation by the Railroad Commission, the railroad request was denied in May 1895. The result was a victory for both oil wholesalers and Californians residing beyond the bay area, particularly for rural dwellers who consumed up to 75 percent of the kerosene used in the state through 1920 and 90 percent by the 1930s.

Between 1881 and 1890, annual kerosene consumption in California climbed from just over five million to eight or nine million gallons. This steady growth in consumption accelerated through the turn of the century. By 1914, it had reached 21.8 million and within four more years climbed to 30.7 million gallons. Annual per capita consumption, which had grown from about 5.8 to 7.4 gallons between 1881 and 1890, and then to 8.9 gallons in 1920, declined by the mid-1930s to around 2.5 gallons. Comparatively, national per capita consumption moved from 3.9 to 7.4 gallons between 1884 and 1894 and declined thereafter. Although Californians consumed a bit more kerosene than the national average, their consumption never amounted to much more than one percent of total national kerosene production. As kerosene lighting decreased, it replaced more expensive gasoline and engine distillate as a fuel for stoves and tractors. These uses, plus a growing population, allowed Standard Oil to double its kerosene sales income between 1911 and 1919, but kerosene use declined after 1920, supplanted by electricity.[20]

Illuminating Gas

By the 1870s, several California cities had adopted what historians Harold F. Williamson and Arnold R. Daum have called the most spectacular American illumination development of the 1830s and 1840s, the manufacture of gas. Illuminating gas could be made through dry distillation of many organic substances—fats and oils, coal and petroleum, rosin and wood. In 1816, the Baltimore Gas Light Company produced a modest amount of coal gas, and, in 1823, the New York Gas Light Company produced gas from whale oil, until prices rose and it switched to rosin. By the 1840s, British scrubber technology for purifying coal of ammonia, cyanogen, carbon dioxide, and sulphur compounds brought wide acceptance for bituminous coal as the principal gas-making substance. Widely employed to light street lamps and thereby secure citizens of life and limb, American cities manufactured gas in fifty-six plants by mid-century, and a household meter invented in 1815 came into general use.[21]

As elsewhere Californians sought safety through street lighting, for protection against crime as well as pitfalls. For example, the City of Santa Barbara, in response to local crime problems, passed an ordinance, in 1855, requiring citizens in the town center to hang lanterns in front of their dwellings between dusk and 10:00 P.M. In San Francisco, oil lamps became the first public street lighting, property owners paying James B.M. Crooks to install them along Merchant Street in October 1850. Two years later,

Crooks had ninety lamps operating in the commercial district, and he received a contract from the city council to expand his oil-lamp system. But gas light was much more effective. Thus, despite the high cost of coal to make the gas, it soon found its way to the Pacific Coast.[22]

In August 1852, Peter and James Donahue, San Francisco's first iron founders, obtained a city franchise and incorporated the San Francisco Gas Company, California's first. Casting twenty-one iron retorts in their foundry and importing eastern cast iron pipe for street mains, they erected their gas plant by the bay and made their first gas from Australian coal in February 1854. Within two years, they sold gas at $15.00 per 1,000 cubic feet to 563 customers, principally businesses and wealthy residents, and they billed the city $46,000 to light some 154 street lamps for an eleven-month period. Though costly, the effort represented the progressive spirit of the West Coast's principal city. Even more important, it captured the imagination of people across the state, encouraging other manufactured gas installations.[23]

Gas plants were opened in nine more communities by 1870, from the new state capital, Sacramento, to the mining town of Grass Valley and the farm market communities of Marysville, Napa, and Vallejo. Los Angeles residents, demanding street lights as an antidote to crime, convinced the city to contract with the Los Angeles Gas Company in 1867. Within two more decades, plants in forty-four communities produced 1.4 billion cubic feet of gas which served 26,397 customers, 2 percent of the state's population. Two-thirds of the gas was made with imported coal—though Los Angeles tried using "brea" or bitumen for a time before switching to coal—and oil and wood accounted for the remainder. With 2 percent of the nation's population, Californians consumed 4.2 percent of the country's gas, and at least a third of that was manufactured and sold to San Francisco residents. There, customer rates declined steadily between 1857 and 1878, from $12.50 to $3.00 per 1,000 cubic feet, and a rate war between the San Francisco Gas Company and the newer City Gas Company drove rates as low as $1.60, in early 1873.[24]

The desire of urban residents to have reasonably priced gas lighting led to battles between consumers and gas companies and eventually persuaded the state legislature, in 1870, to open up competition by permitting cities to give any corporation the privilege of laying gas pipes. For Los Angelenos, whose city was bound by a twenty-year inviolate contract with the Los Angeles Gas Company, the law did no good. The city discontinued street lighting in 1872, unable to pay the asking gas price, and not until 1876 did the company reduce rates enough to entice a new contract. Elsewhere, the hope that competition would lower prices in fact did result in rate wars, but temporarily lower prices disappeared as established companies felt compelled to buy out newer competitors. This drove up prices and prompted calls for greater regulation. The legislature entered this field tentatively, passing a law in 1876 requiring that meters be tested for accuracy by a state inspector. Two years later, legislators took a much stronger step, enacting a regulatory measure affecting all cities with 100,000 or more residents. It established that gas had to provide light equal to sixteen candlepower and that the maximum rate could not exceed $3.00 per 1,000 cubic feet. The following year a new state constitution expanded this regulation to all cities, and public utility regulation in California was born.[25]

The cost of manufactured gas was tied largely to that of imported coal, so it is not surprising that other sources of gas were sought. In 1864, workers drilling for water on the site of the county court house in Stockton tapped natural gas. Although the town had a coal gas plant, established in 1859, natural gas became a principal fuel for many of its residents. The Stockton Natural Gas Company produced eighty thousand cubic feet per day in 1888, and small wells served private residents as well as a variety of other consumers—the court house, Weber Swimming Baths, Saint Mary's Church and Convent, Crown Flour Mill, California Paper Mill, and Solomon Ranch. In all, some thirty-three wells were drilled in the Stockton area, but natural gas was not tapped commercially elsewhere in the state during the nineteenth century.[26]

Californians also turned to noncoal substances for making gas. Pitch pine was used in Nevada City and Marysville, though generally mixed with some coal, and the Gilroy Gas Works experimented with bitumen from nearby deposits at Sargent Station. In Los Angeles, where high prices caused sustained public protest, potential competitors to the Los Angeles Gas Company offered a variety of schemes to provide gas. Among the most fanciful was an 1876 proposal to use garbage, selling enough by-product ammonia and coal tar in the process to provide gas to consumers for free. Oil, however, provided the best alternative to imported coal, and in 1874 southern Californian George J. Luce patented a process to make gas from crude oil. Seven communities, from Ventura north to Red Bluff, installed Luce gas makers, but dirty gas that clogged mains, meters, and burners led to the system's demise in 1877. That same year, a promoter from Pennsylvania, Colonel Zaccur P. Boyer, came to California with the West Coast rights to a new gas system.[27]

The presence of hydrogen in water had prompted chemists to study the decomposition of water, since the late-eighteenth century. A theory of water gas developed, resting in the dissociation of steam by contact with heated carbon, which produced a mixed gas of hydrogen and carbon monoxide. Thaddeus S.C. Lowe, an inventor who had gained notoriety as the chief of the Union Army's balloon corps during the Civil War, made commercial application of the theory possible. In 1872, he invented an internally fired generator used in conjunction with a superheater and built the first water gas plant in Phoenixville, Pennsylvania, in 1874. The Lowe system jetted steam through incandescent anthracite coal resting on the floor of a vertical retort, whence it passed through a superheater loosely filled with white-hot bricks. Since the resulting "blue gas" contained only 322 British thermal units per cubic foot, ineffective for commercial use, Lowe raised the BTU rating to over 500 by spraying petroleum into the top of the generator above the coal bed. The oil mingled and passed with the blue gas into the superheater, where it vaporized and both were fixed into permanent gas.[28]

Colonel Boyer, attracted by two characteristics, moved to San José, just below the southern end of San Francisco Bay. The existing coal-gas company in the town of some twelve thousand people paid a great deal for coal and produced poor gas, dissatisfying many consumers. At the same time, oil prospects in nearby Moody Gulch seemed quite promising. This was ideal turf for a gas firm using the Lowe system, and Boyer successfully lined up local backers for the venture. The Garden City Gas Company started construction, in August 1877, and opened almost a year later. Fierce competition from the established San José Gas Company led immediately to a rate war and ultimate absorption of the new company by the old. But this escape from expensive coal gas and the high quality of Lowe's water gas system caught the attention of other California entrepreneurs.[29]

The Oakland Gas Light Company switched from coal gas to the Lowe system in 1880, and two years later the Central Gas Light Company in San Francisco opened a plant, specifically to serve the new, luxurious Palace Hotel. Lowe systems spread even more quickly after the inventor moved to Pasadena in 1887. He was followed two years later by his son, Leon P. Lowe, who started a water gas company in Los Angeles and went on to convert several plants in northern California to his father's system. More significantly, Leon patented his own system to manufacture gas from oil, in 1889, which began a total transformation in northern California's manufactured gas industry after the turn of the century.[30]

As gas became a significant lighting source in urban communities, efforts were made to bring this new illuminant to rural areas and businesses beyond the service of central stations. The first gas-producing system that could be applied to individual buildings, a gasoline system, was born out of technical developments by A.C. Rand, a New York oil marketer and refiner. In 1867, he introduced a "pneumatic gas machine," a portable gasoline carbureting unit. Two years later, he acquired two patents from Dr. Leonard K. Gale for vaporizing hydrocarbons and fixing them into stable gas. In the East, the Gale-Rand process was employed somewhat successfully to provide gas light for cities without using coal, and through the 1880s commercial systems spread.[31]

The idea was introduced to California, in 1869, by the Pacific Pneumatic Gas Company of San Francisco, which held the West Coast rights to early Rand patents. The company organized to install carburetors using imported eastern gasoline for use in commercial buildings, country homes, and communities that were too small to support a coal-gas works. A second firm, the Maxim Gas Company, also used gasoline machines, contracting with small California towns to provide lighting. The Santa Barbara Gas Company, for example, became a Maxim firm in 1871. Selling gas at $6.00 per 1,000 cubic feet and lighting six lamps along the main street, it was unprofitable and short-lived, as were other Maxim adventures.[32]

Both the Pacific Pneumatic and Maxim companies disappeared during the 1870s, the former briefly rejuvenated as San Francisco's Metropolitan

Gas Works in 1871. Hoping to capture the urban market from coal gas firms with the Gale-Rand system, it constructed a large plant. But it was too tightly financed and had troubles with its gas-making process. When an explosion in the retort room injured seven workers and blew off the roof, in April 1873, the catastrophe ended the company as well as other community gasoline lighting system efforts in California. Nevertheless, systems for use in buildings continued to develop.[33]

Gasoline lighting, because of the fuel's volatility, was much more dangerous than kerosene or coal gas. Lamps, like those used in urban gas lighting systems, brought fuel through a pinhole to feed the burner. Gasoline flowed to the burner by gravity pressure through a tiny feeder line from a fuel container located a few inches above the burner, or it was forced by several pounds of air pressure gained from a hand pump attached to the storage container. A truly convenient house lighting system could be attained by using a main pressurized storage tank that fed tiny hollow wires conveying gasoline to several lamps in various locations throughout the house. Sixteen pounds of pressure produced around 145 candlepower, and thirty-four pounds produced 300 candlepower, a remarkable improvement over kerosene. However, pinhole feeders had to be kept very clean to prevent clogging, and some people could not shake the fear that gasoline lines passing through the house were unsafe.[34]

Gasoline light was employed by a few rural Californians of means. In 1891, San Francisco's Badlam Brothers introduced a home gas producer plant designed to provide both gasoline lighting and heating. A compressor, driven by clockwork-like weights that needed but occasional winding, forced air through gasoline in a generator tank buried underground outside the house. The resulting vapors flowed back to a small basement tank, where more fresh air was injected before the gas vapor moved through hollow wires to lamps and heaters. In 1910, an Oakland company sold "the Stewart Gas Lamp," a safe and steady burner with an overhead gas storage container, which the company advertised as a must for every farmhouse. Similarly, in Stockton, "Gould the Light Man" offered "The Light of the Future," gasoline lamps with separate individual storage containers or a hollow wire system for the house.[35]

Another illuminant, especially popular among midwestern farmers,

Figure 4.1. Acetylene gas provided one of many resources for domestic lighting during the nineteenth century. (Courtesy of the San Jose Historical Museum Archives, San Jose, California.)

also found its way to some rural California homes. During the early 1890s, the J.B. Colt Company in New York introduced acetylene gas lighting, produced from combining calcium carbide with water. Some acetylene advocates hoped the gas could be used in enriching manufactured coal or oil gas to stave off competition from newly introduced electric light systems. Experiments revealed that it was, in fact, more expensive than electric lighting, but it did have a potential market as an illuminant for railways, carriage headlamps, country homes, and even small communities where other means were not feasible.[36]

Acetylene gas, however, experienced a period of controversy. In 1896, the New York Board of fire underwriters condemned it. Apparently explosive when contacted by air, a problem exacerbated by burning at a low 480 degree centigrade versus 600 degrees for other gases, the Board refused to insure buildings in which it was used. In California, *The Pacific Electrician* warned that, in spite of the "wild talk afloat as to the wonderful virtues of acetylene gas," it was extremely dangerous under pressure, even at atmospheric pressure. Severe explosions had occurred in New Haven, Connecticut, and Paris, France; Germany prohibited its use. "The ordinary man," concluded the *Electrician,* "had better get his light in some other way for the present."[37]

But the illuminating properties of acetylene were too great for fear to allow its abandonment. It produced almost 240 candlepower, and it did not vitiate the air as did coal gas. In 1898, San Francisco's *Journal of Electricity, Power and Gas* described acetylene at length, noting that it gave fifteen times as much light as coal gas and that its explosive character could be reduced effectively by using it with carbonic acid. Later that year, the editors ran a four-part series abstracting a set of London Society of Arts lectures delivered on the subject of acetylene generation, storage, danger, and use. The series concluded by suggesting it would be used best where coal gas was unavailable and that it was an up-and-coming illuminant.[38]

By the early 1900s, acetylene found fairly wide use as an illuminant in rural America, and perhaps a few thousand Californians adopted it. The common generator produced about one-half cubic foot of gas per hour, which greatly diminished the danger of explosion. Safety was increased by locating generators outside buildings, as recommended by the National Board of Fire Underwriters. In 1912, the Iowa Agricultural Experiment Station suggested in 1912 that acetylene was safer and more sanitary than gasoline lighting, its white light "easy on the eyes and . . . very desirable for domestic use." The earliest generators, in which water dripped on to the carbide, gave way to newer ones that fed carbide into water, provided fuller chemical action, and washed the gas as it passed through the water. Systems cost between $150 and $300, plus $50 to $70 annually for carbide, and offered rural dwellers a good illuminant.[39]

Summary

Californians, then, illuminated their homes, businesses, and streets in several ways during the nineteenth century. In towns and cities, people readily abandoned candle light for camphene lamps, then kerosene, and finally gas lighting distributed from central manufacturing plants. Kerosene also provided an alternative to fuelwood and coal for heating and cooking, particularly in fuel-starved, semiarid southern California.

Although manufactured gas never reached most rural areas, change came with kerosene lamps, and some rural residents installed independent gasoline lighting systems or adopted acetylene gas systems. In the end, Californians' choice of illuminants involved a variety of factors: product and resource availability, comparative price, geographic circumstances, inventiveness, and individual perceptions of efficiency and convenience. Furthermore, although a sense of progress imbued Californians as they adopted newer and newer methods of illumination, they remained shackled by prices generally dictated by eastern business interests.

CHAPTER V Wind, Tide, and Sun

AMONG THE VARIOUS energy resources that contributed to California's energy budget, nature provided the potentially free and ubiquitous power of the wind, tides, and sun. A temperate climate with abundant plant, animal, and maritime life had attracted to the region the largest density of Native Americans found anywhere in North America. When the Spanish arrived, they introduced passive solar buildings, constructing their missions and homes with thick adobe walls that retained the sun's heat at night and remained cool through long, hot summer days. They often oriented their structures so that they faced south with an overhang extending from the front that permitted the winter sun, low in the sky, to reach into the building while the high summer sun was deflected. Although a few Americans also built with adobe, the rapid population growth that accompanied the gold rush brought with it construction techniques from the East. Turning away from passive solar energy, Americans utilized California's wind, tidal, and solar energy in an active way.

Wind power provided a major energy asset for Californians. The age-old resource was employed in three tasks: raising ground water, powering small machines, and propelling ships. The prevailing Pacific breezes reached well inland through most of the state, and windmills soon provided power for small shops and competed strongly with hand pumps and

Photo 5.1. Two Dutch windmills were erected in Golden Gate Park during the 1890s to raise water for irrigation. (Postcard courtesy of the California History Center Foundation, De Anza College, Cupertino, California.)

horse whims in raising water for domestic use, stock, and irrigation. Wind machines appeared in ever increasing numbers around San Francisco Bay, Sacramento, and Stockton in 1860. *Scientific American* reported that

> Windmills are becoming great institutions in San Francisco. They are being extensively employed for pumping of water, propelling the shaft of the machine shop, turning the burr-stones of the flouring mill, &c. The weather here is particularly adapted to the wind mill business, a large supply of wind being constantly in the market and obtainable without money and without price.[1]

The first "effective windmill for supplying water" built in California was probably that of W.I. Tustin in the town of Benecia on the Sacramento River Delta. By 1859, Tustin sold his mills for prices ranging from $75 to $250, and several other windmill builders entered the business. Prizes for windmill designs awarded at the 1860 San Joaquin Valley Fair spurred a lively competition among windmill builders. Tustin attacked the efficiency of the special premium-winning Philips's mill, built by Hyde and Storer of Benecia, challenging them to a trial of their mills on San Francisco's Telegraph Hill. He argued that only two of Philips's mills actually were operating, versus many of his, and that they did not run smoothly in gusts. For their part, Hyde and Storer claimed that many of the mills to which they had Pacific Coast patent rights were in use in and around Benecia. Good for driving grist mills, tanneries, turning shops, and circular saws, plus churning and pumping, they were advertised for $100 for one with a twelve-foot diameter fan and $350 for an eighteen-foot fan.[2]

The widespread success of windmills stemmed from Daniel Halladay's 1854 invention, in Connecticut, of a mill which governed its own speed and turned automatically into the wind. Soon windmill makers throughout the United States sought to perfect self-governing windmills, and San Francisco's James B. Johnson won the 1857 Mechanics Institute Fair highest premium for his "self-regulating" mill. It may or may not have worked. As the *California Culturist* editor observed later, "a great variety of wind wheels, with their appliances, all of them termed self-regulating, have, from time to time, resulted from mechanical effort in this direction; but many of them upon trial, have not come up to the standard of excellence desired." He suggested that Jacob Dickerson's mill, which took First Premium at Sacramento's 1859 State Agricultural Fair, truly met "every point of excellence, except that of running with no wind at all." In an accompanying advertisement, Dickerson boasted that over one hundred of his Sacramento-manufactured mills were in use in mid-1860.[3]

Regardless of who best solved the dilemma of self-governing windmills, the careful attention they received at agricultural fairs signaled the importance of them to Californians. Besides prizes granted to Hyde, Johnson, and Dickerson, the mills of Balthis and Cozzens won prizes in 1860 and 1861 at Stockton's San Joaquin Agricultural Society Fair. At the 1861 fair, John W. Hart also displayed his iron windmill, and the next year Stockton's J.E. Roberts entered a mill for competition. An independent category, with its own judging committee, was formed for windmills. "The

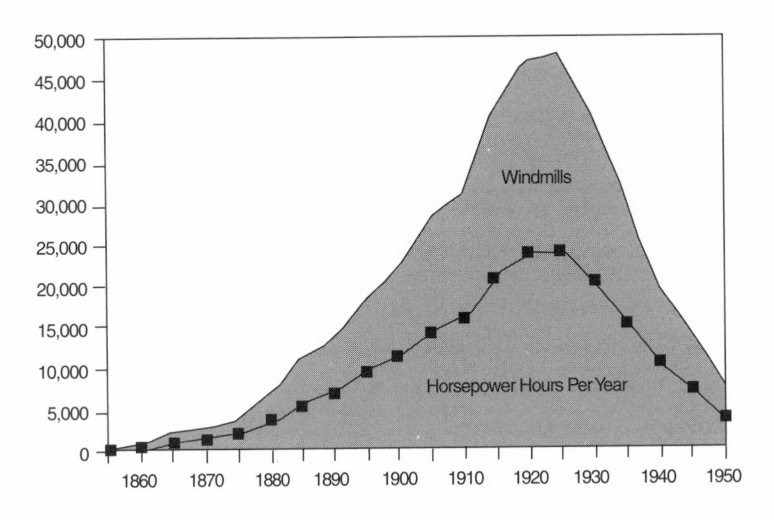

Figure 5.1. Estimated California Windmill Power, 1855–1950. Source: *Supra*, Table C-4.

immense importance of the wind power of California to the agriculturist and mechanic," observed the *Culturist,* "is just now beginning to be properly appreciated."[4]

The next two decades saw windmills erected throughout California. William Brewer wrote, in 1861, that hundreds of them could be seen along the west side of the Central Valley down to today's Pacheco Pass region, and another traveler wrote that one appeared every four or five miles between San Francisco and Sacramento. The number of windmills increased from perhaps eighteen hundred in 1860 to twenty-eight hundred in 1870, and better than doubled in number during the next decade (figure 5.1). By 1884, the state mineralogist reported that "these machines, the most of which are made here, are of diverse patterns—some simple and cheap, costing not over $50, others large and complicated, costing as much as a thousand or fifteen hundred dollars—average cost, about $200."[5]

The average fourteen-foot fan windmill used for irrigation took about ten days to apply four acre-inches of water to 5.41 acres, working a ten-inch piston, thirty strokes a minute, eight hours a day. Larger fans could handle larger pumps and greater flow. Most of the deep wells in the Salinas valley were pumped by windmills in 1890, and Sonoma County mills lifted water from wells twenty to sixty feet deep. In San Mateo, Alameda, Santa

Clara, and Los Angeles counties, artesian wells brought the water level within a couple of feet from the surface. Thus, some windmills had but a slight lift, unless raising water to an elevated storage tank. Other locales, such as Sacramento, could depend on the wind from April to August, but found it light and capricious during the latter part of the summer.[6]

In the early 1880s, over a dozen different firms manufactured wooden windmills in northern California, four in Stockton, eight in San Francisco, and two in Oakland. While some firms produced mills developed in the East, California-designed wooden mills were common. Oblique vanes, radiating from a redwood shaft, formed the fans. Redwood was used because it was light and deteriorated less in the rain and sun than other woods. Tough Oregon pine was preferred for braces. Iron castings and brass bearings came from San Francisco foundries, but mill builders in Stockton preferred to use hard maple bearings. Metal bearings had to be oiled weekly, whereas wooden ones, once saturated with oil, required oiling only monthly. Wooden mills imported from the East, as with so many other products, sometimes sold more cheaply than locally made ones, but purchasers discovered California's dry climate revealed that they often were made of partially seasoned wood. In general, mills with fans from six to sixty feet in diameter could be purchased, a good wooden one lasting thirty years if properly maintained. Some farmers made their own from purchased plans.[7]

Stockton probably used more windmills per capita than any other community in the state. Although data is not available to confirm such a claim, the harnessing of steady winds permitted irrigated garden crops to thrive in long, dry Stockton summers. John Hittell observed, in 1879, that the town's nickname, "The City of Windmills," appeared "appropriate to the traveler who approaches the town on a windy day and, at a distance, sees little save a multitude of great arms revolving furiously above and among the trees and house-tops." Three years later, he observed that within a radius of a few hundred yards of Stockton's Yosemite Hotel, "one may count more windmills than are to be seen, in the same area, elsewhere in the State." Mills pumped water for homes, commercial establishments, and surrounding farms and ranches. John Rich, for example, employed eight windmills to irrigate nine hundred fruit trees, in 1861, and, later in the century, J.B. Haggin's ranch used no less than sixty windmills.[8]

Photo 5.2. Wooden windmills on a hundred-acre strawberry field near Florin. From *Sunset Magazine* (September 1906): 284.

Figure 5.2. "E.J. Marsters, Inventor & Manufacturer of Feeders, Elevators, and Wind Mills," was one of several Stockton-based wooden windmill manufacturers. From *History of San Joaquin County, California* (Oakland, CA: Thompson and West, 1879), plate XXV.

To supply the local demand, as well as that of surrounding communities, Stockton windmill manufacturers maintained heavy production. Abbott and Stowell, producers of the Relief Windmill, occupied a 10,000-square-foot building at Market and California Streets and employed twenty laborers. The factory had grown from a small blacksmith shop opened, in 1871, by Americus Miller Abbott, who had immigrated ten years earlier to California. In the mid-1870s, he began devoting his full attention to windmill construction and soon formed a partnership with John Stowell. In 1889, they incorporated under the name of the Relief Windmill Company. His principal competitor through the 1880s and 1890s, Richard F. Wilson, produced wooden mills under the brands Improved Davis and San Joaquin, plus the iron and steel mill, Hercules. Wilson's firm, originally started in 1858 by John Davis, manufactured about one hundred mills each year at a factory near the steamboat landing on Commerce Street. In all, some eight windmill, pump, and tank dealers were found in the Stockton area in the mid-1880s.[9]

Yet, by 1890, the U.S. Census reported only eight California windmill companies. Two operated in Oakland and one in Los Angeles, while others advertised from San José, San Francisco, and Stockton. Eastern and midwestern manufacturers had worked actively and successfully to gain a share of the California market. As early as 1882, the midwestern brands U.S. Star Windmill and Enterprise had agents in the state, and, in 1889, Indiana's National Wind Engine and Illinois's Challenge Windmill were advertised in the *Pacific Rural Press.* In San José, E.B. Saunders offered the imported Eureka Improved Windmill, which could "be controlled by a boy or a woman" and needed "oiling only occasionally." Even in Stockton, William B. Prahser sold the imported all-steel Peerless "IXL" Windmill. The Aeromotor Company of Chicago, destined virtually to dominate the market during the 1920s, boasted in California advertisements, during 1892, that they would sell sixty thousand of their galvanized steel windmills and towers across the nation versus the 20,049 sold the year before.[10]

The eastern competition spurred some California mechanics to move away from wooden machines. R.B. Sinclair introduced his Alameda Steel Windmill in 1891, seeking agents to sell the product and advertising it as the only steel mill made in the state. The next year, the California State

Agricultural Society awarded a premium to the Crane Company of San Francisco for the best windmill at the state fair. Offered with eight-and-one-half and twelve-foot fans, the Crane was painted and galvanized, a common technique to assure skeptical buyers the metal mill would not rust. But competition continued to be fierce, as mail-order houses, such as Sears Roebuck and Montgomery Ward, joined Aeromotor and other out-of-state firms in offering cheaply priced mills. In 1905, only four California firms employing nineteen workers remained in business, two in San José and two elsewhere in the state. By 1914, only one firm persisted.[11]

Besides providing pumping water for farmers, households, and commercial establishments, windmills also functioned at railroad water stops and served small town water systems. In 1904, the Salinas Water, Light and Power Company, for example, used two fourteen-foot Aeromotors on opposite corners of the town to pump water into fifty-thousand-gallon storage tanks. As an auxiliary to the main water supply system, the mills provided 7 percent of the town's water needs, drawing on steady ocean breezes from about 10:00 A.M. to 9:00 P.M. during the spring, summer, and fall. In southern California during the 1880s, the Inyo Development Company, a manufacturer of carbonate of soda from the waters of Owens Lake, used a windmill to feed water to its evaporating vats. It could lift seventy-five thousand gallons of water ten to twelve feet in an hour.[12]

In addition to pumping water, windmills could provide mechanical power. In 1889, the California journal, *Industry,* suggested more development needed to be done in this field: "We need here some Dutch engineers and millwrights to build wind powers of large size for raising water and other purposes. In some of the large valleys, especially in Southern California, the natural circumstances are especially favorable for wind power, and the time will no doubt come when such power will be made available for many purposes now served by steam engines." In fact, two Dutch mills were installed in San Francisco's Golden Gate Park to raise irrigation water, and midwestern products, such as the Challenge Double-Header Geared Windmill, offered Californians mechanical power mills in a variety of sizes.[13]

In 1890, for example, Duncan McKinnon, a Canadian immigrant to California, purchased eleven thousand acres of the El Sausal Rancho in the

Salinas Valley and installed a Challenge Double-Header windmill atop a three-story building. Using steel worm gears, a main vertical drive shaft, and an overhead belting system, the mill ran a myriad of machines on the three floors of the building, and in an adjacent single-story workshop. On the ground floor, McKinnon located two heavy planers, on the second floor a corn grinder and barley cleaner, and on the top floor a walking beam to a ground-level water pump. Through all three floors he built a grain-processing system not unlike that designed for waterpowered grist mills. An elevator raised corn and barley to the third floor and delivered the grain to the proper feeder bins by an endless screw. One bin fed corn to the second-floor grinder and the other fed barley to the crusher. The processed corn and barley then dropped through a chute to a first-floor bagging bin, while chaff dropped to a separate first-floor compartment. In the adjacent workshop, an overhead transmission shaft brought wind power to grindstones, two drill presses, a wood lathe, a band saw, a circular saw, and a metal lathe.[14]

Shortly after the turn of the century, windmills began giving way to steam, gasoline, and electric pumps. A 1902 survey in portions of Riverside and San Bernardino counties revealed the inroads made on wind power by gasoline and electricity: 227 farms used windmills, fourteen exclusively for irrigation, 113 for irrigation or stock watering plus domestic use, and 100 for domestic use. But 113 gasoline engines, twenty-four electric motors, and sixteen steam engines also pumped water. About 49 percent of the mechanical water systems relied on wind power, gasoline and electricity accounted for 29 percent of the total, steam provided for less than 3 percent, and human and animal muscle powered the rest. Throughout the state, windmills faced competition from gasoline engines at the turn of the century, because motors and fuel were relatively cheap and offered the convenience of pumping at any time. Similarly, wherever electricity became available from transmission lines, the windmill was threatened. Even McKinnon removed the blades from his Challenge Double-Header in order to power his equipment with a Westinghouse ten-horsepower, Type "C," induction motor.[15]

Nevertheless, the windmill continued to serve rural Californians. In 1917, the *Pacific Rural Press* described a modern indoor plumbing system in a San Luis Obispo country home where water was pumped from a seventy-seven-foot-deep well by a windmill, a one-and-a-half horsepower gasoline engine serving only as an auxiliary power source. Elsewhere, old mills still operated and midwestern firms continued sales. While use of windmills in towns waned early, serious farm use declined only toward the end of the 1920s. Indeed, a revival in windmill sales, particularly of Aeromotors, came during the post-World War I recession years and persisted into the Great Depression. Several hundreds of these mills continued to operate in California during the 1980s.[16]

Wind Power in Transportation

However many windmills may have existed in the Golden State, wind power played its greatest role in transportation. In this capacity, wind contributed more energy to nineteenth-century Californians than any other resource save human muscle and wood fuel, over 20 percent of the state's total energy consumption. Sailing ships of 43.2 million total tonnage arrived in San Francisco between 1850 and 1900. Assuming they spent half their sea time in trade with California, their use of wind power equaled the energy produced in burning fifty-four million tons of coal—50 percent more than the actual amount of coal burned by Californians and 25 percent more than equivalent energy generated by work animals. Additionally, hundreds of other sailing vessels plying coastal waters, the bay, and other inland waterways drew on wind energy, using an equivalent of another 36.6 million tons of coal (figure 5.3).[17]

American and other merchant sailing ships came to California waters in steadily increasing numbers after 1800, where they hunted whales, otters, and fur seals. After Mexican independence from Spain, in 1821, the new Mexican government legalized foreign trade with California, and even more merchant ships arrived to trade for cattle hides and tallow. Discovery of gold in the Sierra Nevada thirty years later inundated the Pacific Coast with ships. Within a year after U.S. President James Polk verified California's treasure trove in his December 1848 message to Congress, 775 sailing vessels had departed Atlantic ports on the six-to-eight-month voyage to Eldorado. In 1849, these and hundreds of foreign ships deposited 91,405 passengers in San Francisco. The next year, some five hundred ships were

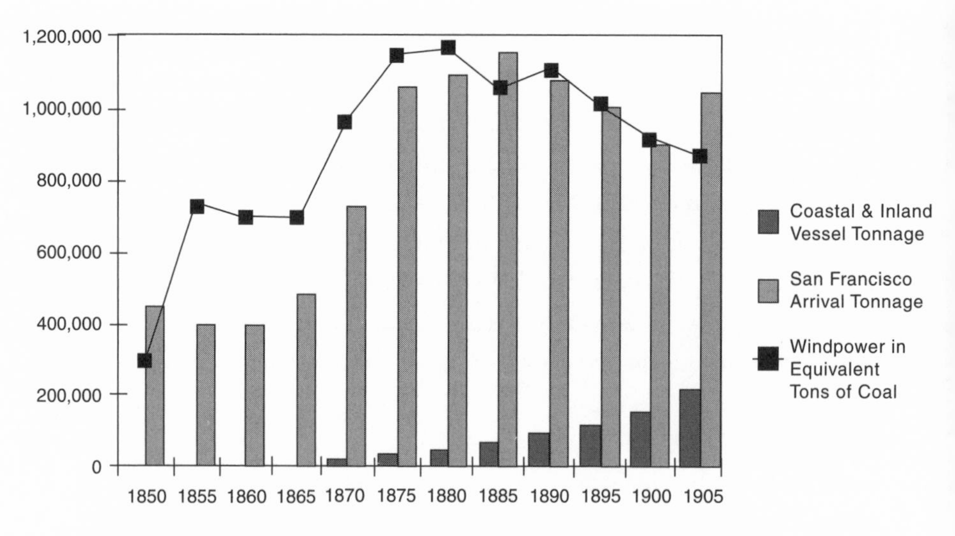

Figure 5.3. Sailing Ships in California and Energy Equivalent in Coal, 1850–1905. Source: *Supra,* Table C-4.

abandoned in the harbor, one hundred of them permanently becoming hotels, store ships, or other establishments.[18]

The gold rush began an epochal era in American maritime history, spurring construction of fast clipper ships destined for the California trade. During the 1850s, American shipbuilders launched 445 clippers, producing 120 vessels in 1853 alone. Almost all the clippers passed westward once around Cape Horn, eighty-one of them making 359 of a total nine hundred passages in the 1850s. They lowered the time from New York to San Francisco to between 100 and 140 days, departing Atlantic ports in the peak year of 1855 at a rate of almost ten ships per month. And eighty-two of the clippers made the trip five or more times in their careers for a total of 671 passages through 1885.[19]

Although fast clippers still govern our image of ships on the Cape Horn passage, they accounted for only a little over one quarter of the 4.1 million sailing ship tonnage that reached San Francisco during the 1850s. The bulk of arriving tonnage was attributable to schooners, barks, brigs, and traditional ships. Such vessels maintained their dominant position in California's maritime trade until the late 1890s. A bumper wheat crop, in 1882 alone, brought 559 vessels to California to pick up 1.1 million tons of wheat and barley and 919,898 barrels of flour. Between 1881 and 1885, only eleven of 1,512 ships participating in the trade were steam-powered. Similarly, over 250 sailing craft dominated the heavy coastal lumber trade between San Francisco and the redwood country of Humboldt and Mendocino counties. Only twenty steamers were involved in the trade before 1888.[20]

Sailing vessels initially also dominated local trade on San Francisco Bay and inland waterways. The bay, inlets, sloughs, and rivers provided essential transportation corridors before railroads spanned the region in the 1870s. Schooners, barks, and brigs first carried passengers and goods between Sacramento, Stockton, and San Francisco. As they often took a week or ten days to make the trip, smaller and handier vessels soon replaced the larger ships. As a part of the river and bay commerce, European-American sailors and boat builders in San Francisco developed a square-toed packet or scow schooner, a modification of scows which they had known on the Great Lakes and elsewhere. Cheap, shallow-drafted, maneuverable craft that could carry a heavy load, over two hundred sailing scows were built between 1850 and 1900. They transported hay, oyster shells, lumber, bricks, wheat, and a variety of other goods and did well in competition with steamers. "The rivers are dotted with sails," wrote Horace Davis in the *Overland Monthly*, in 1868, "and the winding channels among the tules are clouded with the smoke of numberless steamers . . . loaded with the rich produce of the land."[21]

The difficulty and time spent kedging or using horses to pull sailing craft upriver, through narrows where winds did not cooperate, accentuated the value of steamers. They soon supplemented sailing vessels. Sixteen engaged in the river trade in 1850 and accounted for some 15 percent of the total tonnage. Twenty-five plied the waters by 1853, and seventy-one were active by 1888. Steadily, river traffic gave way to steamers and, after 1890, to gasoline-powered craft. By 1915, only seventy-three sailing craft worked the Sacramento and San Joaquin rivers, just 18 percent of the total number of vessels. Similarly, ocean-going sailing ships gave way to steamships, dropping from 70 percent of the gross tonnage docking at San Francisco in 1870 to 50 percent in the early 1880s and 10 percent by 1911.[22]

On 4 July 1867, sailors staged a regatta for local and coastal sailing workboats on San Francisco Bay. In succeeding years, thousands of bay residents lined the docks and climbed San Francisco's hills to watch the annual race sponsored by the Master Mariners' Benevolent Association. Between thirty and forty freshly painted and decorated schooners, barks, scows, and sloops challenged each other over a twenty-mile course. Combining keen competition with the celebration of Independence Day, they sought the treasured "Champion" banners given the winner of each class, and as many as one hundred revelers might ride the decks of a large competing schooner to cheer on the captain and crew.[23]

The regatta, symbolic of the importance of sailing craft to Californians, remained an important event through the 1870s and was staged erratically thereafter. For a time, it regularized previously informal contests between local sailing craft, and it engendered as much excitement among bay residents as had earlier competition between clippers to set the fastest time between New York and the Pacific Coast. The meeting of sailing craft in the Master Mariners Regatta was not unlike the competition at mechanics and agricultural fairs between makers of windmills. In a sense, each celebrated the importance of wind power for Californians.

Wave Motor Development

Already drawing heavily on Pacific breezes, some Californians also looked to the energy potential of the ocean itself. For several hundred years, European millers had trapped high tidewater in reservoirs, releasing it at low tide through waterwheels. Americans built a few tide mills on the East Coast, and, in 1851, the Weatherly Company, a logging and saw mill operation, built a tide mill on the Mendocino coast. The firm was probably the only one in California to try tidal power, and Californians did not seek again to harness the ocean's power until 1889. Then a San Francisco entrepreneur sought to use ocean wave motion as power for pumping water. Near the Cliff House, he threw a wooden truss frame across a notch in the cliff. From it he swung a large pendulum, a vane at the bottom entering the ocean's surface. The pendulum was pushed back and forth by ocean waves, and linkages transferred the motion to a pump that lifted water to the top of the cliff. Skeptical observers watched the wave motor fail, the crush of Pacific waves battering the machine apart.[24]

Other experiments with wave motors followed, and half of thirty-two

American patents issued for the devices, in 1892, went to Californians. In 1894, successful experiments off a pier at Long Beach led to the formation of a company in Capitola two years later. The firm obtained a twenty-year lease on the 750-foot long Capitola wharf and added a 200-foot extension. They installed a Gerlach wave motor that consisted of two sets of three nine-by-twelve-foot vanes slung from an overhead shaft. The vanes swung eight times per minute in a radius of thirty-two feet, the motion carried by chain and sprocket gearing to a flywheel twenty-four-feet in diameter. Built by San Francisco's Atlas Iron Works, the entire apparatus weighed thirty-nine tons and produced about 135 horsepower. Redondo Beach entrepreneurs installed a similar motor a year later, and *The Pacific Electrician*'s editor observed: "There seems to be always something more needed to make these wave motors completely successful."[25]

Californians continued to hold interest in wave motors, particularly as they might be applied to electric power generation, but it was felt that such generated energy had to be stored to be commercially useful. In a paper presented at the 1892 meeting of the American Society of Mechanical Engineers, Albert Stahl, a U.S. Navy engineer based in San Francisco, described a wave machine and suggested that it would take as much as ten days to reservoir enough energy for practical use. While electric batteries would be the best storage method, their high cost eliminated them. Therefore, storing the wave-generated energy by means of water in elevated reservoirs seemed to be the reasonable solution. San Francisco engineer George W. Dickie observed that any apparatus producing one thousand horsepower must cost less than $400,000, the present cost of the equivalent amount of coal in the city. If that could be done, he supported the concept.[26]

But E.T. Molera, a well-known West Coast electrical engineer, was skeptical: "I think that the apparatus which Mr. Stahl has described is exceedingly ingenious; . . . but, as Mr. Dickie says, when you have got it, what about it? The wave motion is exceedingly unmanageable, and I have found it exceedingly unprofitable." To store water pumped by a wave motor, one needed high places. San Francisco, Molera conceded, was better adapted than other cities for this, but even there only two spots three hundred feet high existed within a mile of the shore. "Therefore you have to provide from the motor a pipe one mile long to the reservoir. When you have the

Photo 5.3. J.E. Armstrong's wave motor, built on a bluff near Santa Cruz in 1900, raised water from Monterey Bay to sprinkle the town's dusty streets. (Postcard courtesy of the California History Center Foundation, De Anza College, Cupertino, California.)

reservoir there you have to convert the power of the water into another power down at the bottom, 300 feet below. . . . Then you have again to reconvert this power into . . . electrical current, then again conveyed through a conductor six or seven miles long to the centre of the city, and here again conveyed into small electric motors for the purpose of utilizing the power." The whole thing appeared terribly impractical.[27]

Despite skepticism, wave motor development persisted for a time. In 1900, J.E. Armstrong of Santa Cruz constructed a motor located on a bluff outside the town. There his motor pumped salt water 120 feet above to an eight-thousand-gallon water tank, whence it was piped to several points along Pescadero Road where it was used for road sprinkling. Within four years, Armstrong abandoned the machine, corrosion and rust severely damaging the iron structure supporting the motor and tank. In 1907, a Los Angeles Wave Power and Electric Company was formed and requested permission from local officials to build a seven-hundred-foot pier at Redondo to install a wave motor. The scheme collapsed from lack of capital. Finally, a 1911 patent for a wave motor to Telesphore J. Beaudette of Los Angeles seems to have been the last breath among wave motor enthusiasts.[28]

Solar Energy

In addition to harnessing wind power and experimenting with tidal energy, Americans in California turned to solar energy, expanding beyond its passive use via architectural design to harness it in other ways. In 1868, booster Titus Fey Cronise praised solar energy's potential for the sericulture industry, observing that one would not need kilns to destroy the silk worm in its cocoon. Exposing the worms to the sun for two or three days "in troughs with a glass covering" would do the job easily. French immigrant botanist, Louis Prevost, pioneered sericulture near San José, touting California's unique climate, and state legislative bounties spurred other growers as far south as Riverside to plant some 1.8 million mulberry trees. But, despite cloudless skies, episodes of frost and lack of grower organization caused the industry to fail.[29]

The deciduous fruit industry did not fail. Sun-drying prunes, apricots, pears, and other fruits permitted California horticulturists to overcome the perishability and fresh market limitations of their products. They developed national and international markets. Some ten thousand tons of fruit were dried each year during the 1890s, and annual tonnage gradually increased in the early 1900s to about forty thousand. Acres of drying trays and crews of workers became the late summer trademark of Santa Clara County and other horticultural areas. By 1919, California produced 94 percent of the nation's dried fruit, relying entirely on solar energy. Early rains in September 1918, however, caused almost total loss to the drying prune crop, resulting in earnest federally supported experiments with artificial dehydration. Within five years, a fifth of the prune crop was dehydrated, and, by 1946, the sun dried only a quarter of the state's prunes. A similar shift occurred in the walnut industry, which also had long relied on solar energy for drying.[30]

In southern California, some energy entrepreneurs looked to solar energy not for drying but for powering engines to pump water and run machinery. In 1901, Pasadena's A.G. Eneas built a 642-square-foot cone-shaped solar radiation reflector. Focused on a boiler, his thirty-three-foot, six-inch-diameter collector raised steam to a pressure of about 150 PSI, enough to pump 1,400 hundred gallons of water per minute a distance of twelve feet. Eneas's four-horsepower-rated, eighty-three-hundred-pound solar engine pumped water for the birds on a Pasadena ostrich farm, and he went on to experiment with two other plants in Arizona, in 1903 and 1904. At the same time, Albert Carter opened the Solar Furnace and Power Company in Serena, near Carpenteria, California. There he built a concave reflector to raise steam for electrical generation, using storage batteries for cloudy day backup. Finally, beginning in 1904, H.E. Willsie and John Boyle, Jr. built five solar engines, one in Needles rated at twenty horsepower. Their solar devices heated water to volatize either ammonia or sulphur dioxide, which then powered an engine. As business fortune would have it, theirs and the solar engines of others reached a marketable stage just as the natural gas industry appeared in the Southwest.[31]

Clarence M. Kemp, on the other hand, utilized the heating power of the sun more modestly. In 1891, he patented a solar water heater. Consisting of four galvanized, iron tanks, painted black and placed in a glass-covered pine box insulated with felt paper, his Climax Solar Water Heater could be placed on a roof of a building or attached to an outside wall at an angle. It did not work well in Baltimore, but it did function successfully in California. In 1895, E.F. Brooks and W.H. Congers of Pasadena purchased from Kemp the manufacturing rights for the device. Since heating water alone could consume three-quarters of a ton of coal or a cord of wood annually, the $25.00 Climax heater marketed by Brooks and Congers could pay for itself in about three years. Although some other fuel source had to provide backup heating during periods of continued cloud cover, rain, or for late night or early morning hot water, the heater permitted those people who had water piped into their homes to avoid having to strike a match every time hot water was desired.[32]

The solar heaters did not sell rapidly, although some sixteen hundred were operating in Pasadena and other areas of southern California by 1900. Residents remained skeptical, and press coverage as well as advertising was limited. But, in 1898, Los Angeles contractor and realtor Frank Walker improved on the Climax with a new design that included plumbing into already-existing stove or space heater water-backs. Entrepreneur Charles Haskell acquired rights to both the Climax and Walker heaters by 1905 and developed a shallower tank design that quickly could achieve water temperatures up to 120 degrees. He marketed it as the Improved Climax heater. In 1907, J.J. Backus, Superintendent of Buildings in Los Angeles, wrote Haskell an open letter in *The Architect and Engineer of Southern California:* "All your company needs is a little judicious advertising to increase your business."[33]

Three years later, William J. Bailey, from the Los Angeles suburb of Monrovia, received a patent on a revolutionary new solar water heater. According to Ken Butti and John Perlin, "Bailey's solar heat collector was a shallow glass-covered box, only four inches deep, lined with felt paper. It contained copper tubing coiled back and forth across a copper sheet to which it was soldered. The copper sheet and tubing were painted black."[34] It was connected to a storage tank that stood above the collector in the house attic. Bailey's Day and Night Solar Heater Company, established in 1909, soon became the world's leading solar heater manufacturer, advertising widely and demonstrating models at local fairs.

Susan Swaysgood wrote the *Pacific Rural Press,* in 1913, with her praises of solar water heating, a step "in the ladder of progressive living." Her forty-gallon system provided hot water from 10:00 A.M. to 8:00 P.M. "without one cent of expense for fuel or one minute's labor to carry the fuel in." It cost $65.00, plus the wages of a plumber to install it, and she used her stove as a backup. "I think it is safe to say we have 265 days in the year," she wrote, and "that old Sol will furnish the heating power for the house. . . .

No California home no matter how small can afford not to have a solar heater." Furthermore, the supply tank on the roof, if painted brick color, easily could be made to resemble a chimney.[35]

The original Climax heater cut fuel consumption for water heating in southern California by 40 percent, and Bailey's Day and Nite reduced it a full 75 percent. By 1918, Bailey had sold over four thousand heaters in southern California and Arizona and introduced a nonfreezing heat exchange system. In the peak sales year for the company, 1920, he sold over one thousand Day and Nite heaters, and through the coming decade sales averaged perhaps three hundred per year. Unfortunately for Bailey, intensive marketing and adoption of natural gas heating by southern California gas companies undermined the solar water heater market. Although he continued to manufacture the Day and Nite heater, as a good businessman he also turned to building thermostatically controlled, insulated gas-heated water tanks.

Other efforts to promote solar heating continued, but only for a time. During the 1920s, Mary Seacrest and other University of California home management extension specialists worked with rural housewives to demonstrate modern electric appliances. In addition, they provided assistance in making and installing solar water heaters and fireless cookers. Their work led to inquiries about solar heaters to the U.C. Agricultural Experiment Station in Davis, which released a 1929 bulletin giving California solar radiation data, test results of various solar heaters, and installation information. Since several thousand home solar water heaters operated in the state during the 1930s, Professor F.A. Brooks argued, in a second solar energy bulletin, that solar heating also would be beneficial to schools, dairies, and manufacturers using hot water at temperatures below 140 degrees. "Every square mile of ground in California," he observed, "receives during each clear summer day about as much energy as can be produced by all the power plants of one of the largest electric utility systems in the state."[36]

Summary

Energy from the wind, tides, and sun played roles of varying significance for nineteenth-century economic development in California, but, despite the region's environmental endowment of these natural energy resources, their use eventually languished. Before the midtwentieth century, windmills, sailing vessels, ill-fated wave motors, sun-drying of fruits, and solar devices disappeared as important contributors to California's energy budget. Technologies exploiting solar and wind energy proved convenient and efficient only when they were substantially more cost-effective in competition with other energy resources. With the exception of sailing vessels, their virtues were short-lived in all but the most isolated corners of the state. As the twentieth century unfolded and newer energy resources became available, consumers yielded to their penchant for easier living, letting utility companies handle the production, maintenance, and distribution of power. Yet, certainly it would have pleased W.I. Tustin, Albert Stahl, William Bailey, Susan Swaysgood, and other users of wind and solar energy to know that ever-changing factors in the interplay between technology and the environment would lead Californians once again to tap these natural energy resources during the last half of the twentieth century.

CHAPTER VI

Waterpower and Steam Engines

EUROPEAN-AMERICANS MIGRATING to California brought with them the United States' developing tradition of economic growth through intensive energy use, obtained in part from mechanization and the exploitation of environmental resources. In the East, the New World's labor and capital scarcities had been offset not only through tools fashioned to extend the power of human and animal muscle, but through power from abundant flowing rivers, endless forests, and rich bituminous coal deposits. Eastern immigrants to California knew well both the immense waterpowered textile mill towns that had transformed the rivers and landscapes of New England and the thousands of individual mills dotting rivers elsewhere. They understood how to burn wood and coal to raise steam for engines, a revolutionary technique born in England during the eighteenth century and since employed wherever waterpower fell short. They were the forebearers of Mark Twain's Connecticut Yankee, who bragged "I could make anything a body wanted—anything in the world, it didn't make any difference what; and if there wasn't any quick, new-fangled way to make a thing, I could invent one—and do it as easily as rolling off a log."[1] If manufacturing was to be planted in California, it would be done by ingenious settlers effectively utilizing waterpower and steam en-

Photo 6.1. Hydraulic mining at the Malakoff Diggins near the Sierra Nevada town of North Boomfield forever scarred the landscape. Photo by author.

gines. Whether manufacturing arose or not, waterpower and steam engines would reshape important parts of California's landscape.

Overland immigrants made their way through the Sierra Nevada to John August Sutter's New Helvetia at the confluence of the American and Sacramento rivers, and they easily assessed the potential of flowing streams and vast forests. So, too, did the settlers who found their way to the coastal mountains and valleys. They soon began constructing waterpower mills. Isaac Graham arrived in Santa Cruz from Tennessee, and he built a sawmill in 1841. Trapper George Yount settled in the Napa Valley in 1836, where he later erected a grist mill and a sawmill. John Read, the earliest settler in Sausalito, across the bay from San Francisco, built a grist mill, followed by a sawmill in 1843. On the eve of John Marshall's discovery of gold in the tailrace of the sawmill he was constructing for Sutter on the south fork of the American River, water drove six mills in the Central Valley. And, six miles up the American River, a group of Mormon immigrants, also employed by Sutter, were working on a dam, a four-mile millrace, and a flour mill with four sets of stones.[2]

Although the gold rush temporarily halted the construction of mills, and thieves destroyed Sutter's new sawmill and unfinished grist mill, the demand for lumber and flour from tens of thousands of argonauts and new settlers soon led to renewed efforts to tap waterpower. In the 1850s, rude sash sawmills, driven by traditional overshot waterwheels, sprouted along streams flowing out of the Sierra Nevada, and pioneer John Bidwell built a successful grist mill in Chico. South of Sacramento, near Stockton, J.H. and D.M. Locke, immigrants from New Hampshire, built a large flour and sawmill at Knight's Ferry on the Stanislaus River. Near the shoreline, in San Francisco's North Beach area, "Honest" Harry Meiggs opened a successful lumber business by erecting a sawmill along a brook fed by two springs. Fifty miles south of the city, in the Santa Clara valley, J.A. Forbes harnessed Los Gatos Creek, in 1850, to run a flour mill.[3]

Throughout northern California, mills of traditional eastern design appeared during the 1860s. A three-story flour mill served Stockton, while in Marin County the Pacific Powder Mill and Pacific Paper Mill drew power from Tokeluma Creek. In Santa Clara County, Forbes's mill was followed by several grist mills along the Guadalupe and Coyote rivers, and waterpower in Santa Cruz County made it second only to San Francisco as a manufacturing center. There, tanneries, sawmills, and powder works received power from the San Lorenzo and Pajaro rivers and several creeks. Even a few individual farms, fortunate enough to have a good running stream on their property, installed small waterwheel systems. A dairy farmer in the San Bruno hills south of San Francisco, for example, built two reservoirs to provide water for irrigation and domestic use; he saw that it was "well fitted with machinery run by water-power," including a cream separator, revolving brushes to wash milk cans, and equipment in a general farm workshop.[4] Meanwhile, a considerable number of waterpower projects were undertaken near Sacramento.

A large flour mill on the American River at Folsom boasted a five-hundred-barrel daily capacity, and the owners, Stockton and Coover, were preparing the way to lease or to sell to other manufacturing interests. They envisioned woolen and paper mills, a machine shop, and a winery. The *California Farmer* reported confidently that the venture would surely "awaken attention to the value of our 'natural waterpower,'" and, in 1863, a San Francisco newspaper attributed ten waterwheels to Folsom's five-story Granite Flour mill. Horatio Gates Livermore, a native of Maine, also dreamed of transforming the American River and creating a waterpowered, industrial city along it. With two sons, Charles Edward and Horatio Putnam, he gained financial control of the Natoma Water and Mining Company and acquired several thousand acres of land in the vicinity of Folsom. In 1868, the three men proposed to give land and waterpower access to the State of California for a new prison. In exchange, the state would provide them with convict labor to construct a dam and two canals along either side of the river. The canal from the north side of the dam would provide irrigation water; the southern canal would provide industrial waterpower with a total fall of eighty feet.[5]

By the 1870s, although trivial by eastern standards, waterpower in California generated sixty-five hundred horsepower to process grain, saw lumber, and run other manufacturing equipment (figure 6.1). Transplanted New Englanders, such as the Livermores, pushed forward various waterpower schemes, and boosters sang praises of California's potential. "The time is at hand," bragged John Powell, "when her villages, towns, and cities shall hum with the wheels and spindles of her own manufactures." In California's mild climate, frost never halted the mill wheel and winter never required mill operators to unhealthfully shut up factory doors and windows. Even earthquakes, of some concern to potential immigrants, seemed to offer a blessing: "The manufacturers of [Santa Cruz and Santa Clara counties] derived an important advantage from the great earthquake of 1865. . . . That shaking increased the waters of all the creeks and rivers to nearly double their previous volume, during the dry season." Beyond the coastal rivers and streams, Sierra Nevada waterpower, boasted Titus Fey Cronise, equaled "the force exerted by five hundred thousand horses. . . . If all the waterpower existing in the New England States were added to that

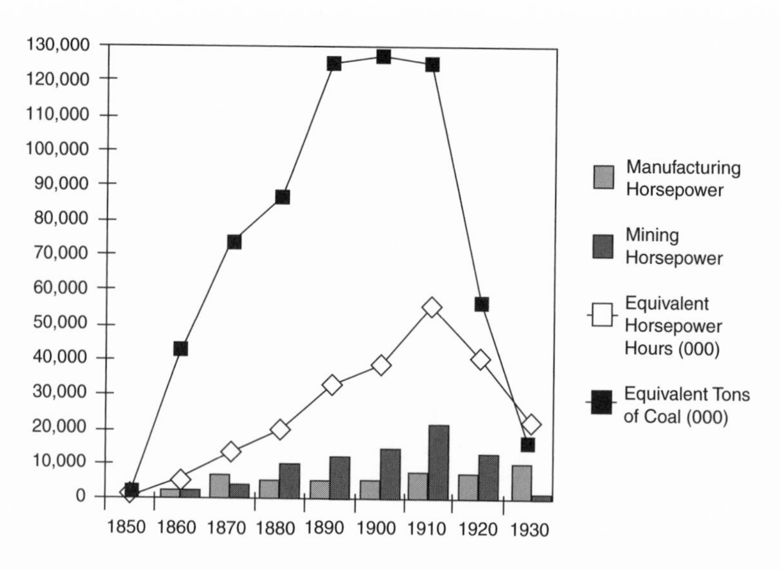

Figure 6.1. California Waterpower, 1850–1930. Source: *Supra,* Table C-6.

of New York, New Jersey, and Delaware, it would scarcely exceed that still running to waste down the side of the Sierra."[6]

Cronise and others may have slightly exaggerated the potential of Sierra Nevada waterpower, but its use by gold miners hinted at its possibilities. Using water to wash gold from dirt and gravel, they became familiar with Sierra Nevada stream characteristics. The snow-laden mountain watersheds reached heights of 14,000 feet and fed scores of rushing streams. Unlike rivers in the East, such as the Hudson, with a heavy water volume and a gradual fall of fourteen feet per mile, Sierra Nevada streams dropped a small volume of water abruptly in elevation, as much as 133 feet in a mile on the north fork of the Stanislaus. The small volume, observed historian Louis Hunter, permitted "the conveyance of high-potential water for miles across rugged mountain and Piedmont terrain by ditching, fluming, and wooden or iron piping." Placer miners first tapped this possibility, followed by hardrock and underground quartz mining companies. They introduced two permanent hydraulic technology innovations, "the rude but effective hurdy-gurdy waterwheel and the ditch system of high-head water supply."[7]

Before the 1870s, miners built several hundred miles of ditches and flumes. By 1870, large mining firms and water companies had commodified Sierra Nevada water and, reshaping the stream pattern to fit their needs, created hundreds of energy delivery systems. They supplied water through a network of 874 main ditches some five-thousand-miles long and, perhaps, as many miles of subsidiary conduits. The 1880 census reported that nationwide such ditch systems totaled almost eleven thousand miles, California accounting for three-fifths of the mileage. Typically smaller than transportation canals, four to six feet at the bottom and five to nine feet at the top with a two- to four-foot depth, ditches in the Sierra Nevada had grades ranging from nine to twenty feet per mile. Construction costs extended from $2,000 to $3,800 per mile. Almost $20 million was spent for all the state's systems, but those who financed ditches generally earned back their investments handily, through the support of their own mining operations and/or by selling water to miners and quartz mill operators. In 1880, the state engineer advised the legislature that "the world probably does not present elsewhere an example of canal building where so many large works, constructed through such rugged country, may be found within an equal scope of territory."[8]

Integrated into the ditch systems, builders installed timber flumes where they had to cross irregular ground, traverse canyons or narrow valleys, or follow the sides of precipitous abysses. In 1859, the Golden Rock Ditch Company in Tuolumne County erected the Big Gap suspension flume. Supported by wire cable from eleven sugar pine towers—two reaching upwards almost 265 feet—the 2,200-foot-long flume carried water between portions of the ditch two hundred feet above the valley. In the same year, the Eureka Lake Ditch Company completed its 1,200-foot-long Magneta Aqueduct, which crossed the Cherry Hill Gap, a gorge in Nevada County, on a 225-foot-high trestle. The firm celebrated its opening with flags, fifty feet in length, raised above the trestle's highest point, a band playing the "Star Spangled Banner," and appropriate orations; among the guests was Governor John B. Weller. In Butte County, an effort to avoid building a trestle led the Miocene Ditch Company to build a 350-foot-long, four-by-three-foot flume on the sheer walls around a bluff, anchoring it with by iron L brackets drilled into the escarpment and supporting it with wire cable.[9]

To avoid wooden trestles, which inevitably decayed and had to be replaced, ditch builders sought other materials. In July 1868, the Big Gap suspension flume collapsed. Rather than replace it, the company returned to a plan it had foregone a decade earlier, and installed a 2,262-foot, 22-inch-diameter iron inverted siphon across the valley. A year later, Butte County's Spring Valley Canal and Mining Company began the Cherokee inverted siphon, which the *Mining and Scientific Press* predicted, "if successful will stir up the engineering world." In December, the siphon, 13,000 feet long and 30 inches in diameter, was opened. With a drop of 837 feet, it marked the first use of iron pipe under high head.[10]

Initially, miners simply built small diversion dams to direct water into ditches, but water and mining companies turned to storage dams so as to lengthen availability of water through dry, late summer and early fall months. At first, they built storage dams from a series of log cribs filled with rock and waste and faced with timber, a technique familiar to them from the East. But canyons in the Sierra Nevada pushed them to build crib dams far higher than the typical five to twenty-five feet. The English dam, constructed between 1856 and 1859, was perhaps the largest, reaching over one hundred feet high with a fifty-foot-wide base and 331-foot crest. But the crib dam was inadequate to the task, and from it a distinctive, new rock-fill dam evolved. The first came in 1876 with an enlargement of the Bowman crib dam, in which the log crib was left in place and covered with loose rock-fill and a dry rubble wall on each face. Built by the North Bloomfield Gravel Mining Company, it was followed by two other rock-fill dams along the Yuba River, which by 1890 had eleven principal reservoirs with a capacity of over 2.1 billion cubic feet of water. By 1900, the Yuba, Bear, and American rivers had sixteen reservoirs holding back almost fourteen billion gallons.[11]

As Louis Hunter observed, miners and mechanics exploited the waterpowers of the gold regions with skill and imagination, although not without environmental impact. Many of the big water companies served hydraulic mining operations, using water's kinetic energy in a devastating way to "mechanize the prospector's manual methods of washing out gold by pan and rocker." Using monitors, cannon-like iron nozzles that fed water under extreme pressure through iron pipes and were mounted so

Figure 6.2. Cutaway of a forty-stamp water-powered stamp mill. From California State Mining Bureau, *Eighth Annual Report of the State Mineralogist for the Year Ending October 1, 1888* (Sacramento, 1888), 548.

they could be moved in any direction, hydraulic miners literally washed away mountain sides. In terms of the environment, placer mining already had damaged fisheries, altering water flow and silting river bottoms. The use of direct water pressure, called hydraulicking, washed so much gravel, sand, and silt into the Sacramento River and its tributaries that, by the 1880s, it had destroyed the salmon fisheries in three rivers and smothered some oyster beds around San Francisco Bay. Additionally, it ended upriver navigation by filling up riverbeds, and it narrowed and permanently raised, as much as fifteen feet, the channels in the delta. The rivers could carry less and less of the rainy season run-off. Spring floods, normal anyway in the valley, became progressively worse. Debris and silt left by receding floodwaters began leaving the land ruined for farming, and outright war threatened to erupt between valley residents and miners. In 1884, a permanent federal injunction put an end to the discharge of any debris into Sierra Nevada streams.[12]

Water pressure, however, also provided miners with power for the hoists, pumps, drills, and stamp mills necessary in hardrock and quartz mining. The cost of providing fuelwood for steam engines, which had first powered much mining and milling equipment, accounted for 30 to 40 percent of most mining and milling expense. As easily accessible timber disappeared, and as transporting fuelwood over poor roads and trails became increasingly expensive, steam engines lost their magic. In fact, the high initial expense required to haul heavy steam engines and equipment into mine sites, coupled with fuel costs, led many miners to utilize waterpower from the start.[13]

Miners soon discovered that eastern overshot and undershot wheels, plus reaction or pressure turbines, were not suited to the high head, low volume water of the Sierra Nevada. One got little power with standard wheels, and the tremendous water impact caused excessive wear on turbine gates, guides, and wheel shaft bearings. Although they later were modified for successful use with medium-high heads, turbines suffered other drawbacks. They had to be geared down to lower speeds for direct drive hoisting, pumping, and stamp milling; sand and silt in the water quickly wore down the turbine blades. California mining engineer Hamilton Smith observed, at the 1884 American Society of Civil Engineers meeting, that "this wear was so objectionable, that with perhaps nearly 800 wheels now in operation, there is not, to my knowledge, a single turbine at work." Fellow California engineer, E.B. Dorsey, agreed he "did not know of a single one."[14]

Miners improvised to harness waterpower, developing the hurdy-gurdy waterwheel. A narrow, vertical wheel, ten feet or more in diameter and less than a foot wide, it had numerous buckets, a few inches in depth, that resembled saw teeth in profile and were enshrouded on either side by wood. High head water was directed tangentially against the wheel, striking the buckets in rapid succession; the wheel turned as a result of impact. In time, iron replaced early wooden and canvas pipes and wooden nozzles, heads increased, and more power was attained by adding nozzles directed at different points on the wheel or by mounting additional wheels on the same shaft. Simply constructed and maintained, the hurdy-gurdy provided miners with effective direct drive waterpower. The rotating speed was controlled by the size of the wheel.[15]

Improvised, on-site hurdy-gurdys gave way gradually to factory-built wheels, and improvements by California mechanics led to the true impulse or tangential waterwheel. First, attention was given to the wheel buckets; according to Louis Hunter, the goal was "to secure the greatest advantage both from impact by the entering stream and from reactive force in leaving the wheel bucket, with minimal interference in the process."[16] In 1866, a San Francisco foundry constructed the first cast-iron wheel, the performance of which proved superior to the rough-hewn hurdy-gurdy. Through the 1870s and 1880s, several California mechanics built successful, inexpensive wheels. All shared the key characteristics of simplicity in construction, installation, and maintenance, flexibility in location and operation, and ease of transport. With all the wheels, an inexpensive timber frame and wooden casing to stop splashing sufficed in place of masonry foundations and wheel pits. Bolted-on buckets could be easily replaced, and access to all parts of the wheel for inspection and adjustment was easy and immediate.

The critical element in successful wheels was the "splitter," or ridge cutting through the bucket center. It divided the entering water jet into two streams which, in following the bucket contour, reversed direction and discharged on each side of the wheel. The water thus spent its energy first in impulse and then through reaction. Several mechanics may have independently developed the split bucket, but Lester Allen Pelton, a builder and amateur mechanic in Comptonville, was the first successfully to develop it commercially. He patented his wheel in 1880 and, eight years later, incorporated and located the Pelton Water Wheel Company in San Francisco. With the split bucket design, tangential wheels developed high efficiency. Although no scientific testing supported these claims, the state mineralogist, in 1888, believed the wheels had an efficiency of 85 to 90 percent, and William A. Doble, chief engineer at Pelton's firm, claimed 80 percent, in 1899. The tangential wheel came close, noted Hunter, "to realizing the long-proclaimed goal of waterwheel design, a wheel in which the water would enter without shock and leave without velocity."[17]

In the nineteenth century's last decades, the hurdy-gurdy and its successor, the tangential waterwheel, provided the majority of the power for mining and milling. Employment of the tangential wheel was aided by the increasing cost of fuel for steam engines and by high head waterpower available from ditch companies seeking new markets, while legal restrictions in the 1880s caused a rapid decline of hydraulic mining. According to an 1870 U.S. Treasury report, just over 50 percent of California's 417 quartz mills employed waterpower: 196 exclusively and sixteen with auxiliary steam power; 205 used only steam. Two decades later, the state's gold mining industry used 370 waterwheels rated at 12,039 horsepower and 266 steam engines rated at 9,813 horsepower (59 percent of the nation's total number of mining waterwheels, but only 20 percent of its mining steam engines). Furthermore, the state mineralogist's report suggested over sixty of the steam engines served only as auxiliaries to waterpower. Waterpower ran elevators, air compressors, hoists, pumps, and quartz mill stamps. "It is a matter of astonishment," noted *Industry* magazine, "how large a share of the quartz mines in California employ water power."[18]

In 1896, the state mineralogist inventoried mining mills in twenty-eight counties, finding that water powered a majority of them (figure 6.3). Siskiyou and the eight Mother Lode counties of Tuolumne, Calaveras, Amador, El Dorado, Placer, Nevada, Sierra, and Plumas—all served by water ditches—had 82 percent of all waterpowered stamp mills, 71 percent of all mills using both water and steam power, and 42 percent of all mills with steam power. Although about 32 percent of the total number of mills inventoried did not specify power source, water likely still powered 60 percent of California's mining mills. However, by 1913, mining and manufacturing tangential wheels supplied only 36 percent of the available horsepower and, by 1919, only 17 percent. Nevertheless, California mining districts provided, notes Hunter, "the apprenticeship of a new and basic wheel type with a large future in the hydroelectricity field" of the next century.[19]

Hydraulic power in the Sierra Nevada combined with what waterpower was used in manufacturing to raise California's aggregate available waterpower rating to almost 20,000 horsepower. Even though, if used regularly, it still only equaled 3 percent of the state's actual energy budget in 1890, during the previous thirty years waterpower had increased as a percentage of total energy consumption in California, whereas in the nation it had declined. Yet California manufacturing centers were not located in the Sierra

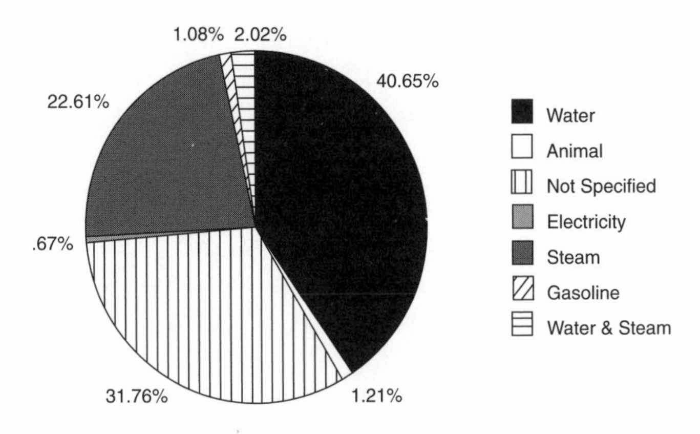

Figure 6.3. Power in California Mining Mills, 1896. Source: *Supra,* Table C-7.

Nevada or its foothills and, therefore, had to use steam power. Furthermore, as transportation improvements and lower fuel costs made competitive large, modern, steam-powered flour mills in the cities, smaller waterpowered grist mills serving local communities disappeared. By 1880, steam-driven flour mills outnumbered waterpowered mills almost three to one.[20]

Despite the fact that waterpowered manufacturing centers like those of New England were becoming obsolete in the face of steam-power technology, boosters such as Titus Fey Cronise still argued that the Sierra Nevada foothills were "designated by nature as the great manufacturing field of California." For them, waterpower never lost its magic. Water from miners' aqueducts, said Cronise, had to "be made subservient to the propulsion of machinery." John Powell remained hopeful that "the great waterpower of the Sierra" would some day replace steam, which was dependent on expensive fuel, as the motive power for California's principal manufacturers. W.W. Morrow entered the 1880s telling a Mechanic's Fair audience that the time was near when the mining industry's water system would supply California's manufacturing power needs. Even mining engineer Hamilton Smith predicted that, "when manufacturing in California assumes large proportions, doubtless most of the motive power will be obtained from these mining ditches, which in aggregate will afford several hundred thousand horse powers."[21]

But without moving population centers to the Sierra Nevada, an impractical though suggested alternative, the main source of California's waterpower was limited to the mining industry and a few grist mills and sawmills. Nor could hydraulic mining technology be moved intact to the Central Valley and coastal regions. Only small waterwheels and current wheels found use elsewhere. The high efficiency of the former made them useful for small applications on farms. The Pelton Company produced a twelve-inch wheel for this market, in 1889, as well as an eight-inch wheel or water motor for use in urban water supply systems. The latter, first used by placer miners to raise water from streams into sluices, was a simple paddle wheel, several feet in diameter with buckets attached to the rim. As the stream current turned it, the buckets filled with water and spilled it out as they reached the top of the wheel's rotation. A few farmers, from Reedley to Visalia and Fresno, built current wheels to water fields from streams or irrigation canals.[22]

Use of Steam Power in Manufacturing and Agriculture

While boosters touted the potential of Sierra Nevada waterpower, steam engines—fueled by wood, coal, and eventually oil—came into use. Called by Thomas Aston "the pivot on which industry swung into the modern age," steam engines attracted every Californian eager for the advanced civilization they heralded. Many Pacific Coast residents already had experienced widespread use of steam engines in mills and manufacturing in the Midwest. They understood that steam power released manufacturers from the locational restraints of waterpower, an advantage that would permit firms to establish themselves in the population centers of California's major entrepôts—San Francisco, Sacramento, and Stockton. This benefit outweighed the high cost of fuel, but, in opting for steam power, Californians had to put great importance on the discovery of reliable local fuel deposits or on imported supplies of coal, the latter being the preferred fuel. As we have seen, fuelwood was often expensive in urban areas, California coal was scarce and poor in quality, and imported coal was expensive.[23]

Stephen Smith installed the state's first stationary steam engine in the mid-1840s, running a combined grist mill and sawmill at Bodega Bay, north of San Francisco. Not until 1849, just after the discovery of gold, was another such engine employed. But engines from the East arrived on the first ships responding to the gold rush and soon were employed in a variety of enterprises. Although limited wood and coal fuel supplies, initially, may have slowed adoption of steam engines, regular coal imports in the 1870s and 1880s brought steady growth in steam power. By 1870, at least 604 steam engines ran mills and manufacturing machinery, their rated horsepower totaling 73 percent of the state's water and steam power (compared to 52 percent nationally). Within a decade, steam horsepower increased by half to account for 85 percent of water and steam power (compared to 64 percent nationally). By 1890, the number of steam engines had more than doubled and horsepower rating climbed 140 percent. Despite high fuel costs, manufacturing steam power grew more quickly in California than it did nationwide.[24]

Lumber and flour mills used half of California's steam power, the former benefiting from cheap fuelwood. The remainder of the state's steam engines powered machinery for a variety of manufactures. Stockton, Sacramento, Marysville, and other communities had steam flour mills, while the Gladding-McBean Company used steam power to produce pottery in Placer County, and Garrett's mill, recently relocated from Alvarado to Santa Cruz County, processed sugar beets with steam heat and power. In part because imported coal unloaded at San Francisco, the city became the state's principal manufacturing center. By 1855, it boasted eleven flour mills, five coffee and spice mills, four sawmills, six candle and soap works, several foundries and iron works, and cracker and candy factories. The city's factories grew in number as the century progressed, and, by 1890, its 4,059 companies expended $1.8 million on fuel for steam power and process heating.[25]

The expense of coal led some firms, such as Garrett's sugar beet mill, to relocate near cheaper fuelwood supplies. But others chose to remain in population centers, which forced them to reduce fuel consumption through better furnace and boiler efficiencies. In Sacramento, the Phoenix Mill modified its furnaces with special air chambers to permit use of low grade lignite coal from mines in nearby Ione. The fresh supply of air ignited gases arising from burning coal before they escaped from beneath the boiler, increasing fuel efficiency by 5 percent and eliminating black stack smoke. Sacramento's Capital Packing Company similarly modified its furnace; and, in San Francisco, Reese Llewellyn, proprietor of the Columbia Foundry, sought to economize coal consumption by inventing an exhaust steam condenser to preheat boiler water. Soon other firms adopted similar condensers, which reduced fuel consumption by an estimated 8 to 10 percent. By the 1880s, the Pacific Rolling-Mill Company had achieved significant fuel saving by employing its eighteen reverberatory furnaces to fire forty steam boilers, as well as to heat tons of iron each day. Its steam engines then ran iron rollers, trip hammers, spike cutters, and bolt and nut machines.[26]

The high cost of fuel slowed adoption of steam power particularly in agriculture. The first portable farm steam engine had been manufactured in Philadelphia, in 1849, but few saw Pacific Coast use in the 1850s. Wheat farmer demands for threshing machinery, however, prompted Joseph Enright to build a farm steam engine in San José, in 1861. Other mechanics followed his lead, among them the Brown Brothers of Salinas, Vallejo's J.L. Heald, and Hayward's Harvey W. Rice. Californians became farm steam power pioneers, as threshing machines were belted to eight- and twelve-horsepower portable engines, replacing horse-driven wheat threshers on the biggest ranches. Soon imported eastern machines were advertised as "California style" engines of twelve to fifteen horsepower with broad, fluted smokestacks aimed at reducing wheat field fire danger from sparks.[27]

During the 1870s, advertisements for portable steam engines to run feed grinders, threshers, and other farm implements became common in California farm journals. But fuel availability still proved a bottleneck to widespread steam power adoption; high farm labor costs and fuelwood availability offset this problem only slightly.[28] Harvey Rice, Hayward's farm steam engine builder, provided a solution, in 1874, when he patented an engine that burned straw, the one fuel available in superabundance to wheat farmers. As inventors of other important devices had discovered, Rice found his patent did not prevent mechanics, such as Enright, Heald, and others, from producing machines very similar to his. Straw burners soon were used in wheat fields throughout the state and exported to South America, Egypt, Turkey, South Africa, Russia, and other locations where cheap fuel for farm steam engines was a factor. According to John Hittell, California farms used some four hundred straw-burning engines by the start of 1882, 270 of them built by Rice or others using his design.[29]

With straw-burning steam engines, California wheat farmers found that eastern-built threshers were too small for their needs. Now huge steam threshers followed the path blazed by mechanics in the development of giant horse-powered harvesters. Historian Reynold Wik writes that "the large-scale of harvesting and threshing equipment used on the Pacific Coast . . . amazed the farmers living east of the Mississippi River." George Hoag, a blacksmith on Dr. Hugh J. Glenn's huge Sacramento Valley ranch, for example, constructed an immense separator, thirty-five-feet long with a forty-eight-inch cylinder. Driven by a twenty-five-horsepower Enright steam engine, it threshed 5,770 bushels of wheat on 8 August 1874, a time when nine hundred bushels was considered a good day's work in Ohio. In 1877, Glenn ran eight steam threshers to harvest 45,000 acres, and two years later Hoag's record was broken when 6,183 bushels were threshed in one day using a twenty-five-horsepower Garr-Scott engine and a forty-four-inch thresher.[30]

Off the field, stationary steam engines worked irrigation and reclamation pumps as early as the 1850s. Although used sparingly and during drought periods, steam irrigation pumping was discouraged by high fuel costs and canal companies that provided water by force of gravity. It, nevertheless, served many farmers, from the Rio Bonito Company vineyards in Butte County and horticulturists in Marysville to farmers near Needles on the Colorado River, and others in San Bernardino. San Joaquin Valley farmers, particularly, sought to avoid exorbitant canal water costs with their own steam-powered pumping. Steam pumping was especially cost-effective in fruit-tree irrigation, where large water expenses were offset by high per acre product value and tree trimmings could be utilized for fuel. Finally, the Pearson Reclamation District's use of steam power to pump water at 37,907 gallons per minute from eight thousand acres surrounded by a 15½-mile levee started extensive reclamation of thousands of acres in the Sacramento and San Joaquin river deltas region.[31]

Meanwhile, smaller stationary engines provided power for farm workshops and specialized farm operations, such as dairies. Local mechanics, such as E.H. Farmer of Gilroy, produced three- and four-horsepower steam generators for the farm market during the late-nineteenth century. Smaller engines, ranging from two- to ten-horsepower, however, were extremely costly to operate, requiring as much as ten pounds of coal per horsepower hour, compared with two pounds for large engines. Not unexpectedly, gasoline engines undercut small steam engine use beginning in the 1890s, but a California farm power survey, in the 1920s, estimated that 7 percent of the state's farm energy still came from some thirty thousand stationary steam engines, each engine averaging six horsepower and delivering about 450 horsepower hours of work per year.[32]

Use of Steam Power in Transportation

Stationary and portable steam engines fulfilled an essential role in manufacturing and agriculture, but in California, as well as in the nation at large, steam power's greatest visibility was in transportation. In early 1849, three coastal steamers of the Pacific Mail Steamship Company arrived for regular service on the Panama-San Francisco run, and a year later fifteen steamers operated the coastal route. They preferred coal for fuel, the nine ships of the Pacific Mail Steamship line consuming thirty-five thousand tons in a year, but burned fuelwood if necessary.

The rush for gold also meant profitable traffic on the Sacramento and San Joaquin rivers. The first regular steamboat joined schooners, barks, and brigs in carrying passengers and supplies between San Francisco and Sacramento in October 1849. Within a year, sixteen steamers accounted for 15 percent of the tonnage operating on the state's northern inland waterways, and twenty-five plied the waters by 1853. When Mount Diablo lignite coal became available at river's edge, after 1860, the number of boats increased. By 1880, there were over seventy riverboats and twenty-two bay ferries.[33]

As gold fever waned, farm commodities and local passengers became the staple loads on California riverboats. Between June 1883 and June 1884, Stockton received 257,350 tons of goods and shipped 185,600 tons, 93 percent of the outgoing tonnage comprising farm products. Points above Sacramento shipped 70,876 tons of wheat, accounting for almost one-third of the Sacramento River's total traffic, and considerable trade occurred in fruits and vegetables. Since the Sacramento could be navigated to Red Bluff, 262 miles upstream, and the San Joaquin could be traveled 195 miles to Fresno County during spring high water, farmers of the Central Valley were served well by water transport. Additionally, the Feather and Mokelumne rivers, Sacramento River tributaries, served rural residents, and Petaluma Creek and the Napa River provided water transport to portions of Sonoma and Napa counties. In 1880, the state's riverboats, bay steamers and ferries, and ocean steamers carried 1.6 million tons of freight and 6.3 million passengers. California, with 1.8 percent of the nation's population, accounted for 6.1 percent of its steamer freight tonnage and 3.7 percent of its passengers.[34]

Well into the twentieth century, steamboats provided transportation for Californians and their products. In 1900, steamers on the San Joaquin and Sacramento rivers carried 732,201 tons of freight, grain, and livestock; fifteen years later, they carried 1,598,169 tons, plus 381,183 passengers. Traffic, in 1915, on the Mokelumne and Feather rivers equaled 91,530 tons. Although gasoline launches began to replace steamers, the sixty-six steamers traveling California rivers, in 1917, accounted for 79 percent of the total tonnage, excluding barges. In 1911, the State Board of Agriculture explained that "water service is especially valuable to a considerable area of rich farming land situated on islands in the deltas." Not only grain, but potatoes, beans, asparagus, and other vegetables were shipped in large quantities.[35]

Besides steamboats, Californians also built railroads, the nation's Atlas of steam power. The state's first, the Sacramento Valley Railroad, ran twenty-three miles between Folsom and Sacramento and was completed in 1856. It remained the only line in the state for several years, serving as the foundation for the first transcontinental railroad, which commenced construction during the 1860s. But, by 1870, rails linked Sacramento with Chico to the north and Oakland to the west, they ran between San Francisco and San José, and in the south they connected San Pedro to Los Angeles. The following decade saw mileage increase from 925 to 2,201, and, by 1884, rails reached from the Oregon border down through the Central Valley to Los Angeles and on to the state's southern border at Yuma.[36]

As with stationary engines and steamboats, the value of rail transportation outweighed the fuel cost, even though railroads required vast amounts of fuel. The rule-of-thumb was 140 cords of wood per year per mile, and California railroads used wood from the start, burning as much as 160,000 cords per year in the 1890s. But fuelwood's bulk and heating value made it inferior to coal. Therefore, railroad owners went out of their way to find coal supplies for their locomotives, and during the last part of the century they found enough coal to permit consumption of between 600,000 and 800,000 tons of coal annually. In 1864, the Western Pacific laid tracks toward coal finds in Corral Hollow, and the San Francisco and San Joaquin Valley Railroad surveyed a line in 1888. Similarly, the California and Nevada Railroad started a line to Livermore, in 1887, but halted construction to exploit coal found in a cut in Contra Costa County. The Central Pacific, parent of the Southern Pacific, went far afield, opening mines in Tacoma, Washington and shipping the coal to San Francisco, in the 1870s. Along their transcontinental route, they tapped coal fields in Utah, where they used it, as well as shipping back to California over 250,000 tons a year during the 1880s and 1890s.[37]

Wherever the railroad went, it impacted the environment. In addition to cutting wood and extracting coal from the earth for fuel, railroad builders converted vast amounts of timber into cross ties, lumber for bridges, mountain snow sheds, and other structures. To level out gradients, they cut through rolling hills and filled in depressions, arroyos, and even deep ravines. They slashed roadbeds along mountain sides, cleared pathways atop ridges, and drilled tunnels through solid granite. The effect of railroads on the landscape was substantial, yet environmental obstacles, such as mountains too steep for locomotives to climb, also constrained it. Nevertheless, for nineteenth-century Americans, the railroad represented the technological sublime. Its powerful and fiery locomotives were instruments of progress in civilizing the wilderness by transforming it. In much the same way that the mining ditch system reshaped the Sierra Nevada's natural stream pattern, adding to it an immense maze of artificial capillaries that gradually came to be seen as natural themselves, railroads comprised a network of transportation corridors that, as William Cronon suggested in his study of Chicago, "it seemed at once so ordinary and so extraordinary—so second nature—that the landscape became unimaginable without it."[38]

The railroads, like steamboats, played a crucial role in transporting people and commodities, and California agriculture particularly experienced the economic consequences of the railroad. Transportation of perishables became possible by the 1890s and, with improved irrigation, prompted tremendous expansion in the cultivation of fruits and vegetables. The first shipment of oranges to the East occurred in 1886, and almost 19,000 carloads of fruits and nuts were sent out of state by rail in 1891. The development of ventilator and refrigerator cars, plus methods of precooling produce before shipment, opened vast markets for perishable goods.[39] Although great political controversy arose in California over the power of the dominant Southern Pacific Railroad, nineteenth-century boosters gushed praises of the benefits that it brought:

> It is through the railroads the demand for agricultural lands has been rendered active. It is through their agency that farms are being opened; wild, worthless, tangled forests being cleared, where now gardens have been planted, vineyards flourish, and corn-fields wave, and flowers of paradise bloom in place of a wilderness. . . . Through them the wilderness smiles, fruitful trees wave, and flowers blush, and a cloud of fragrance follows their track wherever they go, and they cause the desert to smile as a garden, and be glad as paradise.[40]

Where railroads could not be financed, Californians sought other ways to employ steam traction. To reach the Corral Hollow coalfield, to which railroads finally seemed unwilling to build, the Eureka Coal Mine directors asked San Francisco's Peter Donahue to build a steam traction vehicle with roller chains transmitting power to wide, ribbed steel wheels. A trial run pulling three wagons up a San Francisco hill proved successful, but it had no brakes, relying instead on compression to slow the vehicle. On its trial descent, it went out of control and crashed.[41]

Although Donahue's steam wagon failed on San Francisco's hills, the steam engine seemed an obvious solution to urban transportation systems. Unfortunately, city residents objected to railroad engine exhaust steam, smoke, cinders, and noise. Introduction of small "dummy locomotives" disguising the engine mechanism and condensing exhaust steam, did not fully overcome these objections, although dummy steam lines operated in several California communities. But a steam power answer for urban transport did exist in what historian George Hilton described as "a mechanical connection between the engine and the passengers." Experiments with this concept began in the 1860s, and San Francisco wire rope manufacturer Andrew S. Hallidie moved concept to reality with the Clay Street Cable Railway in 1873. An expensive, energy-inefficient system, it nevertheless proved cost-effective in high-density urban areas. Other cable lines followed Hallidie's—in San Francisco, Los Angeles, and Oakland, and in 1882, lines in Chicago. By 1890, seven San Francisco cable companies operated forty-six miles of line, 72 percent of the state's total but only 16 percent of the nation's total.[42]

Steam Traction

Donahue's steam wagon effort also led to other traction developments. In 1881, for example, John Dolbeer invented a successful steam donkey to replace oxen in skidding fallen trees throughout the logging industry, but steam traction innovations for the farm proved revolutionary. As early as the 1830s, English farmers experimented with cable plowing, using stationary engines to pull plows by cable across fields. Costly, time-consuming, and plowing only twenty acres a day, cable plows were impractical for California's large wheat farms, which needed self-propelled engines. In 1858, Warren P. Miller, of Marysville, built a tracked steam wagon and won a $400 prize at the State Fair, but for unknown reasons he did not manufacture his early forerunner to the Caterpillar tractor. Two years later, the *California Culturist* presented its readers with an engraving and description of the Wadsworth Steam Rotary Digger, inviting "the agriculturists and mechanics of California to test its principles and capacity . . . ." Yet this giant rototiller-like tractor, behind which the maker claimed one could attach any other cultivating device, probably never received a trial in California.[43]

Another inventor, ingenious New England mechanic Philander H. Standish, came closer to solving the steam plowing dilemma. In 1851, along with his father and brothers, he established Standish and Dalton Foundry in Pacheco, a farming town east of San Francisco. With financial backing from Oliver C. Coffin, a ferryboat and flour mill owner in Martinez, Standish embarked on a plow-making venture. In 1867, he built the "Mayflower," an eight-ton, twenty-four-by-twelve-foot plow with two drive wheels, each eight feet high by thirty-two inches wide. In front, a single eighteen-inch-wide, four-foot-high wheel guided the vehicle, which used an old threshing machine boiler to raise steam. The plowing mechanism resided in four vertical shafts at the rear of the vehicle. At the bottom of each shaft was a three-foot-diameter wheel to which were attached six vertical knives or coulters. As the plow moved across a field, the four wheels turned rapidly, the knives cutting four three-foot swaths in the earth to a depth of as much as six inches.

In 1868, Standish won prizes at the California State Fair and San Francisco's Mechanics Institute Fair, and, in November, he gave his plow its first real trial. At the Kiliehor farm near Mount Diablo, the "Mayflower" suffered breakdowns while plowing several hundred acres but justified itself. Plowing almost twenty acres in a day, Standish's fees netted him $2.00 per acre, while Kiliehor paid him only $.50 more than the cost for draft horse gang plowing. The next year, Standish improved the "Mayflower" and rented it out at $5.00 an acre in the Petaluma area. But his production costs at the foundry reached nearly $7,000, so he moved to Boston to find cheaper labor. There he constructed a second plow and demonstrated it in Ohio, but his improvements raised costs even higher, and his backers gave up. Standish finally settled in Ohio, and his plow disappeared after briefly exciting California farmers.[44]

Reports of steam plowing experiments continued in California through the 1870s, but mechanics turned their attention to the development of a general purpose traction machine. The Thompson Road Steamer, designed in Scotland and built by the Great Locomotive Works in Paterson, New Jersey, was tried in Stockton. It traveled at ten miles an hour pulling over twenty tons of freight, but cost stopped its adoption. In the late 1870s, Riley R. Doan of Sacramento tentatively embarked on building several fifteen-ton steam traction engines to haul grain and lumber, but the Southern Pacific Railroad allegedly bought him off to prevent competition. Meantime, the state legislature encouraged mechanics, in 1876, by offering a $15,000 bounty to any California citizen building a machine powered by steam to take the place of horses on farms or roads.[45]

In 1887, George Berry of Tulare County became the first California farmer to build and to use a self-propelled steam harvester. Burning straw, he received traction power from an Oakland-built twenty-five-horsepower

Mitchell-Fisher engine. A smaller six-horsepower Westinghouse engine received steam from the same boiler to operate the thresher. Berry claimed a season harvest average of ninety-two acres per day for his combine. At the same time, successful development of steam traction engines had occurred in Ohio, and they proved a timely innovation for Californians. Summer heat could be devastating, killing many of the horses that pulled the state's giant combines. Soon Daniel Best of San Leandro and Benjamin and Charles Holt of Stockton, all builders of harvesters, prepared steam combines for the market. Another San Leandro mechanic, Jacob Price, who had built a traction engine to serve as a road roller in 1887, joined the J.I. Case Threshing Machine Company in Racine, Wisconsin, which later built several plowing engines designed by him.[46]

In 1890, traction engines were a major feature at the State Fair. Daniel Best's "Pathfinder," weighing thirteen tons, easily pulled twelve plows, while his smaller seven-ton "Native Son" also proved its worth. Jacob Price's field locomotive weighed eight and one-half tons, and the Benecia Agricultural Works displayed a thirteen-ton engine. During the year, the *Pacific Rural Press* carried steam plowing success stories, and Best's "Pathfinder" received good notices for hauling gravel in the City of Sacramento. Logging companies placed orders with Best, and a party in King City reported plowing eight acres an hour with one of Best's twenty-two operating California engines. As the decade unfolded, traction engines from Best, Holt, and other mechanics could be found in most large California wheat fields and logging regions. Combines harvested 125 acres a day with cutting bars ranging from thirty-six to fifty feet in length, immense compared to the common twelve-foot bars used in the Midwest. Some models had 110-horsepower engines powering eight-foot drivewheels, and one Holt "Monarch in the Field" near Stockton, in 1900, weighed twenty tons. To prevent it from sinking into the soil, much of which was soft in the delta region, it had eighteen-foot-wide drivewheels.[47]

By the end of the century, hundreds of straw-burning traction engines pulled plows and combines through California, Washington, and Idaho wheat fields. The Holt Manufacturing Company alone accounted for over nine hundred of them in 1898. Steam plowing became cheaper in California, by 1910, than anywhere else in the nation, averaging $.85 an acre versus

$1.92 in the Pacific Northwest, $1.90 in Canada, and $1.31 in the Southwest. A farmer plowed fifty-one acres a day with a sixty-horsepower California machine that cost $6,000, compared to twenty-five acres with smaller, out-of-state, twenty-six to twenty-nine-horsepower machines, costing up to $4,000. Steam power relieved thousands of draft animals from plowing and belt work. Reynold Wik has suggested that, by 1913, steam engines nationwide produced as much power as could seven million horses and mules, theoretically freeing from feed grain cultivation some thirty million acres of land. Equally significant, steam traction laid the technical and experiential foundation for the gasoline tractor, pioneered by the same manufacturers.[48]

Summary

During the nineteenth century, Californians were both empowered and constrained by technology and the environment in fulfilling their energy needs. They made energy use choices by roughly calculating advantages. They weighed resource availability, environmental and geographic possibilities, costs of capital and fuel, convenience and suitability of technology, impressions of efficiency and time savings, their own sense of progress, and their material appetites. They lighted their homes with imported candles, turpentine, and kerosene, all the while seeking to develop local resources. They employed animal power and their own muscle to mine gold, cultivate and harvest crops, and settle the region. Faced with expensive, scarce human labor, they imported labor-saving machinery and invented their own. They burned fuelwood and harnessed wind power and solar energy. They innovatively tapped the waterpower potential in high mountain streams and, for use in wheat farming, they invented straw-burning steam engines.

Since waterpower and steam engines were essential to the development of industrial manufacturing, they possessed special significance to Californians who were seeking to replicate the eastern industrial revolution in the West. The location and character of California rivers nullified dreams of waterpowered industrial communities. Gold miners, mechanics, and entrepreneurs, nevertheless, pioneered various hydraulic technologies, commodified Sierra Nevada water, and, with hydraulicking, precipitated tragic environmental change. Meanwhile, other Californians, enthralled by steam power as much as any easterner, employed steam engines, but found that the absence of good regional coal deposits was a bottleneck to steam-powered industrial development. Despite high fuel costs, caused by having to import coal from the Pacific Northwest and trade wheat for it from nations thousands of miles away, a number of manufacturers produced mining equipment and processed agricultural products, innovations in steam traction actually helped break through the worldwide bottleneck in farm power technology, and steamboats and railroads provided transportation.

By 1900, although it may not have developed as an industrial region like parts of the Midwest and East, California was no longer a halcyon agricultural frontier. In fifty years its population had grown more than tenfold, from 127,000 to 1.4 million. Instant gold-rush cities created a lasting urban and commercial lifestyle, grain farming introduced industrialized agriculture, residents began overlaying the natural environment with their own second nature, and California prospered, even as an economic colony of the East. Its isolation from the rest of the country diminished over time, as the telegraph and railroad lines linked it with the nation. Moreover, its society mirrored the movement toward centralization and specialization that characterized the national experience. The Southern Pacific Company's monopoly of rail and maritime transportation became as much a national as a regional symbol of the new corporate domination that heralded the birth of urban/industrial society. In farming, bonanza grain producers overshadowed subsistence agriculture, auguring imminent development of large-scale, single crop, commercial farming. Thus, even though not an industrial region itself, California appeared at the edge of the twentieth century thoroughly sharing in the rise of urban industrialism and in the breakdown of the nation's rural, agricultural, island communities.

CHAPTER VII

Oil and Natural Gas

FOLLOWING THE GOLD RUSH, California's fast-growing population steadily consumed more nonrenewable than renewable energy resources. Energy demand outstripped the capability of available technologies to utilize waterpower, wind power, and solar energy, and it overreached the carrying capacity of accessible subregional fuelwood. By 1900, renewables provided only half the energy needs of Californians. For the other half, they relied on nonrenewable coal and petroleum. Despite having to import 90 percent of their coal and over half their refined petroleum, Californians found that their consumption of nonrenewables nevertheless surged ahead of renewable resources. If Californians were to prosper in the twentieth century, they needed to achieve independence from imported energy resources. They did so in large part by developing regional petroleum deposits. In the process, they metamorphosed their lives and the world around them.

Mechanics enhanced the value of California's asphaltic petroleum, discovering ways to use it as a substitute for coal in raising steam and manufacturing gas. New discoveries of petroleum in the Los Angeles basin and petroleum and natural gas in the Central Valley freed Californians from reliance on imported coal for industrial and domestic fuel. Automobiles and trucks replaced horse-drawn vehicles in towns and cities, and provid-

Photo 7.1. The McKittrick oil field in Kern County helped transform the southern San Joaquin Valley into a major North American oil area. From *Sunset Magazine* (December 1908): 794.

ed people with an entirely new interurban transportation technology. Similarly, the automobile extended rural mobility, while tractors hastened the industrialization of farming. Californians led the nation in adopting the automobile, designing around it their fastest growing new city, Los Angeles. And the Federal Trade Commission reported, in the 1920s, that California agriculturists relied more on gasoline and other motor fuels than farmers in any other part of the nation.[1]

Developing and employing petroleum and natural gas, along with the working out of the political, economic, and social issues that accompanied the rise of urban/industrial society, dramatically changed people's lives in a number of arenas. Growth of large oil and gas corporations, plus parallel development in the electric power industry, led the California State Railroad Commission and the state legislature to craft public utility regulation. In 1911, state adoption of direct democracy—the initiative, referendum, and recall—provided the public and corporations alike with powerful tools for directly influencing the political and social environment. Prosperity, economic growth, and events well outside California caused concern over use and depletion of natural resources; control over pumping oil and wasting natural gas became matters of serious debate at the state and national levels. California's energy experience became more and more complex as its population continued growing, people used new technologies more successfully to exploit resources, the environment was altered, and people's values about the natural world began to change.

Domestic Fuel Resources

When petroleum began to light Pacific Coast homes and businesses in the form of kerosene soon after the opening of Pennsylvania oil fields, Californians participated in an ill-fated oil rush during the 1860s. The collapse of the oil bubble prompted even some regional boosters, like John Hittell, to doubt that the state would ever produce much oil.[2] But the demand for kerosene, plus experiments with oil as steam fuel, provided thrilling possibilities for petroleum developers. If oil could be burned as fuel, the entrepreneurs would be rewarded handsomely, and California would be liberated from reliance on expensive, imported coal by an abundant domestic fuel resource. This prospect kept entrepreneurs going, drilling at promising sites and searching for methods to burn oil as fuel. They could not have known the extent of California's petroleum potential, but they held grandiose images of oil flowing from rocks like rivers. Their persistence was finally rewarded, not just with immense quantities of oil, but also with natural gas and natural gasoline (map 7.1 and figure 7.1).

In mid-1867, oil developers were tantalized by the prospect of facilitating Pacific Ocean trade with a steam fuel that required less ship storage

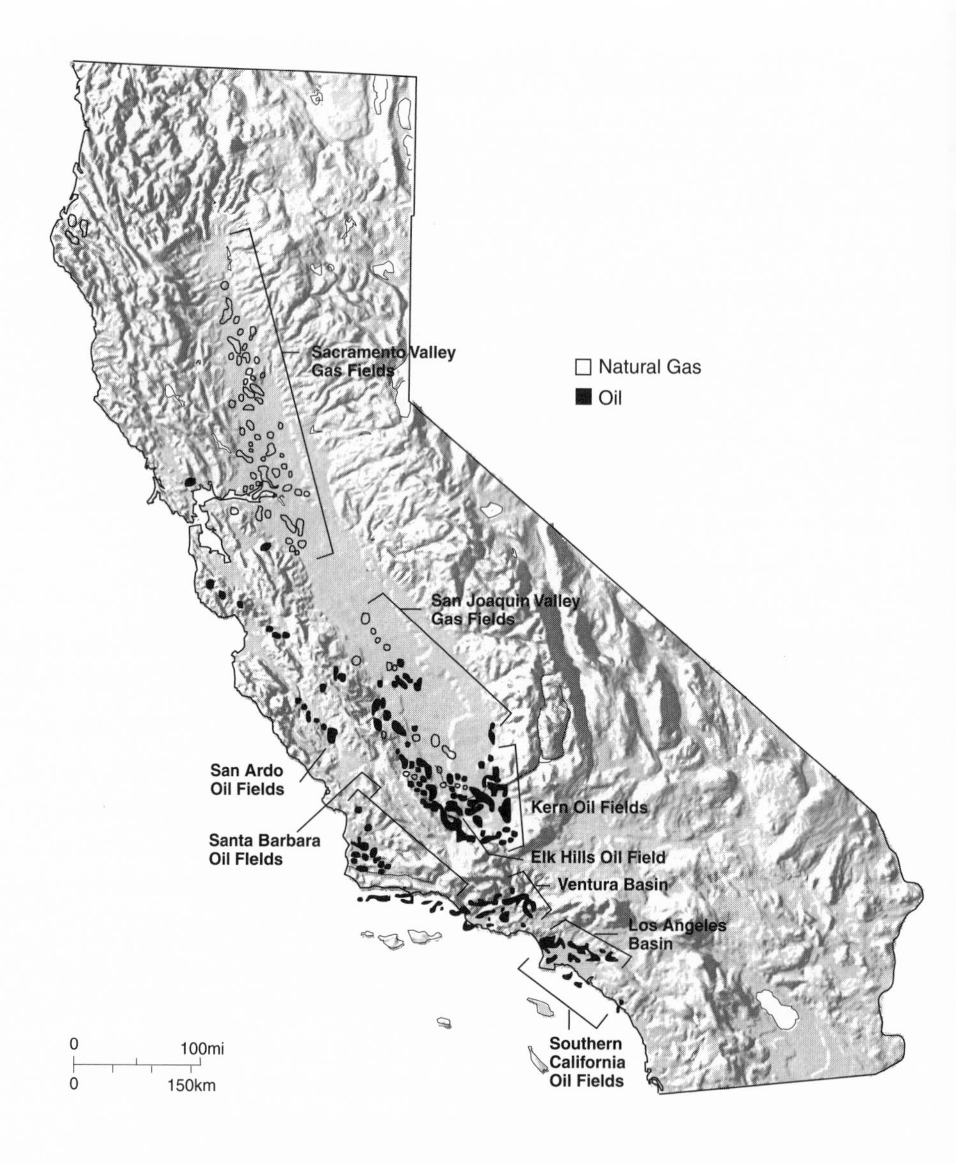

Map 7.1. California Oil and Gas Fields. Adpated from California Energy Commission, *Energy Tomorrow, Challenges and Opportunities for California, 1981 Biennial Report* (Sacramento, 1981), 12; and David Hornbeck *et al., California Patterns: A Geographical and Historical Atlas* (Palo Alto, CA: Mayfield Publishing, 1983), 115.

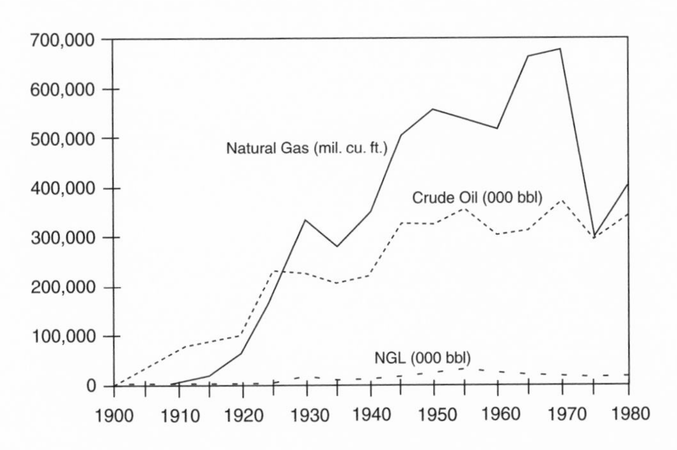

Figure 7.1. California Oil, Natural Gas, and NGL Production, 1900–1980. Source: *Supra,* Table C-8.

space than coal and less labor in stoking boiler fires. In Boston Harbor, the U.S. Naval gunboat *Palos* completed a successful oil-burning trial. California's oil entrepreneurs and steamship companies responded excitedly. In April 1868, the California Steam Navigation Company burned oil in a seven-mile trial with the ferry *Amelia,* but the effort was sullied by an imperfect oil burner and oil contaminated by dirt. In fact, had the *Amelia* test been a complete success, California's greatest annual oil production to date—twelve thousand barrels in 1866—would not have been enough to sustain the potential market. But the *Amelia*'s failed test and limited oil production did not dampen all spirits, even though oil prospecting dwindled to minor efforts near San Buenaventura. Hopes about petroleum assumed for many Californians the mythic qualities that so often characterized their optimism over other energy resources. Even as oil activities ceased, Titus Fey Cronise wrote of petroleum as fuel in the Pacific steamship trade, predicting that "should the result anticipated from the experiments now being made with this new fuel be ultimately realized, the coast region of California will be rendered quite independent of other sources of fuel supply."[3]

William Hutchinson described California's oil industry as "a shirt-tail and starve-out game for some years to come." Production depended largely on fluctuations in eastern kerosene prices, and an economic depression in the late 1870s hurt the industry. The situation seemed to change when San Francisco entrepreneurs Charles N. Felton and Lloyd Tevis entered the business. They formed the Pacific Coast Oil Company in 1879, bought out some older firms, constructed a refinery at Alameda, drilled briefly in Santa Clara County, and began successful drilling in the Pico oil district. Their efforts increased state oil production from 13,543 barrels, in 1879, to 128,665 barrels, in 1882. Meanwhile, other oil people prepared the state's first pipeline from Santa Paula Canyon to storage tanks near the San Buenaventura wharf. By late 1883, two vessels in San Francisco were being fitted up with iron tanks to carry oil in bulk between the southern coast and the Bay Area.[4]

The Price Current, a weekly in San Francisco, praised the Pacific Coast Oil Company's "energy, perseverance, and pluck" in staying out the hard times and forecasted that "the day is not far distant when oil from the oil fields of California will illuminate every home on the Pacific Coast." Plans to ship oil by water to the city, plus new discoveries in the southern part of the state, enlivened the industry. *Resources of California* predicted that in a few short months "a forest of derricks will stand like sentries on the Santa Ana mountains . . ." and observed, in a second article, that "the oil trade of California promises to attain gigantic proportions in the course of a few years. . . . We have good reason to expect that we will stand at least second in America as an oil producing region, and by consequence second in the world."[5]

Seeking an Effective Oil Burner

Because of California petroleum's heavy viscosity or stickiness, oil people soon found out that only a small percentage of it could be refined easily into kerosene. Crude oil, of course, could be used to make manufactured gas, and petroleum producers found this an attractive market. But their largest trade opportunity rested in oil's use as fuel for steam generation. Although fuel experiments in the 1860s had not proven this potential, difficulties distilling good kerosene, plus the high cost of coal, motivated both oil entrepreneurs and potential fuel oil users to find an effective oil burner. Early efforts to burn the state's asphaltic oil by mixing it with blasts of air proved ineffective. Insufficient air produced incomplete combustion and dense clouds of black smoke. Greater amounts of air, while eliminating the smoke, produced a short flame which, according to a report by California's state mineralogist, concentrated heat at the boiler's front "to such an extent as to rapidly burn out and destroy the iron."[6]

Experiments with various burners had been carried out in England, France, Russia, and the East, but Evan A. Edwards, a San Buenaventura storekeeper and oil businessman, discovered California's own answer to the burner problem. In 1876, he joined two other oil enthusiasts to revive drilling on known nearby oil lands, and he experimented with oil as fuel at his small San Buenaventura refinery. In 1882, not long after moving to Los Angeles to join the Continental Oil and Transport Company, he patented a successful oil burner. It used a jet of steam to atomize the petroleum and directed a fine spray of fuel into the furnace. The steam retarded the oil's combustion, resulting in a steady, long, smokeless flame that spread evenly across the boiler. Edwards, who later joined Union Oil's fuel oil staff, turned over distribution of his burner to Sutherland Hutton and former congressman H.H. Markham. They formed the Los Angeles Oil Burning and Supply Company in 1885, and built up a trade with foundries, brick kilns, and other steam users in southern California.[7]

Although Edwards's was the first successful burner, others were put into use by the mid-1890s. A.J. Stevens, master mechanic at the Southern Pacific Railroad's Sacramento shops, worked with oil as fuel as early as 1879. He continued his development and testing, particularly after seeing plans for a burner devised by Scotsman Thomas Urquhart, who was Superintendent of Motive Power for the Garzi-Tsaritzin Railroad in southeast Russia in the early 1880s. Another railroad master mechanic, W. Booth of the Peruvian Central Railroad, made a burner that the Southern California Railroad acquired. They put it to use and improved it during the late 1890s. Still another inventor, speaking before the Technical Society of the Pacific Coast, claimed his steam-type burner provided a six-to-one efficiency over burning coal, but members were skeptical that a burner could provide more than a two-to-one efficiency ratio.[8]

Because California railroads consumed more steam fuel than any other western industry, their high operating costs attracted them to the potentiality of oil as inexpensive fuel, and they assumed an essential role in

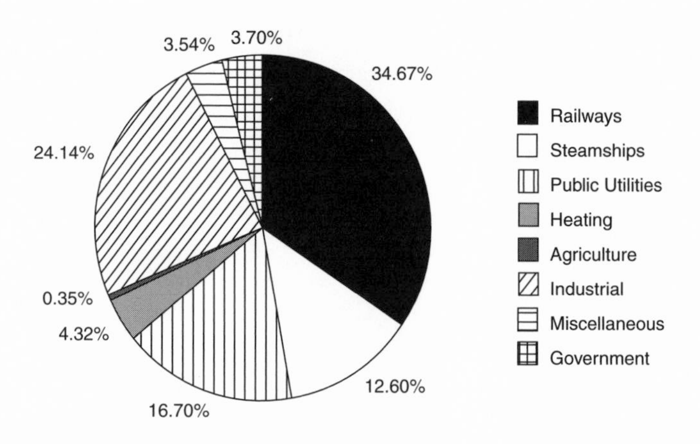

Figure 7.2. Consumption of Fuel Oil in California, 1919. Source: *Supra*, Table C-9.

proving its effectiveness. By 1895, the Southern California Railroad, soon part of the Santa Fe system, had three steam locomotives operating on oil, and Southern Pacific put its first into service. Six years later, Southern Pacific's Los Angeles division discontinued coal, and a year later, in 1902, the Sacramento division adopted oil. Its Salt Lake division continued burning coal from company-owned Utah mines until 1912, when it also switched to oil. In addition to being cheaper than coal, railroads found oil to be cleaner, more efficient, and easier to load into tenders. By 1919, oil provided good service at half the cost of coal, and railroads became the state's principal petroleum guzzlers (figure 7.2). To ensure continued cheap oil supplies, both the Southern Pacific and Santa Fe railroads went into the oil production business in California during the late 1890s, expanding their holdings in the early 1900s.[9]

In the maritime industry, Southern Pacific Railroad's San Francisco Bay ferries built on earlier experiments with fuel oil. Between 1885 and 1887, the company carried out a series of tests on several ferries, achieving an average 20 percent or more cost savings over coal. A tragedy in 1888, however, raised serious questions about oil as fuel. The double-end ferry *Julia* had been converted to oil for seven months. On the morning of February 28, 1888, steaming near Vallejo, she exploded and caught fire, causing the loss of twenty-eight lives. The coroner's jury concluded that fuel oil played a contributing role in the disaster and recommended it be banned on passenger vessels. Captain H.S. Lubbock, the U.S. Department of Treasury's Supervising Inspector for Steam Vessels, held hearings and concluded the "prime cause of the destruction . . . was the explosion of petroleum gas within the furnaces," despite evidence that the boiler was the fault. On his recommendation, the Secretary of the Treasury banned fuel oil on all American vessels.[10]

The year before the disaster, fire had destroyed the warehouses of three San Francisco kerosene merchants, and another fire had struck the Fulton Iron Works. All were blamed on oil, and, in late 1887, the Pacific Insurance Union urged the San Francisco Board of Supervisors to pass an ordinance curbing the use of fuel oil. The Board refused, but in the wake of the *Julia* tragedy and subsequent Treasury Department action, it passed a law raising the burning test for fuel oil from eighty to ninety degrees Fahrenheit. Although the law did not stop use of fuel oil, the petroleum industry and local fuel consumers felt threatened. They lobbied unsuccessfully for its repeal, while Union Oil's Thomas Bard got Congressman William Vandever to introduce successful national legislation permitting fuel oil on nonpassenger vessels.[11]

By the end of 1888, the storm over *Julia* had subsided, but steamer fuel oil adoptions proceeded slowly. Through the next decade, uncertain oil supplies slowed adoption, as did insecurity about safety. Treasury Department inspectors insisted that ships equipped to burn oil have strengthened hulls and fuel storage areas as a precaution against volatile and explosive oil gases. By 1904, the industry—except the Navy, which continued to have concerns about oil on warships—had dealt with safety problems and accepted oil's benefits over coal. One hundred thirty-seven coastal and inland vessels had converted, and bunkers held fuel oil in Los Angeles, San Francisco, Port Hartford, and Portland. Since burners using steam atomization required additional supplies of fresh water, at first only coastal vessels, tugs, ferries, and river steamers converted. But newer, improved, air-atomization burners that utilized steam-powered air compressors or rotary blowers soon made it feasible to equip ocean ships with fuel oil. By 1915, the space- and labor-saving benefits of fuel oil over coal had seized all who were progressive in the maritime industry. Pacific-based merchant ships generally burned oil, the Navy was converting, and the opening of the Panama Canal stimulated oil's use in the Atlantic.[12]

Beyond rail and maritime transportation, oil also meant cheap and efficient fuel for industry. With the introduction of Edwards's burner, petroleum companies sought industrial users. They convinced a variety of customers to try oil: the California Sugar Refinery, Standard Soap Company, San Diego Rapid Transit Company, and the Del Coronado Hotel all tested it in the late 1880s. Within a few years, a Los Angeles iron mill, a brick and terra cotta company, an electric railway, and several other firms had adopted fuel oil, and among its biggest early industrial users were agricultural processing plants. The Chino Valley Beet Sugar Company refinery, Watsonville's Western Beet Sugar Company refinery, and the Union Sugar Company refinery near Santa Maria all became heavy oil consumers in the 1890s. The Chino refinery used thirty thousand barrels in 1893, and the Western plant contracted for sixteen thousand barrels the following year. Opened in 1899, the Union refinery began with some thirty thousand barrels a year, and, according to a 1900 oil promotional booklet, two of these refineries used approximately one hundred thousand barrels each during their four-month seasons.[13]

The viscosity of California petroleum also affected the transport of oil fuel to consumers. The San Francisco Bay Area, the state's most industrialized and populated region, provided the earliest major fuel oil market. In 1896, after using ships to carry oil in barrels from the southern coast to the bay area, Union Oil built the Pacific Coast's first true tanker, *George Loomis.* Soon other tankers comprised a small fleet that carried crude oil in bulk to refineries in Alameda County, and the Southern Pacific Railroad put tanker cars into service. But transferring oil to tankers cost both money and time, so companies turned to pipelines. Short lines carried oil from coastal fields to ocean and rail shipping points, and long-distance pipeline service developed after the opening of the Kern River and other southern San Joaquin Valley fields. In 1902, Standard Oil began construction of a 275-mile line from Kern County to its Bay Area refinery in Richmond, and two years later the Coalinga Oil Transportation Company began a second 110-mile line from Coalinga to Monterey.[14]

Because much of the state's oil was too sticky for easy pumping, howev-

Photo 7.2. Deposits of lignite coal gave Coalinga its name, but, after 1900, oil made the Fresno County town prosperous. From an advertisment in *Sunset Magazine* (December 1912): n.p.

er, California pipelines required operational innovations and greater investments than elsewhere in the country. Engineers on Standard's Kern-to-Richmond line first tried mixing the oil with water but got an emulsion even harder to pump, so they brought civil engineer Forrest M. Towl from their New York headquarters to investigate the problem. He recommended reducing the oil's viscosity by heating it, which local engineers accepted although it had not been tried on long-distance pipelines. They insulated the eight-inch pipe by wrapping it with an asbestos blanket and burying it three feet underground. Then they installed four boilers and two high-pressure compound pumps at each pump house (spaced twenty-eight to thirty miles apart), from which exhaust steam fed into a heater and raised the oil temperature to 180 degrees Fahrenheit. Finishing the line in 1903, they began pumping oil on March 19th, but because it cooled by the time it moved ten to fifteen miles through the line, it did not reach Richmond until July 18th.[15]

Next, Standard unsuccessfully tried to solve the viscosity problem by mixing the oil with hot water. When this failed, they mixed the Kern River oil with lighter crude from Coalinga, but this only increased the flow from

Photo 7.3. The heart of the Bakersfield oil field in Kern County. From *Sunset Magazine* (December 1908): 796.

one thousand to three thousand barrels a day. In early 1904, they added nine new pump stations between the original ten. Although construction cost $18,800 per mile, $8,550 more than that of mid-continental lines, the flow increased to twenty thousand barrels daily. Meanwhile, other companies continued to experiment, some using an auxiliary steam line to heat the oil pipe and others partially refining the crude oil before pumping it. The Associated Oil Company built a rifled pipeline to San Francisco into which they injected water to form a film between the oil and pipe, but separating the water and oil at the terminus proved difficult, so they abandoned the system. Most companies finally turned to Standard's technique, and 2,019 miles of trunk pipelines were built by 1919.[16]

Oil Consumption

The 1895 opening of the Los Angeles oil field glutted the fuel market. J.A. Graves recalled that at first "people were loath to change from coal to oil, as none of them knew how long the oil being produced in the city would last and there were no storage facilities." But the field kept producing, entrepreneurs built storage facilities, and an enduring and growing demand for the product developed. By 1900, new fields at Coalinga, Sunset, McKittrick, Kern River, Fullerton, Brea-Olinda, and offshore at Summerland seemed to assure a bountiful, steady oil supply; wells went in at twice the national rate and successful ones exceeded the national average by 20 percent. California became the leading oil-producing region in the world. Whereas state per capita coal consumption collapsed to less than one-half ton versus the nation's 5.3 tons, Californians burned 80 percent of their oil for fuel, compared to 20 percent nationally.[17]

But the industry continued to experience rocky development. New fields led production to outstrip demand, while capped wells in 1904, and again in 1912, reversed the situation. Unstable periods of supply and demand adjustment led to business combinations, pools, and horizontally integrated firms for the storing and marketing of oil, which were soon followed by highly developed, vertically integrated companies. By 1920, seven firms, which had produced under 50 percent of the state's crude oil in 1914, controlled 68 percent of the state's proven oil land and accounted for 71.5 percent of its crude production. Perhaps worse, inconsistent production led to fears that oil reserves might not last. In 1908, the United States Geological Survey estimated California's total reserves at 8.5 billion barrels, enough for 113 years at an annual consumption rate of 75 million barrels. A much more optimistic 1911 estimate doubled this, predicting a 230-year supply with a reserve of 17.2 billion barrels. But the experts did not fully agree and consumption steadily increased. Between 1900 and 1910, state per capita oil consumption climbed from three to twenty-eight barrels, compared to one-half to two barrels nationally.[18]

Both national and state government saw the oil supply question in terms of natural resource conservation, an important part of America's progressive reform agenda. Therefore, in 1908, the Geological Survey recommended to the Secretary of the Interior that filing of claims on publicly owned oil lands, in California and elsewhere, should be suspended. If this was not done, argued the agency's petroleum experts, an adequate supply of oil would not last for "the remainder of the century or even for the next fifty years."[19] Moreover, since the Navy had determined fuel oil superior to coal and was dependent on petroleum products for lubrication, oil conservation assumed added national importance. In 1909 and 1910, President William Howard Taft issued a number of land withdrawal orders which closed some three million acres in California to oil exploration and production, and, in 1912, he ordered creation of two naval petroleum reserves in the withdrawn lands. Additionally, the government undertook litigation to invalidate allegedly fraudulent filings on some of the withdrawn lands, including 172,000 acres owned directly or indirectly by the Southern Pacific Railroad.[20]

Californians varied in their response to government land withdrawals, as well as to the meaning of conservation. The state legislature created a Conservation Commission, in 1911, to study natural resource conservation, and the Commonwealth Club, formed in San Francisco in 1904 to discuss state affairs, focused on the issue in 1912. The club's "Section on Conservation" supported the withdrawals, noting "there is at the present time no need for the oil contained in this territory, and it should be husbanded. . . ."[21] But club members in the oil industry, such as S.C. Graham of Los Angeles's "Oil Conservation Association," opposed arbitrary land withdrawal as a conservation measure. He blamed government for causing waste of oil by requiring entrepreneurs who sought title to oil land to immediately develop the oil on it. Club member A.L. Weil added, if conservation meant getting "a maximum of use with a minimum of waste," existing government policy "brought about the diametrically opposite result."[22]

During the next several years, California's petroleum consumption continued to grow. After 1916, it generally exceeded production, and World War I aggravated the problem. The state's oil industry and the Southern Pacific used the war as grist for their continuing effort to persuade Congress to reverse land withdrawals, end litigation, and enact a leasing bill.[23] On 12 May 1917, State Oil and Gas Supervisor R. P. McLaughlin observed that "a serious shortage of oil is apt to occur . . . [and] such a shortage would cripple transportation facilities . . . , and would also seriously disarrange most other industries on the Pacific Coast." That same month, Governor William D. Stephens created within the State Council of Defense a Committee on Petroleum to study and recommend what to do about the oil situation.[24] At the national level, a wartime fuel administration was established to ensure that critical petroleum needs were met.

State and federal petroleum experts were sympathetic to oil interests. During the summer and fall after completing its report, the state petroleum committee worked successfully with industry representatives and the U.S. Department of Justice to restart production on some of the lands under litigation. It also gained military draft exemptions for oil workers and helped to secure delivery of scarce oil well material and supplies. But despite the wartime emergency and intense lobbying, Congress neither reversed land withdrawals nor passed a leasing law. The federal government still placed petroleum conservation as its highest priority, even though this did not seem to bode well for the Pacific Coast where oil provided some 70 percent of the energy for heat, light, and power. California's petroleum committee concluded that production and transportation difficulties made it infeasible to substitute coal or powdered coal for fuel oil in railroad, utility, and industrial use; and the Southern Pacific provided both the committee and the United States Fuel Administration with a variety of reasons why it should not reconvert to coal-burning locomotives on its Pacific Northwest division or, for that matter, anywhere else in its system.[25] Oil remained the Pacific Coast's fuel of choice.

World War I did not create petroleum shortages on the Pacific Coast. Priority classifications for petroleum deliveries and, during 1918, "Gasless Sundays," were experienced only east of the Mississippi River. But the war did make more acute the awareness that oil was a nonrenewable resource. Stored stocks from earlier periods of flush production declined. Gasoline, which was refined only with difficulty from California crude, had to be imported from mid-continental fields. Standard Oil, for example, brought twenty million gallons to the Pacific Coast in 1919 and eighty-four million gallons in 1920. Californians consumed 114 million barrels in 1920, establishing an annual consumption rate that reduced the Geological Survey's most optimistic estimate about long-term oil supplies from 230 to 150 years. This deeply concerned experts. In December, David M. Folsom, professor at Stanford University's School of Mines and Federal Oil Director for the Pacific Coast during the war, told San Francisco's Commonwealth Club that, "so far, we have been fortunate in discovering new fields at critical times" but that declining older fields, limited untapped reserves, and the digging of more and more dry wells did not bode well.[26]

The scientific search for oil, however, was just beginning. Companies only began permanently employing geologists during World War I and afterwards, and seismographic exploration for oil was not adopted widely in the state until the 1930s. Several new discoveries remained in the future, beyond Folsom's vision, the first coming between 1921 and 1923 in the Los Angeles basin. At the end of the decade, new discoveries came near Taft and in the Midway-Sunset field which had been opened shortly after 1900. Several other pools also were opened in Kern County during the 1930s. The last big single discovery came at Wilmington, near Los Angeles, in 1936. Furthermore, the war had prompted work on the problem of refining California oil, and by 1920, notes Mansel Blackford, "California possessed the most technologically advanced oil plants in the United States." Through new fields and improved refining plants, California maintained its fuel independence and became the only self-contained refining region in the nation through the mid-1940s. Only after World War II did the state once again have to import fuel.[27]

Toward Centralized Manufactured Gas Production

In conjunction with oil, as well as independent of it, Californians also discovered natural gas. First used in Fredonia, New York, in 1821, natural gas later was found in connection with oil in Titusville, Pennsylvania. Oil operators there saw it as a nuisance, and introduced the industry practice of flaring it at the well. Nevertheless, by the 1870s, it was being used for lighting and heating in parts of Pennsylvania, and by 1884 some 150 natural gas companies had been chartered in the Midwest. In California, Stockton residents tapped natural gas during the 1860s, and two decades later reports came of other northern state gas wells, plus gas associated with oil near Santa Barbara and Los Angeles. The state mineralogist, probably reflecting the oil industry's initial attitudes about gas as a annoyance, did not speak of it with optimism: "It is doubtful whether natural gas will be found in our petroleum sections . . . in quantities approaching anywhere near the amount found in the Eastern States. . . . Under the existing geological conditions, it would seem folly to expect that prospecting for gas, along our petroleum belt, would pay interest on waste of time and money."[28]

Beyond the Stockton area, few people used natural gas, but elsewhere the manufactured gas industry, which first had developed using coal and then oil to make gas, expanded with long-distance, high-pressure pipeline technology. Pacific Gas and Electric Company, which, after 1900, consolidated the operations of several independent providers to become the principal northern California gas manufacturer, saw a good market for gas in local industrial use and, as electricity took away their lighting customers, for domestic cooking and heating. This perception, plus a vision of economies of scale and service to suburban communities in the San Francisco Bay Area, led the firm to become a pioneer in high-pressure gas distribution. Its engineers abandoned some common practices developed in the East, such as giving up cement in cast iron pipe sealing in favor of lead wool, which gave greater flexibility. They also adopted steel pipe and sleeve couplings, in 1902, installing a sixteen-and-eight-tenths-mile two-inch line that operated under pressures up to one hundred PSI.[29]

Between 1902 and 1920, the company doubled its gas production facilities, more than doubled the mileage of its distribution pipelines, and almost trebled its customers. A thirty-mile high-pressure extension from San Francisco's Martin Station south to Palo Alto, in 1907, signaled an extensive effort to reach suburban communities from large central production plants. Such an effort necessitated improved pipelines, so they adopted oxy-acetylene welding for improved steel pipe joints. They first tried it, in 1912, on a one-mile stretch of forty-foot-by-eight-inch welded sections, and two years later installed a welded sixteen-inch high-pressure steel pipe loop stretching half way around San Francisco. The project gained some notoriety in supplying gas to the 1915 Panama Pacific International Exposition, the first such exposition to have full gas service.[30]

Other California gas companies saw the same opportunities and sought similar operating economies and expansion through high-pressure service. In 1911, the Southern Counties Gas Company, serving seventeen towns near Los Angeles, adopted high-pressure trunk lines, which permitted retirement of three small production plants. It eventually hoped to have its "entire system served by two plants, operating at each end of the through trunk line, a distance of about seventy-five miles." Farther south, the San Diego Consolidated Gas and Electric Company brought high-pressure service to much of their 344 miles of gas mains, while north of Los Angeles the Central California Gas Company was organized, in 1912, to consolidate two smaller San Joaquin Valley companies and serve seven communities with high-pressure operations.[31]

California gas companies employed long-distance, high-pressure transmission lines and centralized gas production facilities, which bore striking similarities to the electric power industry's interconnected production and distribution system. Pacific Gas and Electric led the trend toward centralized manufactured gas production and high-pressure distribution. By 1923, its five northern California subregional systems, some with more than one large central gas manufacturing plant, delivered 12.8 trillion cubic feet of gas to 320,000 customers in several counties and fifty-eight cities. The "Super-Gas Systems of Pacific Gas and Electric Company," said the firm, paralleled the long-distance transmission and interconnection of its electrical system. In southern California, Pacific Lighting consolidated gas companies serving Los Angeles and other areas and accomplished similar interconnections through the 1930s.[32]

While such gas company consolidations brought efficiency and economies of scale to the industry, some members of the public never really believed that they were receiving the lowest rates and the best service to which they felt entitled. In Los Angeles, early concerns about gas service helped pave the way for legislation, in the 1870s, that brought state government into the field of public utility regulation. State regulation of large utility companies eventually evolved into the most widely accepted solution to rate and service problems; however, another response appeared in the form of community ownership. Toward the end of the nineteenth century, municipal gas and other utility system ownership became a part of the progressive reform movement, the struggle to adapt to a fast-growing urban/industrial society. Public ownership seemed the only remedy to get rid of what its proponents characterized as greedy utility owners who would not provide decent service. From Pasadena to Gilroy and Palo Alto, California cities bought out private firms. Los Angelenos formed the Municipal Ownership Party, in 1905, to respond to the problem, and their mayoral candidate Stanley B. Wilson proclaimed: "The private ownership of public utilities is the cause of all political corruption."[33]

Private gas companies fought municipal ownership at every juncture, relishing the opportunity to undermine it in the eyes of the public. When the City of Gilroy, which purchased its system from a private provider in 1905, turned its back on municipal operation four years later, the industry's private sector immediately portrayed the city's action as confirmation that public entities could not operate a gas utility successfully. This prompted no little outrage from municipal ownership advocates across the state, and Gilroy's Mayor Walter G. Fitzgerald appeared before the annual convention of the League of California Municipalities to defend his city's action. He denied that leasing out operation of Gilroy's gas plant was "any admission or proof" that municipal ownership was a failure. Yet, he did not agree with Los Angeles's Stanley Wilson that the roots of political corruption rested in private utility ownership. "Municipal ownership," he told his fellow public officials, "is not a panacea for all the trials a city is heir to."[34]

Fitzgerald argued that "municipal ownership has its own deficiencies," the principal one involving the structure of local government itself. Like most California cities, Gilroy's government was one in which an elected and frequently changing part-time council served as both the legislative and executive body. Fitzgerald felt that few council members knew "anything about the manufacture of gas, or electricity or the management of a gas works or electric light plant." They had to "depend solely and helplessly upon the engineer or superintendent" to manage the utility operation. Here was the rub: "If the superintendent does not know his duty or perform it properly how is [the] Councilman to know?" A council of amateurs, thought Fitzgerald, could run a simple gravity-feed water system, but before gas and electric light plants could "be operated with entire success by municipalities, some form of government, possibly through some form of civil service, or through boards of commissioners wherein the term of office shall be long and where the terms of all officers do not all expire at the same time, must be established."[35]

Many, but not all, of California's community leaders agreed with Fitzgerald, and local governments gave little more than token opposition to gas companies, which continued consolidating and tying central gas manufacturing plants into high-pressure lines that served a number of communities. In fact, Fitzgerald's analysis of the drawbacks inherent in local government structure was precisely right. City councils found it extremely difficult to oversee public works and utility operations in their communities. Civil engineers in the nascent field of municipal engineering in California tried to help educate local officials, and Ernest McCullogh, one of California's first fulltime city engineers, even published and distributed free to local officials his little book, *Public Works: A Treatise on Subjects of Interest to Municipal Officers.*[36]

But even as Fitzgerald offered his analysis, the Board of Trustees of Ukiah City, one hundred miles north of San Francisco, had taken action to solve the problem. In 1904, they passed an ordinance creating a new municipal position, that of Executive Officer. Four years later the California State Supreme Court upheld Ukiah's action, and by 1920, twenty-one California cities had adopted what became known as the city manager form of government. In 1925, with a city manager in place, Long Beach boasted complete success in operating a gas system; however, by this time most Californians had accepted regulation of gas companies through a state public utilities commission.[37]

Regulating Natural Gas

The issue of natural resource conservation directly affected natural gas development. Even after adoption of gas for domestic and industrial use in the Midwest, flaring it at the well remained the industry's standard practice. During the early decades of the twentieth century, close to 90 percent of the nation's oil-related natural gas (trillions of cubic feet) was wasted in this manner, and virtually all Pacific Coast oil producers flared wellhead gas. Some members of the industry argued that there was little choice. It was economically and technologically infeasible to do otherwise. But resource conservation proponents nationwide mounted strong opposition to natural gas wastage. Californians attacked the issue in 1911, when that year's session of the state legislature enacted a generation's backlog of progressive reform measures. Following the lead of eastern states, which already had passed laws prohibiting flaring or otherwise wasting gas, California state legislators enacted a similar ban.[38]

New gas discoveries in the San Joaquin Valley put the industry in an awkward position. Either they could seek ways around the newly mandated conservation law, or they could invest in developing gas as a usable and, possibly, marketable product. Since the legislature had worded its conservation law to prohibit "willful" wastage, many oil operators chose to ignore the ban but never to admit or concede that they were willfully wasting natural gas. In response to this tactic, Mark Requa, later head of the Oil Division of the United States Fuel Administration during World War I, argued before the Commonwealth Club, in December 1912, that government had to enact stronger laws to curb wastage and regulate the transportation and sale of gas. The state legislature, however, did not move to strengthen its prohibition, nor did it enact related measures.[39]

Meanwhile, some oil companies saw profit in marketing rather than flaring or otherwise wasting natural gas. The first large find of oil-related gas to be marketed in California was discovered, in 1907, near the central coast town of Santa Maria, and a local gas company arranged with oil producers to market it. In Kern County in 1909, Standard Oil tapped a single gas well in its Midway oil field that yielded seven million cubic feet per day. This and a second equally productive well led Standard to form the California Natural Gas Company and to sell their product to manufactured gas firms operating in southern California. "In the spirit of conservation," said Standard's Fredrick H. Hillman, the company sought to market as much gas as possible, hoping "that through education as to its use we could develop a valuable product and one that would eventually be profitable to us."[40]

Standard's Midway gas field developed quickly. By the end of 1912, its fifteen wells yielded fifty million cubic feet of gas per day. The opening of additional wells, such as McNee No. 9, which alone produced thirty-five million cubic feet per day, thrust California into third place as the nation's largest gas-producing state, just behind West Virginia and Pennsylvania. Pipeline construction began in earnest. In 1912, one pipeline connected the Midway field with Bakersfield, where the San Joaquin Light and Power Company switched its manufactured gas system to natural gas. The utility's owners then allied themselves with one of the founders of Pacific Gas and Electric to form the Midway Gas Company to lay a twelve-inch, 112-mile pipeline to Glendale, near Los Angeles. They completed it within a year. Although some observers feared natural gas production would soon collapse and its uses be replaced by electricity, new gas fields and new wells, such as Elk Hills Hay No. 7, which came in with an estimated 187 million cubic feet per day in 1919, dispelled pessimism. By the early 1920s, Los Angeles and four score more southern California cities were well supplied with natural gas.[41]

When the Midway Gas Company's pipeline opened in 1913, the editor of the *Journal of Electricity, Power and Gas* observed that "electric power no longer has a monopoly on the claim for long distance transmission at high pressures." Like hydroelectric sites, the location of natural gas fields were often far from consumer markets and required substantial transmission developments. Improvements in steel pipeline construction permitted the building of pipelines that stretched several hundred miles between San Diego and San Francisco (map 7.2). Pacific Gas and Electric's 287-mile gas line from Buttonwillow to the San Francisco Bay Area and Standard-Pacific's line from Kettleman Hills to Richmond were the longest in the West. And discoveries of dry natural gas fields in northern California near Marysville, in 1933, and at Rio Vista, in 1936, meant more pipelines. By World War II, a statewide natural gas line network comprising close to twenty thousand miles of pipe reached from the Mexican border north to Oregon, mirroring the region's high-tension electric power grid.[42]

Besides simply delivering gas to fulfill the fuel needs of California's industrial and domestic consumers, oil producers discovered other useful gas products. For example, natural gas also yielded an additional important product, gasoline. Refiners could produce asphalt, road oils, and lubricating oils from the state's high-viscosity crude petroleum, but they could not refine good kerosene or gasoline. To help with this problem, both the Union Oil and the Pacific Coast Oil companies had hired chemists in 1892. Pacific Coast's chemist, Walter Price, became acquainted with a second chemist, University of California graduate Eric A. Starke, who worked at a San Francisco gunpowder company. In 1894, Starke developed a "hot-treat" method for refining kerosene, which he turned over to Price. Two years later, he made another discovery that totally eliminated carbonaceous properties from refined kerosene, and Pacific Coast hired him. Starke's work put Pacific Coast technically ahead of other refiners in

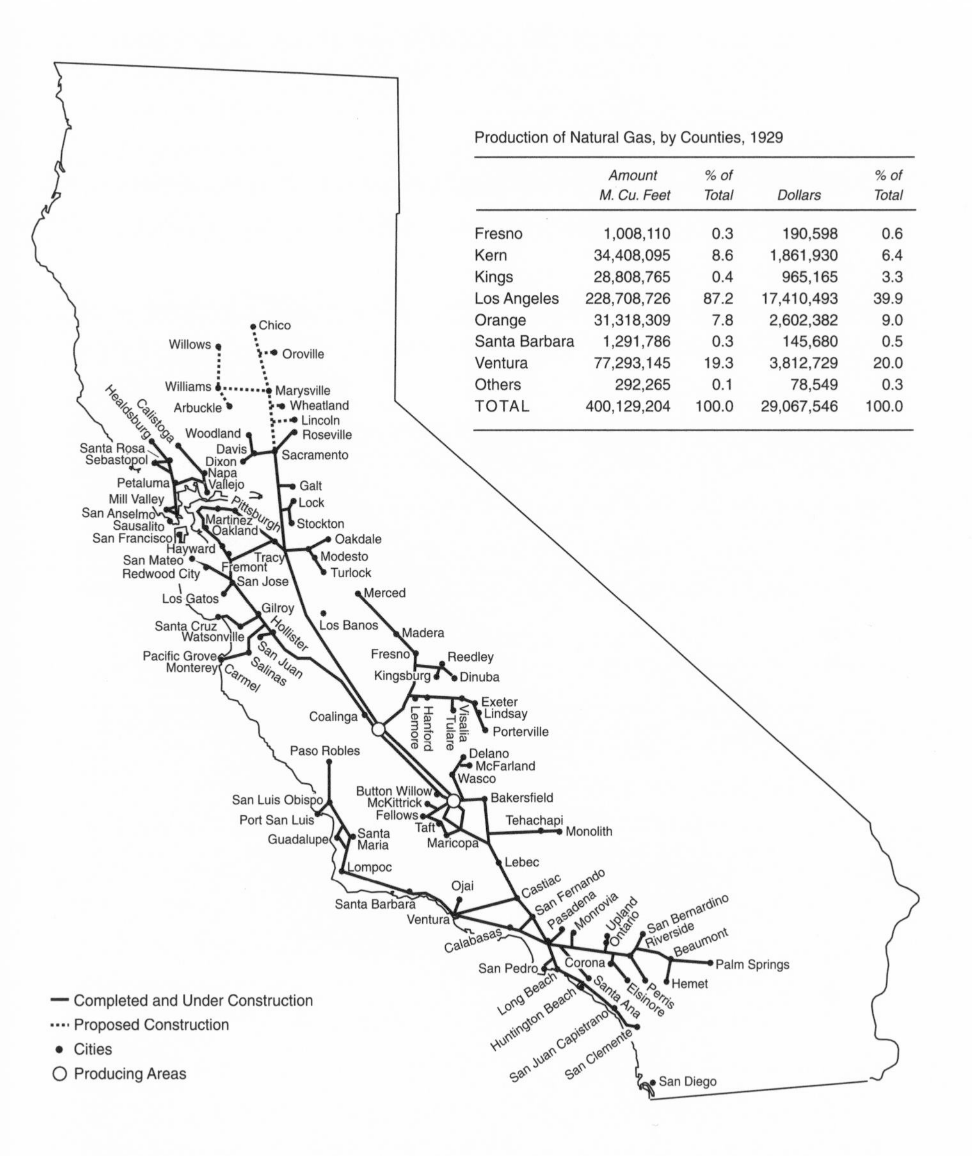

Production of Natural Gas, by Counties, 1929

	Amount M. Cu. Feet	% of Total	Dollars	% of Total
Fresno	1,008,110	0.3	190,598	0.6
Kern	34,408,095	8.6	1,861,930	6.4
Kings	28,808,765	0.4	965,165	3.3
Los Angeles	228,708,726	87.2	17,410,493	39.9
Orange	31,318,309	7.8	2,602,382	9.0
Santa Barbara	1,291,786	0.3	145,680	0.5
Ventura	77,293,145	19.3	3,812,729	20.0
Others	292,265	0.1	78,549	0.3
TOTAL	400,129,204	100.0	29,067,546	100.0

Map 7.2. California Natural Gas Pipelines, 1930. Source: *California Journal of Development*, 20 (December 1930): 25.

treating California oils, and when Standard Oil bought Pacific Coast in 1900, it put Starke's process into commercial practice. But they could still not refine gasoline.[43]

Between 1900 and 1917, the newly introduced automobile increased demand for gasoline. At its Richmond and El Segundo refineries, Standard Oil tried cracking experiments—using heat to break down the complex hydrocarbons of heavy crude into lighter fractions—and it built new facilities to use a cracking process developed by William Burton at its Indiana refineries. But efforts by Standard and other refiners did not produce much or very good gasoline. Natural gas associated with oil, sometimes called "wet" gas, provided an alternative. Through a compression process first used in West Virginia and Pennsylvania and commercialized by 1910, one could produce natural gasoline. Oklahoma soon became the leading producer, and compression plants appeared in California oil and gas fields. Production increased quickly, and Standard's chemist, Eric Starke, pushed it along by developing a gas trap, which separated gasoline from gas coming out of high-pressure wells. Then, in 1912, state gas operators adopted a new absorption process that extracted natural gasoline from "dry" gas.[44]

With early natural gas production far exceeding pipeline and fuel market capacities, natural gasoline proved to be a most useful product. In fact, California's natural gasoline became highly prized, for it possessed a high proportion of iso-octane hydrocarbons that gave it a natural high-octane rating for smooth engine performance and superior power. Demand for it grew so quickly that production could not keep up. At the same time, gasoline refined from crude oil became increasingly important, particularly during World War I, and during the 1920s a process with its roots in California finally provided a solution for refining the state's asphaltic crude oil. The Universal Oil Products Company introduced a cracking process that it developed from a pipe still demulsifier invented, in 1909, by Jesse Dubbs, a Santa Maria refiner. Improved by his son, with the appropriate if eccentric name of Carbon Petroleum Dubbs, it successfully cracked California crude into gasoline, a product that became fully acceptable when blended with natural gasoline.[45]

Through the 1920s, California narrowly trailed Oklahoma in natural gasoline production and surpassed it during the following decade. As natural gas output increased, natural gas companies also followed the national industry in developing and marketing other natural gas products. Liquefied petroleum gases, such as propane or "bottled gas" that was self-vaporizing above forty-four degrees Fahrenheit, were introduced to consumers as fuel options. Propane was aimed at the domestic market, butane found service in industry, and pentane was used in rural central station manufactured gas systems. Because of the continuing importance of fuels to the state, Standard Oil and Shell Oil companies in California led the nation during the 1930s in marketing liquid petroleum gases under the trade names Flamo and Shellane. They particularly brought new fuel options to rural residents of the state, to small towns that did not have piped natural gas service, and to industry.[46]

In 1928, Shell Oil experienced a production boom at its Signal Hill oil field near Los Angeles. Not wanting to waste the gas associated with the field's flush production, and already selling gas to the Southern California Gas Company, Shell arranged to pump the excess back into the ground at its Dominguez Hill oil field. The experiment proved fruitful. The gas repressurized the oil field, helping Shell to retrieve oil more easily from its Dominguez Hill wells, and the gas itself was stored until consumers required it. Union Oil Company joined Shell in this endeavor, and, by midyear, the two firms were putting into underground storage 14.5 million cubic feet of gas per day.[47] For oil companies that saw the waste of gas as inefficient business practice, and for conservationists who saw it as inefficient environmental practice, storing gas through repressurization of oil fields proved a most favorable development.

Despite the waste reductions gained through marketing gas and otherwise using it, enactment of additional oil and gas conservation laws became an issue during the period of flush oil production extending from the 1920s into the 1930s. Virtually all producer states legislated stronger oil conservation programs and, in 1935, formed "An Interstate Compact to Conserve Oil and Gas." As a by-product of control over oil production, gas wastage also was lessened. But, in California, just the opposite occurred. Despite a fast-growing natural gas industry, almost 30 percent of the state's oil-related gas was being flared by producers who claimed they had no other course of action. In 1927, the governor created a Gas Conservation Board to seek voluntary production curtailments and recommend other actions. Two years later, the legislature passed the Lyons Act, which went beyond the state's 1911 prohibition of willful wastage to halt "unreasonable" waste. The state supreme court upheld the new law, and as a consequence, gas utilities wrote new contracts with oil producers who previously had been unwilling to complicate their oil work with gas operations. Electric utilities, among other users, were quick to take advantage of large new supplies of cheap gas to fuel steam-generating plants. By the mid-1930s, wastage dropped to 7 percent. It declined to less than 3 percent during the 1940s and, indirectly, aided in oil conservation.[48]

In 1931, when the legislature took what seemed the next logical step by passing the so-called Sharkey Oil Conservation Act to control the rate of oil production, the handful of major oil companies that controlled over three-quarters of the state's production launched a referendum campaign against it. They created an "association" that quickly gathered sufficient voter signatures to place the Sharkey Act on the ballot. The antioil conservation campaign reached out to voters with a cartoon depicting a shark devouring the consumer, and Californians responded by defeating the Sharkey Act in April 1932. The major California oil producers opposed state-mandated conservation, but most of them understood the necessity for regulating production. Therefore, in the wake of their victory over the Sharkey Act, they formed their own conservation committee of buyers and sellers and established a workable voluntary production restraint program. In 1939, the legislature tried again with the Atkinson Oil Control Bill, but in another referendum campaign, oil producers convinced Californians to defeat it at the polls.[49]

Summary

Although California's oil and gas production was surpassed by that of other states and foreign nations during the 1930s, capture of these prodigious natural energy resources freed Californians from reliance on imported fossil fuels and provided them, in large measure, with energy independence. Novel technological developments allowed for the manufacture of high-quality oil gas, whereas other inventions made oil's direct use as fuel almost universal on the Pacific Coast. Oil provided domestic and commercial light and heat, and also facilitated Pacific Ocean shipping, western rail transportation, and industrial and commercial expansion. Further technical advances enabled effective refinement of California oil into gasoline and other distillates, thus aiding the rise of the automobile and ensuring even greater economic and social change.

As a part of progressive reform responses to the strains of urban/industrialism, how to develop energy resources in ways that would provide the greatest benefit for society became an intense issue. The rise of large oil and gas corporations and the increasingly integral role oil and gas played in people's day-to-day lives brought Californians face-to-face with this. Collision between supporters of public versus private utility ownership was part of the working out of who should control gas production and distribution, and it touched issues from service and efficiency to economic philosophy and the structure of local government. On another level, the rising natural resource conservation movement and World War I greatly influenced California's oil and gas industry, prompting federal oil land withdrawals and widespread discussion about oil conservation. Attempts to attain government regulation of oil production failed, but conservation concerns did result in state legislation prohibiting natural gas wastage. Combined with new gas discoveries, improvements in gas delivery technology, and market opportunities resulting from rapid population growth, the end of gas wastage made Californians the nation's leading consumers of natural gas. Seventy percent of the state's homes used it by 1940, a considerably higher percentage than the national average, and most of the state's electric power companies used it as fuel in their steam-generating plants.

As the twentieth century unfolded, California's oil and gas resources helped to establish a strong regional economic base, which, in turn, helped override the restricted economic development that had stemmed from California's geographic isolation and its colonial relationship with the eastern United States. But the energy independence and economic prosperity state residents attained from oil and gas, and the enormity of the state's consequent population growth, exacted a high environmental price. The very environment that had endowed Californians with oil and gas was transformed comprehensively. The energy resources were essential ingredients in the creation of the Pacific Coast's artificial landscape. From them stemmed a visible, as well as invisible, environment of oil and gas rigs, storage tanks, pipelines, fouled air, and polluted water that many Californians came to think of as natural indeed. The use of oil and gas, however, had an even greater impact on the natural, as well as the human, environment, especially by means of the internal combustion engine.

CHAPTER VIII

Fueling California

IN 1900, Union Oil's Wallace L. Hardison predicted that California's petroleum resources meant "the day is not far distant when there will be built up here communities surpassing in wealth, comfort and advanced civilization those of any part of the world."[1] His hyperbole aside, oil and natural gas did, indeed, change California. Within two decades, railroads, ships, ferries, manufacturers, and gas utilities relied entirely on oil, and gasoline had begun taking on an importance unimaginable to Hardison or any other early oil entrepreneurs. Oil and gas, joined by hydroelectricity, formed an energy triad that modernized California and markedly altered its landscape, economy, demography, and citizens' lifestyles.

During the last two decades of the nineteenth century, Southern California left behind its literary image of "cattle on a thousand hills" to begin a transformation that made it a paradigm of modernity. Less than sixty-five thousand people lived in the southern part of the state in 1880, but by 1900, a dramatic shift in the state's population growth pattern was revealed: in twenty years, southern California had grown 372 percent, compared with the state's overall 47 percent growth rate. Moreover, this trend continued after 1900, with 2.4 million more people settling in the south's eight counties by 1930, Los Angeles County attracting two-thirds of them.

Photo 8.1. The automobile led to construction of many scenic boulevards, such as Foothill Boulevard on the San Francisco Peninsula. From *Sunset Magazine* (May 1908): n.p.

In fifty years, southern California's population increased from just less than one-tenth to over half of the state's total.[2]

The roots of this onslaught reached back into the 1870s when southern California's climate inspired a rush of settlers from the East, seeking to improve their health. After 1880, the area's agricultural potential, epitomized by the citrus industry, combined with railroads to attract settlers seeking farmsteads. To extend its lines, the Southern Pacific had been granted, over the years, enormous tracts of public land, much of it in the south. Eager to sell it, the railroad's publicity department advertised California widely through books, magazine articles, and broadsides. In 1885, when the Santa Fe Railroad began service to Los Angeles, a rate war between the two railroads drove down the price of a one-way ticket from Kansas City from $125 to as little as $1. Fewer and fewer settlers bound for California searched out the Golden Gate and San Francisco. Instead, they created new towns in the south, such as Riverside and Pomona. They sought and found health, wealth, and outdoor living, where clouds rarely blocked the sun, and where, with a little water, a person could grow anything. Some of these settlers spawned early electric power developments, but starting in the mid-1890s, oil discoveries in and around the Los Angeles basin supported and, in a global if not specific sense, dictated southern California's stunning growth.[3]

Health attractions, agriculture, land speculation, and even the motion picture industry helped determine where people lived in southern California, but most historians agree that the greatest influence on the demographic character of the area, particularly the Los Angeles basin, came first with the extension of the Pacific Electric Railway's lines and evolved into its modern form with the automobile. In the first case, suburbs appeared alongside Henry E. Huntington's railway extensions, and new arrivals into the 1920s dispersed according to the pattern proscribed by the interurban system. In the second case, widespread adoption of the automobile influenced population dispersal after 1920 and led to decentralization of the basin. A third aspect of Los Angeles basin development is found in the development of the oil industry itself: "Industrial and residential suburbs appeared at either an oil field or a refinery site. . . ."[4] Additional oil field discoveries through the mid-1920s led to a series of new towns which, when connected by the interurban railways and automobile roadways, contributed enormously to the Los Angeles region's general industrialism.

To the south and southwest of Los Angeles, oil pipelines connected new communities whose individual identities grew and were reinforced by economic links to oil fields and refineries. Economic links created between oil communities ensured their independent political survival, whereas in other areas Los Angeles annexed those suburbs along Pacific Railway lines which possessed no direct oil industry ties. Historian Fred Veihe contends that a wide range of consequences stemmed from the oil suburbs, from suburban working class residents holding political and other views that were more "urban" than those of affluent Midwesterners who settled in Los Angeles proper, to an explanation of Los Angeles's reputation as a "cultural desert" and its development as an "administrative-residential core" surrounded by industrial suburbs. By 1930, Veihe suggests "a suburban industrial network founded by the oil industry . . . occupied much of the . . . the Southland." Furthermore, even the suburbs and small towns north, northwest, and west of Los Angeles—those beyond the immediate area of the oil fields and pipelines, and often owing their existence to the interurban railway system—flourished in large part because the petroleum industry provided an economic foundation for the whole region.[5]

New Uses for Oil: Gasoline

The most profound early impact of the petroleum industry came through the use of oil by manufacturers and individuals. By the 1920s, virtually all the state's steam engines used fuel oil. Engines used by railroads and maritime shipping consumed the most fuel oil, followed by the gas and electric industry. About 25 percent of the horsepower generated in manufacturing came from oil-fired steam engines, and the petroleum industry itself used fuel oil in its operations. Petroleum companies marketed between seventy-five and ninety million barrels of fuel oil annually in California during the 1920s, and fuel oil continued to be a basic industrial fuel into World War II. Meanwhile, natural gas became an almost universal commercial and domestic fuel, and, after 1930, replaced fuel oil in most California manufacturing establishments (figure 8.1). The electric power industry, which had its roots in hydroelectricity, also became a heavy user of fuel oil and natural gas. Steam-operated central power stations at first supplemented waterpower, but, by the end of World War II, as electric motors replaced industrial steam power, giant steam turbine plants became the basic power generators, augmented by hydroelectricity.[6]

Oil's greatest impact on individual's lives, on the other hand, came in the form of petroleum distillates. Kerosene provided lighting for many Californians well into the twentieth century, as well as fueling heaters and stoves. Napthas, of which gasoline became the most important, soon surpassed kerosene as the petroleum fuel most used by individuals. Before the 1890s, refineries burned napthas occasionally, and these distillates had a limited market as cleaning solvents and paint thinners. Their use expanded rapidly during the 1890s. Cooking with gasoline, which Standard Oil Company and others encouraged, was partly responsible, but the introduction of small stationary internal combustion engines launched gasoline's greatest application.

In June 1890, the Pacific Gas Engine Company put the "Pacific Gas or Gasoline Engine" on the market in San Francisco. By December, the firm

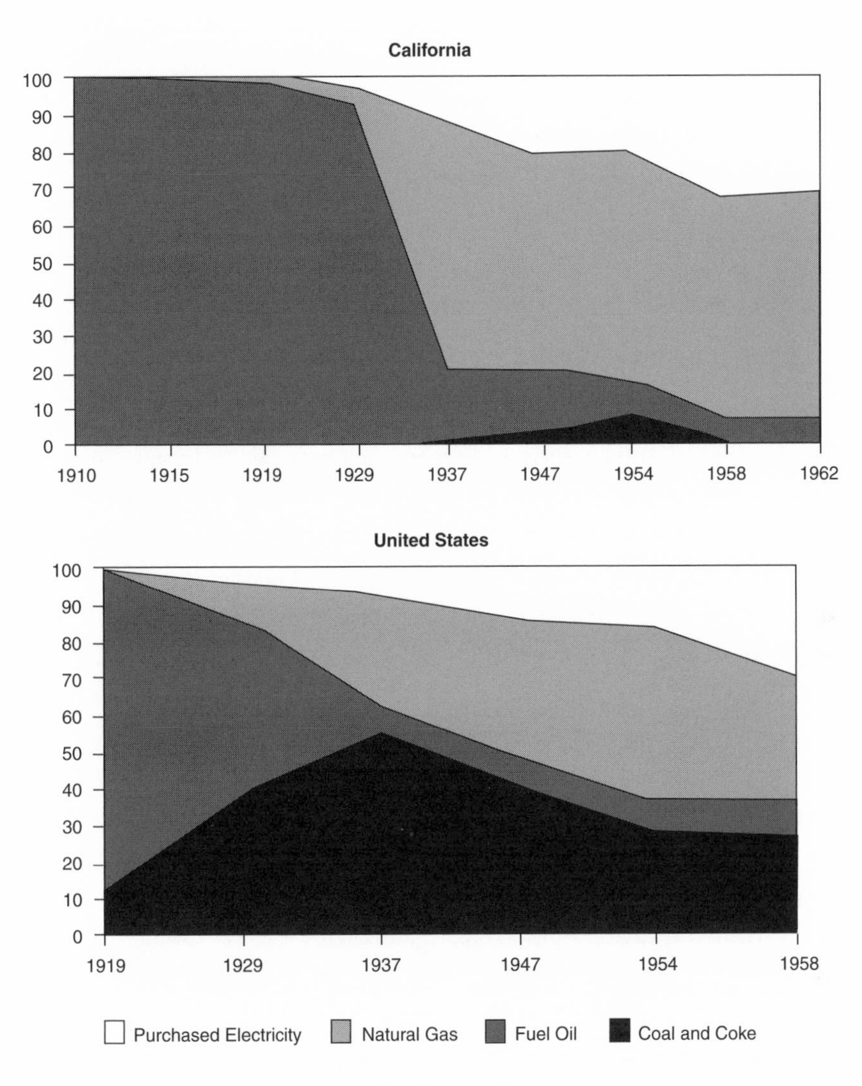

Figure 8.1. Energy Resources Used in California and United States Manufacturing by Percentage, Selected Years. Source: *Supra,* Table C-10.

advertised that over 150 of their engines, designed to run on coal gas or gasoline, were in operation. Four months later, they claimed 221 users of their one-half to ten-horsepower engines. The motors operated grinders, printing presses, cream separators, sewing machines, spice mills, grape crushers, fruit driers, coffee mills, furnace blowers, pipe cutters, small boats, and pumps, from Olympia, Washington to as far south as Central

America. The bulk of Pacific's engine users resided in California, many employing the motors for pumping. A competitor, San Francisco's Regan Vapor Engine Company, claimed over four hundred operating motors in June 1891, and yet another firm, the Electric Vapor Engine Company, sold "Single, Double and Quadruple Acting" motors from "1/3 H.P. to 100 H.P."

The adaptable stationary gasoline engine found many uses. Smaller boats on San Francisco Bay and along the coast and rivers of California adopted them, a variety of small manufacturers used them, and a San José street railway, in 1892, unsuccessfully employed a gasoline-powered streetcar designed by Daniel Best of San Leandro. Gasoline engine manufacturers, however, perceived farmers as offering a particularly important market for their product. Their quarter-page advertisements in the *Pacific Rural Press* touted farm use: "Cheaper Than Windmills for Farmers!" headlined the Electric Vapor Engine Company, while the Regan Vapor Engine Company emphasized "Our Engines are especially adapted for Pumping and Irrigating and Spraying Fruit Trees; in fact, for any use where power is required."[7]

Since the wind turning farmers' mills could not be summoned on demand and, more often than not, was capricious, farmers adopted gas engines. L.K. Sherman, a member of the Western Society of Engineers, observed in 1900 that "the 'bark' of the gas engine can be heard now on almost any ranch in southern California." The California Board of Trade noticed a widespread adoption of engines following a severe drought in 1898. Citing numerous examples of gasoline-pumped wells and irrigation projects, the Board of Trade offered suggestions for improved use and provided testimonials from farmers using the engines. In northern California's Santa Clara Valley, it noted, most of the eight to nine hundred new irrigation plants installed between 1895 and 1898 employed gasoline engines, only the largest plants being driven by steam engines. Along the Sacramento River, where only steam power previously had been used for reclamation and irrigation, fifty- and eighty-horsepower gas engines pumped up to ten thousand gallons of water per minute.[8]

Between 1899 and 1910, two- and four-cycle engines pumped water for irrigation and domestic use and powered various farm tools. W.H. Wright of Los Angeles County, for example, irrigated two acres of celery and kept a three-thousand gallon redwood storage tank full with a four-horsepower engine. He paid for his $696 system by selling water to his neighbors. A California poultry farmer, C.H. Dangers, ran his thirteen thousand White Leghorn operation with a five-horsepower engine, pumping as much as twelve thousand gallons per hour, plus operating an alfalfa cutter, feed grinder, and bone cutter. Additionally, his engine provided heat for incubators and gas lighting for his home. L.S. Wing, an engineer for the California Farm Bureau Federation, suggested the gasoline engine dominated irrigation pumping between 1899 and 1910. Another report indicated that almost all of the some ten thousand portable, orchard-spray rigs used in 1920 were powered by gasoline engines, and a mid-1920s farm power survey reported that thirty thousand stationary engines, averaging six horsepower each, delivered 7 percent of all the horsepower used on California farms. But the arrival of electric power and simpler, maintenance-free motors already was pushing aside the stationary gas engine. No doubt, many of the engines reported in the 1920s survey were maintained principally as emergency back-up for service from the still new central-station electric power companies.[9]

Gasoline and Transportation

As important as the use of petroleum distillates and gas engines were in early twentieth century farm, small shop, domestic, and manufacturing settings, they found their greatest value in transportation and traction. Gasoline and the internal combustion engine made possible individualized transportation and transformed agricultural power. Automobiles and trucks affected both short- and long-range individual mobility, gasoline-powered tractors hastened the expansion and industrialization of farming, and California petroleum consumption moved from fuel oil to gasoline (figure 8.2). The result was absolutely stunning, nothing short of revolutionary.

Urban transportation had moved from horse-drawn omnibuses and horsecar railways to cable cars by the 1890s, when the remarkable electric streetcar appeared. California quickly joined the rest of America in adopting the trolley. Between 1890 and 1907, as the state's electric street railway mileage increased from fourteen to 1,974, all but four of its 207 miles of

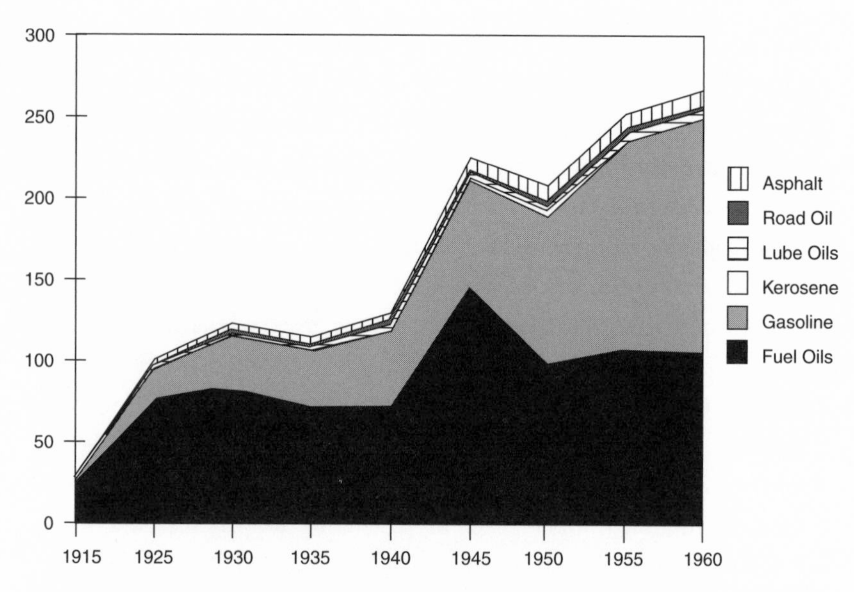

Figure 8.2. California Oil Consumption in Millions of Barrels, 1915–1960. Source: *Supra,* Table C-11.

horsecar and half of its sixty-three miles of cable railways disappeared. By 1917, trolley mileage totaled 2,963. The San Francisco and Los Angeles regions accounted for 90 percent of California's trolley traffic, the Los Angeles basin boasting the most extensive electric interurban and street railway system in the nation. The interurban Pacific Electric Railway Company's "big red cars" tied together forty-two cities and four counties within a thirty-five-mile radius of Los Angeles and operated more than 7,000 cars over 1,164 miles of track. Under the motto "comfort, speed and safety," the Pacific Electric became an institution in southern California, lasting until after the end of World War II. Meanwhile, the Los Angeles Electric Railroad Company provided urban service, carrying nearly 90 percent of the city's rail passengers. As early as 1911, Los Angelenos averaged a ride per day on the system, twice that of other American city dwellers.[10]

The trolley, however, had its drawbacks. It came into American life as a private business, firms receiving franchises to operate in communities, as did gas and electric companies. Their owners found that the real profits to be made were not from passenger fares, even after monopolizing the territory, but from real estate. In Los Angeles, as elsewhere, electric railway entrepreneurs created streetcar suburbs by purchasing undeveloped land, extending a railway line out to it, and selling the property at a tidy profit. At first, service to the new riders was efficient, but as the number of passengers increased, the number of trolleys did not. The cars became more and more crowded, and the transportation system that only a short time before thrilled people and allowed them to move in and out of the city center with ease became oppressive. In Los Angeles, as in other cities, the problem was exacerbated because all streetcar lines converged on the city center and none connected with each other outside the center. To go from one outlying area to another, one had to go into the urban core, which had become so jammed with streetcars, wagons, and other conveyances that pedestrians could walk as quickly as the cars could move. What appeared to be a solid public transportation system turned out to be chaotic, rational only to the real estate speculators who had done well in the new streetcar suburbs.[11]

No easy solution to the inadequate trolley system presented itself. By 1910, because streetcar companies nationwide were running debts amounting to 50 percent or more of their assets, they found it almost impossible to expand. City residents, seeing them in the same light as gas and electric utility monopolies, sought to gain better and cheaper service through regulation. But when local regulatory boards bent to public pressure and kept fares down, the debt-ridden streetcar companies were even less able to invest in new lines or service improvement. Even though regulatory control in California moved to the state level in 1914, where higher fares could be established without bowing to local public pressure, the service problem was so great that fare increases could never be large enough to correct the situation. The tactics of progressive reform did not seem to work. This was a particularly difficult problem in Los Angeles, where the population was doubling every ten years. Moreover, the public had developed such antagonism toward the streetcar companies that they angrily protested even justified fare increases.[12]

The inability to reduce street railway crowding and improve service, observed historian Scott Bottles, "left the individual commuter to fend for himself." The solution in Los Angeles, and soon in the rest of the nation, came with the automobile, which became more than a luxury toy after

1910 . Indeed, by 1920, thanks to Henry Ford, his Model-T, and mass production, the automobile became a fantasy available to almost anyone. As Warren Belasco points out, for urban residents caught in crowded city congestion "the automobile offered pastoral solutions: country drives, wilderness camping, suburban homesteads." Auto clubs formed and newspapers, magazines, and advertising pamphlets from oil companies, automakers, and camping equipment firms boosted the advantages of motorcars. All emphasized the premodern, nostalgic experience of the obviously modern automobile—a curious hallmark of modernity—and lobbied government hard for improved roads and highways. Proponents of the automobile claimed that it reinforced family values by encouraging family outings, vacations, and healthful suburban living: "the family that played together stayed together." It gave individuals control over when one departed, how fast one went, and what route one took: "motoring emancipated travelers from dictatorial railroad timetables." While it eventually doomed rail passenger service, interurbans, and city streetcars, nobody beyond the railway people seemed much concerned.[13]

The environment assisted Los Angelenos' in their running battle with street railways by facilitating use of the automobile. Southern California was ideal for autos: a mild, nearly rainless climate in which muddy roads were an anomaly, population density was low, and life was lived largely on a coastal plain that facilitated automobile travel. Moreover, a well-organized oil lobby and the Automobile Club of Southern California, formed in 1900 by auto aficionados, actively promoted motor vehicles and campaigned exhaustively for better streets and highways. In 1909, Los Angeles County had 1,085 miles of improved public roads, including the first asphalt macadam pavement in the state, and in six years the City of Los Angeles had completed paving all its main streets. At the same time, residents discovered that the automobile could offer an alternative to the streetcar. In 1914, local drivers began offsetting the cost of auto ownership by giving nickel rides to pedestrians. Jitney or taxi service was born, and by year's end taxi companies had been formed, an idea which quickly spread across the nation. Los Angeles's street railways sought to limit jitney competition by persuading the city government to regulate taxis, but they could not hold back the automobile as an alternative to the trolley. By 1920, whereas

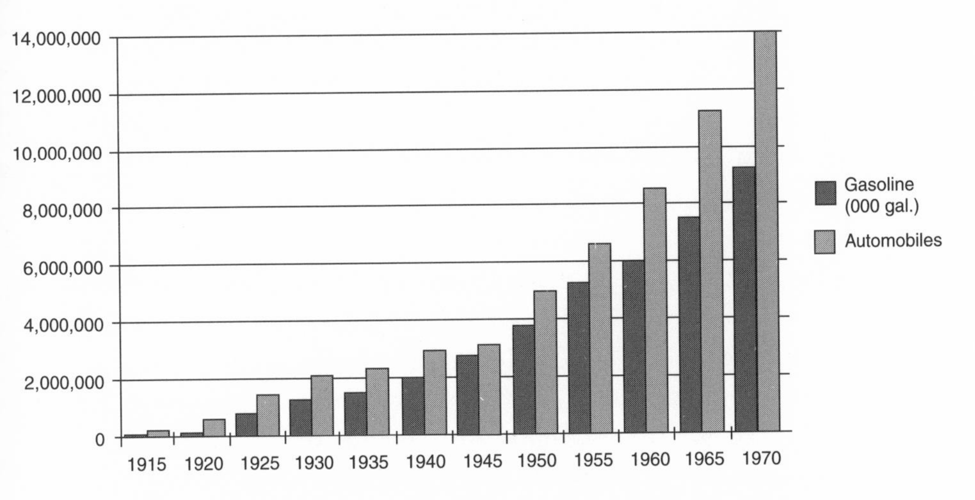

Figure 8.3. California Automobiles and Gasoline Consumption, 1915–1970. Source: *Supra,* Tables C-11 and C-12.

one of every thirteen Americans and one of every six Californians owned a motor vehicle, in southern California, the heart of the oil industry, one in every three and one-half individuals owned one.[14]

"Initially," said Scott Bottles, "the adoption of the automobile was an individualistic response to the failure of progressive reform." Bypassing unresponsive politicians and ineffective regulatory agencies, he continued, "the automobile became, in part, a symbol of democratic technology and civic reform." Fifty-five thousand autos plied the streets of Los Angeles by 1915, increasing to 140,967 in 1919, and to 776,677 ten years later. Residents of Alameda and San Francisco counties owned 29,203 autos in 1915 and, three years later, 60,454. Between 1915 and 1925, statewide motor vehicle ownership climbed from 191,196 to 1,475,852 (figure 8.3), which, at one for every three Californians, was almost twice the national average. In Los Angeles, by 1930, an automobile was owned for every one and one-half residents. More than half the people entering downtown came by car, compared to about 33 percent in Chicago and less than 20 percent in Pittsburgh. Ironically, downtown streets in Los Angeles were extremely narrow and congested. They had been laid out in the 1780s, and with over a hundred years of very slow growth, they had undergone little change. The city had only 21.5 percent of its central business district devoted to streets in

1924, compared to Cleveland with 39.5 percent and even San Francisco with 34.5 percent. Consequently, the growth of individual automobile use proved to be a crushing problem.[15]

The first response by the City of Los Angeles, in 1920, was passage of a stiff no-parking ordinance, but an enraged public soon forced substantial amendments to it. A better solution to traffic congestion came from the Automobile Club and the city public utility board's advisory traffic commission. Both carried out traffic surveys, worked with consultants, and, in 1924, presented a Major Traffic Street Plan to widen and otherwise improve thoroughfares. The city put the plan and a bond issue to pay for between 10 and 20 percent of the work before the voters. Both passed by overwhelming margins. City officials still had to depend on property owners in assessment districts to petition for the work to be done and also to agree to pay for the remaining costs through their property taxes. But the response was so enormous that the improvements moved ahead faster than anyone expected. Consequently, the city council put on the ballot a temporary property tax to support continued work, and voters approved it, even though they defeated every other bond issue on the ballot. "What emerged from these programs," wrote Bottles, "was the public's willingness to tax itself in order to subsidize automobile transit." However, they would not agree to subsidize mass transit, and proposals for elevated railways and subways made no headway during the coming years. Meanwhile, despite obvious traffic congestion, the southland earned an early reputation as a motorist's paradise, and Los Angeles was plainly the most automobile-oriented city in the nation by the mid-1920s.[16]

The auto's social impact was nothing short of phenomenal. Couples in parked cars, for example, became a common evening sight, and, in 1925, Pasadena Police Chief Charles Kelley warned: "the greatest menace now facing the morals of Pasadena youth is the coupe and sedan."[17] But, while the auto granted young people sweeping new freedom and revolutionized relationships between them, its use engendered a plethora of traffic laws that ate away at the sense of freedom drivers first experienced with the auto. Drivers responded by remaining defiant, and a distinctly adversarial relationship developed between them and traffic cops. Upton Sinclair observed in his novel *Oil* that "it must be a dreadful thing to be a 'speed cop' and have the whole human race for your enemy." Drivers used the Automobile Club of Southern California's "Cop-Spotter" rear-view mirror, and came to so loathe police speed traps that they succeeded in getting them outlawed in the early 1920s. Ignoring traffic laws, drivers racked up a terrible injury and death toll. Citations, fines, safety campaigns—nothing seemed to affect the driving carnage. Ultimately, as the auto outnumbered pedestrians and became an essential means of human mobility, people accepted its danger with what historian Ashleigh Brilliant called "a studied indifference."[18]

As the auto found its way elsewhere in California, it brought with it the same consequences evident in Los Angeles: city street congestion, accidents and law enforcement issues, altered lifestyles, and the need to develop an infrastructure. As in Los Angeles, people statewide were willing to pay for the automobile through taxation, and most communities quickly improved their roadways. At the state level, a vehicle registration fee became the first method for underwriting road improvements, but in 1909, under strong pressure from automobile clubs, the legislature authorized an $18 million bond issue for a paved state highway system. The first of many highway bonds over succeeding years, the voters approved it easily in 1910. Four years later, the Federal Highway Aid Act authorized dollar-matching grants to states, and, in the 1920s, California added a gasoline tax, which had been pioneered by Oregon. Within five years, the state built or aided in the construction of 1,124 miles of surfaced highways and, by 1922, had completed 3,264 miles with an additional 820 miles under construction, leading all other states in mileage of paved roads.[19]

Infrastructure built for the automobile brought tremendous change to the landscape, not only in the form of roadways but with establishments that served the auto. Today these enterprises are so familiar that they seem to be an immutable part of the natural world: restaurants, drive-throughs, motels, curio shops, and, most important in terms of energy, gasoline stations. The first gasoline station in California set up on Wilshire Boulevard in Los Angeles, and competition between ever-growing numbers of stations became keen. By 1916, southern Californians "could expect station attendants to check their vehicle's oil and water, clean their windshields, and inflate their tires,"[20] an expectation that drivers soon shared statewide.

Standard Oil lured motorists with a booklet on traffic regulations, the Pinal Dome company began incorporating restrooms in their facilities, many stations opened at 7:00 A.M. and closed at 10:00 P.M., and a station in Anaheim built a sleeping room for a night attendant and offered twenty-four hour service. Stations quickly evolved from "gas" to "service" stations.[21]

Service stations soon appeared along the state's new highways. Standard Oil operated the most stations in 1920, with 150 outlets, followed by Associated Oil Company with seventy-seven. Shell and Union trailed the two leaders, bringing the state's total number of stations to 304, over half located outside southern California. Competition among companies spread everywhere. All of them sold coupon books of one sort or another to encourage return customers, some introduced additional services, such as car washing, and price wars became common. Companies built individual showcase stations, and in the early 1920s, Shell combined attractiveness and standardization in their "cracker box" with a canopy, which they located on landscaped lots connected to the street with a paved driveway. As retailing gasoline became a chore for oil firms, California's companies became the first in the country to lease rather than to operate stations. In 1926, about four hundred were leased to individuals. By 1939, California had 15,218 service stations, and with 5 percent of the nation's population, regularly had consumed through the decade over 8 percent of its gasoline. Service stations—joined by hot dog stands, advertising billboards, and a myriad of other roadside enterprises—reached into every corner of California that people and their autos explored, and with them, visual and trash pollution became an enduring part of the landscape.[22]

In a number of ways, the auto's social consequences and environmental impact merged. Sociologists, noted Ashleigh Brilliant, observed that "the old concept of an urban 'neighborhood' was breaking down," people who owned autos dissociating themselves from the neighborhood in which they lived and finding interests and relationships farther away. For some people, the automobile seemed to offer a solution to social problems that fermented in unpleasant, congested city centers. The auto offered them the opportunity to pursue the perceived moral virtue of suburban, detached, single-family living even farther outside the urban core than streetcar suburbs. Not surprisingly, land developers began building houses, communities, and shopping services around the automobile, going beyond what entrepreneurs had done with the street railway. And, argues historian Mark Foster, Los Angeles planners in the 1920s created a street traffic plan that intentionally promoted urban decentralization.[23]

As people moved farther and farther from the city in which most of them worked, street plans and road construction could not keep up. Consequently, businesses themselves began to relocate. By the early 1930s, fully 88 percent of all the new retail businesses in Los Angeles were being built in the suburbs; by 1940, the market share of downtown stores had dropped to barely more than 50 percent and to less than 25 percent by the mid-1950s.

Starting with fashionable Westwood Village in 1929, developers constructed suburban communities catering to the auto. Within two decades, developers had created Lakewood, north of Long Beach, a mass-produced community of 17,500 homes built around the first regional shopping mall in the West. Meanwhile, new manufacturers located in industrial suburbs, taking advantage of the open space to build roomy, single-story factories best suited for assembly-line production, conveyor-belt operations, and other mass-production techniques made possible, in part, by the employment of electric power. During the 1930s, Los Angeles was well on its way to becoming the nation's first truly decentralized city.[24]

The move to suburbs by those who could afford it brought with it a number of unforeseen problems. For example, minority, quite often poor immigrants to Los Angeles moved into older housing stock left by those who moved outward. By the end of the 1920s, the south-central part of the city, just outside the center, became a solidly black community; east Los Angeles became Latino. Although government centers and business offices remained downtown, the city center and older residential areas experienced increasing economic decay, as stores and public services tended to follow middle-class money in service of the suburbs. Los Angelenos put immense resources into moving commuters in and out of the downtown. Setting the national pace for solving traffic problems, they discovered that mere widening and extension of streets no longer worked. In 1937, the Automobile Club of Southern California proposed a "network of traffic routes for the exclusive use of motor vehicles over which there shall be no crossing at grade and along which there shall be no interference from land use activities." Although not an entirely new concept, the club's plan became the basis for construction of freeways, the 1940 Arroyo Seco Parkway, or Pasadena freeway, becoming the nation's first.[25]

Over time, the automobile's consequences intensified. Traffic congestion got worse and worse both in and out of cities, gasoline consumption increased dramatically, exhaust-polluted air became worse, and parking lots and automobile-related service establishments devoured the landscape. Highways and county roads crisscrossed the state, spanning rivers and gorges with bridges, tearing asunder farmland, leveling hillocks, and filling in arroyos. Worse than the railroad, which novelist Frank Norris dubbed "The Octopus" for its all-consuming impact on people and the environment, the highway and road network decimated the landscape, effecting a carnage as devastating to human society as auto-related loss of life and limb. By adhering to the U.S. Department of Interior General Land Office survey lines set down between 1853 and 1870s, both local and state roads created an artificial, structured grid on the landscape, one still evident in roads throughout California named "Grant Line." The resulting statewide road system, like the survey, showed no relationship to natural environmental diversity. It represented a process, if not an unstated policy, of nature-alienation.[26]

After World War II, the success of Los Angeles freeways led Governor Earl Warren, in 1947, to increase the state gasoline tax in order to finance their construction statewide. Even more of them were made possible after federal highway monies were authorized during the mid-1950s. By the early 1970s, highway building had consumed thousands of acres of land, evicted over 400,000 residents, 44,000 business people, and 13,000 farmers. Reaching into the Sierra Nevada, highway builders took out entire historic community sections; across the state, freeways continued the process of nature-alienation, slashing across valleys and through mountains, forever reshaping the physical environment. Freeways were designed to bring speedy interurban travel and resulted in altered leisure patterns with weekend getaways to mountain and seashore resorts. They improved accessibility to new suburban areas, of which developers took advantage by building

Photo 8.2. Constructing the Hollywood Freeway through Cahunga Pass in 1952. (Used with permission of the Automobile Club of Southern California.)

shopping malls that, while providing easy parking and convenience for shoppers, further contributed to the decay of urban centers. Freeways irreversibly changed the lives of Californians, who owned and operated almost 14 million cars by 1970.[27]

Changes in Rural Life

Motor vehicles were as significant to rural life and agriculture as to urban life. Before the 1920s, railroads and interurbans had provided some rural dwellers access to the outside world, but for the most part they continued to be isolated. Since one could only travel by horse and wagon about twenty-four miles round-trip in a day, this was often just far enough to reach the towns and railroad stations generally planted in farm regions between ten and twenty miles apart. Trips to the city were rare indeed. The automobile, however, emancipated the farm families. It enabled farm women to travel to town easily and regularly, and observers hailed this relief from the hazard to mental health inherent in isolation. The auto saved time for farmers, permitting them to obtain needed supplies and repair parts quickly, and it provided them with a basic power source. It opened the way to postal service rural free delivery and offered migrant laborers a mode of transportation. It would, said some, even stop youth from leaving the farm. In short, suggests historian James Flink, the automobile seemed to be a panacea for the hardships of rural existence: "Best of all, perhaps, the individual did not need to wait for others to act; by buying a car, he could immediately improve his own life appreciably."[28]

In reality, not all farmers agreed, and some frankly opposed the motorcar. City folks had it first, and with it they invaded the countryside, becoming a general nuisance on weekend outings and auto camping trips. Farmers expressed concerns over speeding and reckless driving, and debated whether the horse or the automobile had the right to the road. Some went to extremes, carrying guns to intimidate motorists, placing barbed wire across roads, and even plowing up roads. Near Sacramento, in 1909, notes Reynold Wik, farmers "dug ditches across several roads to block traffic and actually trapped thirteen cars." But opposition was limited to a short period of time. In 1908, the *Pacific Rural Press* announced that the motorcar, no longer an experiment, could be gotten in strong, durable

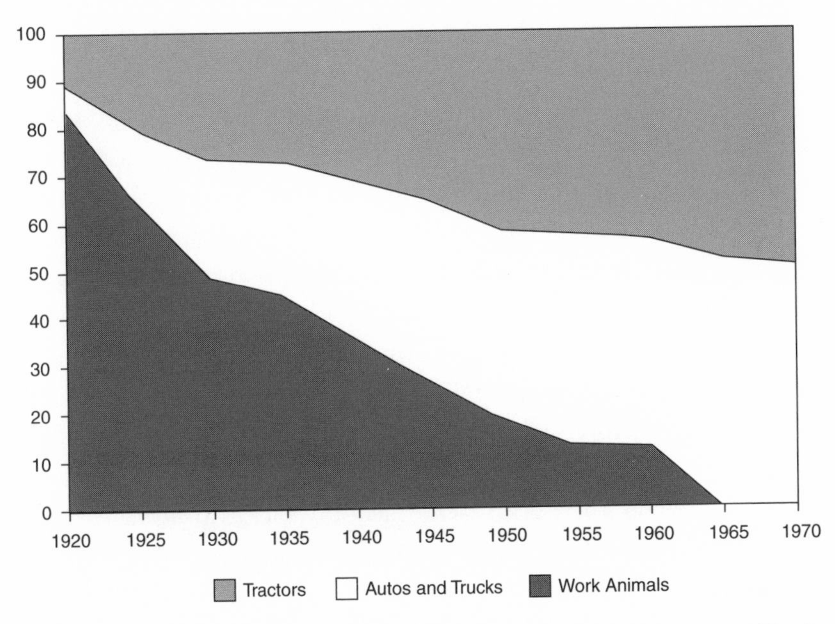

Figure 8.4. Percentage of Tractors, Autos and Trucks, and Work Animals on California Farms, 1920–1970. Source: *Supra*, Tables C-12 and C-13.

models at reasonable prices. And it performed well on the farm, requiring no food when not in use and costing no more to keep up than one would spend for horse shoeing or harness repairs. By 1910, farmers began acquiring motorcars and trucks, and in ten years over 50 percent of California's farms had an automobile (figure 8.4). Motor vehicle growth quickly outstripped that of stationary gasoline engines.[29]

Most farmers used the motorcar in expected ways, hauling goods to and from market and transporting the family, but others put them to a variety of often innovative tasks. In 1914, for example, a Humboldt County farmer jacked up his Garford truck to run a grain separator with a belt harnessed to the vehicle's rear wheels. He used fifteen gallons of gasoline and three gallons of oil a day to shell twenty sacks of oats an hour. At the other end of the state, a San Fernando Valley farmer employed his touring car to clear cactus, yucca, and mesquite. He braced a fourteen-foot beam upright on a small two-wheeled truck, fastened a pair of steel claws to the lower end of the beam, and ran a steel cable from the upper end of the beam to the car. Putting his family in the car for weight over the wheels, he drove forward, causing the beam to pivot and claw the plant from the ground. Rural vehicle owners ran irrigation pumps, wood saws, ensilage cutters, clothes washers, and other machines with their autos and trucks. R.T. Robinson of El Cajon owned four touring cars and two trucks in 1915, no doubt meeting almost any imagined need.[30]

Some farmers tried to use autos as tractors, pulling hay rakes, harrows, mowers, and grain binders, and some used conversion kits, such as the Los Angeles Tractor Company's eight-hundred-pound rear replacement wheel for Model-T's. But these did not make effective field machines. Farmers needed something stronger and more durable. In 1891, a year before Charles Duryea drove the first automobile, Daniel Best had won a State Board of Agriculture award for a gasoline engine that substituted for steam engines on a few giant farm traction vehicles. Meanwhile, others around the country built similar cumbersome machines. The most successful was introduced, in 1893, by Charles W. Hart and Charles H. Parr, who had met as engineering students at the University of Wisconsin. Although two hundred Hart-Parr tractors were being used in 1907, theirs and others were modeled after steam traction machines, and their size and frequent technical problems largely led to failure.[31]

After 1910, mechanics turned to development of smaller vehicles. Minneapolis's Bull Traction Machine Company established the trend, in 1913, with their small "Bull with a Pull" tractor. Sales nationwide reached 14,500 in 1914, climbing to 35,000 in 1916, and when Henry Ford's first Fordson tractor for the domestic market rolled off the assembly line in April 1918, American farmers already were using some ninety thousand tractors. Unlike many other tractors, Ford's immediately found their way to California. Ford gave the first one off the assembly line to his old Santa Rosa friend and famous state agriculturist, Luther Burbank. Ford automobile dealers were pressured to sell them by Ford Motor Company, and soon thousands of people attended Fordson exhibitions across the state. After one in Oakland, on 6 October 1918, the *Oakland Tribune* expressed a reaction probably experienced by many farmers: "They came, they saw, and were convinced. The horse is now the most extravagant motor known."[32]

In 1916, the *Pacific Rural Press* traced the history of the steam plow, thresher, and harvester, tying it into the tractor's development. It noted

Californians had long held "a consuming thirst for tractors" and were hardly backward in their acceptance of new machines. Within a decade, the *U.C. Journal of Agriculture* reported "the California farmer is 'tractor-wise.'" Production of profitable speciality crops and links to powerful marketing cooperatives put farmers in a favorable financial position, one that encouraged them to invest in new farm technologies. By 1925, in the midst of a national postwar slump in farm prices, almost 20 percent of California's farms reported owning tractors, proportionately trailing only North and South Dakota. By 1940, despite the impact of the Depression, 30 percent of farms owned tractors, and the number of farms with horses and mules had declined commensurately (figure 8.4). "The percentage of land broken and disked by tractors in California exceeded that of any other state," according to Robert Ankli and Alan Olmstead, and a custom tractor service market became highly developed as part of the state's agribusiness.[33]

Besides adopting the tractor, Californians offered their fair share of innovation. In 1916, *Farm Engineering* reported three new types of tractors from the state. Farmer Rush Hamilton built a fourteen-hundred-pound, six-foot-long, three-foot-wide machine, with four-foot-high spiked front drive wheels, which worked "somewhat after the manner of a cultivator" or soil aerator but also pulled any implement. It had a turning radius of six feet and could "be driven backward as easily as forward." Another inventor in Pasadena produced a three-cylinder, air-cooled, gasoline engine tractor of equally unique design. The engine was housed inside a three-foot-high, two-foot-wide iron-spoked drive wheel on the front of the tractor's steel frame. Intended for use on small farms, the wheel revolved about the twelve-horsepower engine, the axle serving as the crankshaft.[34]

The principal California contribution to tractor development came with the "caterpillar" tractor. As early as 1858, William P. Miller of Marysville had won a cash prize and medal at the State Fair for his crawler steam plow. It possessed all the basic design features of later crawler tractors, but slipped into obscurity without being marketed. Several other mechanics experimented with tracked steam machines, and, in 1900, Maine millwright Alvin O. Lombard designed the first to be marketed successfully anywhere in the nation. Improved for rough terrain in 1901, the seventeen-ton Lombard steam crawler locomotive became a sensation in New England's logging fields, and, between 1904 and 1915, it also was used in Great Lakes and Pacific Coast logging operations. Meanwhile, Benjamin Holt, Stockton manufacturer of horse-pulled and steam combines, began tracked vehicle development.[35]

Holt and his brother Charles had pioneered steam combines during the 1890s. Seeking ways to prevent their heavy steam traction engines from bogging down in the soft alluvial, peat soils prevalent in the delta region of the San Joaquin and Sacramento rivers, they had added drive wheel extensions that widened the wheels to as much as eighteen feet each. These impossibly costly, cumbersome, and inefficient monstrous machines barely could turn around at the end of a field and could not cross bridges. In 1903, Benjamin Holt pursued solutions to their problem by traveling to England to see what had been done there with crawler tractors. He sent others to locations throughout the United States to observe tracked vehicles. The next year, Holt modified a forty-horsepower steam traction engine, replacing the drive wheels with nine-foot-long tracks. On the Mormon Slough near Stockton, it carried out a successful test, and Holt named it the Caterpillar.[36]

"Fortunately," writes Reynold Wik, "the crawler tractor arrived at the time internal combustion engines were beginning to make gasoline tractors more practical than heavy steam engines." Holt installed one on a Caterpillar in 1906, and three years later he supplied twenty-seven gasoline Caterpillar tractors to help build the 233-mile-long water aqueduct from the Owens Valley to Los Angeles. The project forced Holt to improve his machine and provided him "enough capital for his company to launch the industry on a sound financial basis." It also provided marvelous publicity for the Caterpillar, and, by 1914, Holt had built a second factory in Peoria, Illinois, and filled orders from forty-five countries. Several thousand Caterpillars sold in the next few years, including 5,082 for military use to the governments of France, Great Britain, Russia, and the United States. In 1925, the Holt Manufacturing Company merged with its California rival, the C.L. Best Traction Company, which also had begun to produce tracked machines.

Many California farmers readily adopted the Caterpillar along with other tractors. After 1910, virtually all farm journals praised the new machines as almost universally superior to horses. Tractors worked faster than draft animals, eliminated labor necessary to care for horses and mules, and freed acreage from production of hay and oats, making it easier to change the mix of farm output. More significantly, tractors permitted expanded production and increased the power supply available during periods of peak need and, in this sense, appeared much more practical than the horse. Tractors did not eat after the job was done, their engines did not overheat and require rest, and they did not seriously pack down the soil. Perhaps their biggest drawback was that they substantially increased the amount of petroleum used in farming.[37]

While tractors provided distinct advantages over horses, the early ones lacked versatility, so some farmers held back. In 1922, L.J. Fletcher, head of the Agricultural Engineering Division at the University of California Farm in Davis, both encouraged and cautioned California farmers about tractors. While observing that there were more tractors per improved farm land acres in California than in any other state, he suggested that farmers be careful in considering the purchase of a tractor. Shop around before buying. Ask if the machine would have enough power to pull needed equipment, if it could be controlled easily at turns and driven where necessary, if its wheels would not get mired in soil, if it could go between rows and branches of trees, and if enough work existed to justify its purchase. He also asked the farmer to shop only with a responsible dealer and to ensure the tractor manufacturer had a good reputation and would back his product.

Fletcher particularly argued the tractor's ideal use for nut and fruit farming, noting that Los Angeles County had more tractors than the states of New Mexico, Arizona, Nevada, and Utah, and that Tulare and Fresno counties had more than in all the New England states plus New Jersey and Delaware. Because little or no room for horse feed storage and pasture existed in orchard farming and the ground had to be turned frequently, the tractor had a distinct advantage. Furthermore, because orchard farmers usually ran their own equipment, unlike operators of larger farms and ranches who often left it to hired hands, probably fewer tractor breakdowns would occur. Finally, he thought capital outlay for a tractor would be reasonable, although a retrospective cost analysis done years later by historians Ankli and Olmstead revealed horses remained more cost effective through 1940 on all but the smallest orange groves.[38]

Four years later, Fletcher and his university colleague C.D. Kinsman reported the results of a California farm tractor survey conducted in 1924. They concluded the farm tractor was used an average of 650 horsepower hours per year. It had become the principal power source for California farm field work, and they asserted its value for all farm horsepower hour conditions. Yet they noted that draft animals possessed a better short-period overload capacity and continued to be valuable for corner work and other difficult places. Draft animals, they implied, might have an advantage on large farms and ranches. They appeared to concur with John E. Pickett, who argued in the *Pacific Rural Press* that both the tractor and horse always would have a place on the farm. In the end, observed Fletcher and Kinsman, the tractor's success depended on the ability of the farm operator to use it to best advantage and on the character of mechanical attention given the tractor.[39]

In reality, tractors had low clearance and could not be used for cultivation of such crops as beans, tomatoes, and cotton, until the 1930s. Improved efficiency and flexibility depended on various innovations, which came gradually. Nineteen hundred nineteen brought a power takeoff attachment, and starters and lights followed in the early 1920s. Rubber tires also were introduced and, according to Reynold Wik, "increased the speed of most field operations from 25 to 50 percent." In the 1930s, hydraulic lifting attachments and hitches became available, but not until the 1940s did conditions change strikingly in favor of tractors. Advanced transmissions and other innovations finally brought the tractor real sophistication after World War II.[40]

In a quarter century, petroleum's immense impact on rural life moved from small stationary engines to automobiles, trucks, and tractors. By 1925, the internal combustion engine provided about 34.5 percent of the total annual horsepower hours used on the farm, at a cost of approximately $42.2 million. By midcentury, every state farmer employed some type of gasoline-driven vehicle, either their own or contracted. By 1970, California's 106,052 farms possessed 185,232 motor vehicles, 56,285 farms had 138,914 tractors, and 42,631 farmers spent $246.2 million to hire machines, custom work, and labor. Farmers' direct expenditures for gasoline plus other petroleum fuels and oil climbed from $20.9 million in 1940 to $85.1 million in 1969, and they became as dependent on the oil industry as any urban or suburban dweller.[41]

Summary

Californians refashioned their world with the gasoline-driven engine. They created a new, decentralized modern society. Those for whom the automobile provided a suburban bungalow or ranch-style house and the pleasures of weekend country outings and wilderness explorations, perhaps felt they had arrived at the comfortable civilization that Wallace L. Hardison had predicted in 1900. For those whom demographic changes left behind to tend increasingly neglected urban centers or who lost their land to the automobile's unsatiable demands, life in the new society was less sanguine. In any case, people's appetites for the automobile chewed up old symbols, landscapes, and ways of doing things. They were replaced by new ones, which gradually became second nature: paved roads, highways, freeways, service stations, motels, restaurants, billboards, traffic lights, center lines, parking lots, meters, commuters, traffic cops, wrecking yards, ambulances, air pollution and smog-induced crimson sunsets, and nature-alienation.

For good or ill, Californians eagerly adopted motor vehicles and enjoyed the individual freedom and power provided by them. The Golden State's Mediterranean climate and its geological endowment of oil and gas helped to bring its residents and the motorcar together. Californians produced a large proportion of the nation's petroleum during the first half of the twentieth century, they consumed more of it—increasingly in the form of gasoline—than any other energy resource, and they became dependent on the highly centralized oil and motor vehicle industries for transportation. Yet, oil and gas formed only two legs of Californian's energy budget and experience during the first half of the twentieth century. The third came from hydroelectricity and the electric power system that it spawned. Its development, crucial to understanding the interplay between technology and the environment in California's energy experience, also transformed the lives of Californians.

CHAPTER IX Hydroelectricity

AT THE OUTSET of the nineteenth century, Alessandro Volta's simple primary battery gave birth to the electrical industry. During ensuing years, dozens of electrical pioneers developed it, intermingling science, technology, and entrepreneurship. By 1883, Dr. Wellington Adams announced the presence of utopia brought about by cheap, clean, noiseless, and safe electricity. Although his notice was and remains premature, electricity's magical qualities quickly caught the American whim. "Electricity is destined to be one of the most powerful factors entering our social condition," wrote Joseph Wetzler in 1890. "It must bring forth changes in the social order which are even now hardly realized." Indelicately referring to electricity as "this latest and efficient slave," *Harper's Weekly* underscored positive change for the future, while a writer for *Cassier's Magazine* emphasized that "no bounds can be placed upon this spirit of the nineteenth century. . . ." Electricity's impact on American life was nothing short of phenomenal.[1]

In California, before the midtwentieth century, electric power meant hydroelectricity, and interplay between the environment and technology played a crucial part in shaping the region's electric power industry. California's subregional wood shortages and inadequate coal supplies, plus the

Photo 9.1. Bucks Creek powerhouse on the Feather River, completed by PG&E in 1928, had the highest head of any hydroelectric plant in the western hemisphere until the early 1950s, water dropping 2,561 vertical feet to turn a Pelton wheel. (Courtesy of Larry Shoup.)

lack of a ready substitute for coal, stimulated development of waterpower just as it had propelled oil explorations. Sierra Nevada snowpack and melt-off conditions, plus the experience of miners and, in some cases, the availability of their hydraulic works, steered entrepreneurs and engineers in devising and siting high-head hydropower plants and in creating water conveyance systems; later, broader climatic trends led them to modify some of their initial practices, and they added greater water storage to their systems. Geography also dictated that hydroelectricity be transmitted long distances if it were to serve valley and coastal cities, so engineers had to develop effective electric power transmission technology. Long-distance power transmission, along with high-head hydropower, presented distinctive engineering challenges, and because California was geographically isolated from the rest of the nation, its engineers carried out their work with relatively limited help from the eastern engineering establishment.

Early Efforts

Electricity came to California early. In 1871, Father Joseph M. Neri, a professor at San Francisco's Saint Ignatius College, installed a small battery-powered electric light in his window, and, in 1876, he lit an industrial exhibition at the Mechanics' Institute pavilion with arc lamps. Two years later, the Palace Hotel replaced gas lighting in its public rooms with arc lighting. The *San Francisco Chronicle*'s Charles de Young used a Gramme generator, in 1879, to light five arc lamps outside the newspaper's offices. The *Engineer of the Pacific* suggested that "it is only a question of a little time, when the electric light will be in as common use as gas." Within a year San Francisco's California Electric Light Company incorporated. Operating two dynamos powered by a coal-fired steam engine in a small frame building at Fourth and Market streets, it became the first central arc lighting station in the United States.[2]

During the next decade, several other California communities received and celebrated electric lighting. In 1881, San José citizens constructed a 237-foot light tower supporting six 4,000-candlepower arc lamps, on the theory that the lamps would light the entire community. The project failed both technically and financially, but it gained worldwide publicity and encouraged other communities to seek electric lighting. A year later, the Los

Angeles Electric Company installed a coal-fired, steam-driven generator, erected seven 3,000-candlepower arc lamps on 150-foot-high towers for street lighting, and provided lighting circuits for stores, hotels, and saloons. Meanwhile, wood-fired, steam-driven dynamos lit arc lamps at north coast lumber mills, and lighting soon followed to Eureka and surrounding towns. In 1884, two companies brought arc lamps to Sacramento, a year later the Oakland Gas Light Company installed arc lighting, and soon Berkeley boasted a system. By 1889, Los Angeles and San José ran thirteen and three-quarter miles of electric street railways, and the San Francisco magazine *Industry* wrote that "our city is now well 'sprinkled' over with [electric] motors. . . ." Tall towers also remained an attraction. An elaborate 272-foot-high tower with 3,213 incandescent lamps, a searchlight atop, and a cafe at the eighty-foot level was erected for California's 1894 Midwinter International Exposition in San Francisco.[3]

Although the generators of these early systems were turned by water converted into steam in coal- and wood-fired boilers that could be sited almost anywhere, the cost of fuel in California restricted their development. Instead, falling water turned the generators that resulted in widespread adoption of electricity in early twentieth-century California. The water with power potential was found largely in the Sierra Nevada, and hydroelectric development there really owed its success to the gold mining industry. "What the early miners of the Far West learned about the exploitation of high-head waterpower," observed historian Louis Hunter, "became the foundation for the flourishing hydroelectric industry of the Far West."[4] Unlike the large volumes of water and minimal falls in elevation of eastern rivers, miners discovered that Sierra Nevada streams typically carried a small volume of water that descended rapidly into the Central Valley. They learned that the high-mountain snowpack gradually melted through spring and summer to feed scores of streams and that, although stream flow varied greatly over the year and drought could result in dry years, this hydrological characteristic made waterpower feasible without extensive artificial storage.

To tap this resource, gold miners devised and constructed sophisticated hydraulic-power water-delivery systems. The wooden and iron pipes, ditches, flumes, and dams that comprised them became the basis for hydroelectric power stations, as well as providing a conceptual model for transmitting power long distances; indeed, parts of the mining systems were converted to hydroelectric development throughout the Sierra Nevada. Engineers ascertained an understanding of high-head water delivery from hydraulic mining, where water under enormous pressure was used to wash away gold-bearing hillsides. From the hydraulicker's water cannon, invented in 1852 by Nevada City miner Edward E. Matteson, evolved the forerunner of nozzles used to direct high-head water on the tangential waterwheels adopted for hydroelectric generation. Also from hydraulic mining came the technique of controlling water flow by a needle through the water nozzle, which was developed in 1895 by Andre Chavanne, a French immigrant miner in Grass Valley. Finally, experience in driving mining tunnels through hardrock informed the tunneling work required to construct some of the most complex hydroelectric installations.[5]

During the century's last decades, California miners installed small hydroelectric plants, using existing hydraulic systems and technologies. In April 1879, Nevada County's Excelsior Water and Mining Company became the first California mining operation to employ electricity. Using a water-driven Brush dynamo to light three 3,000-candlepower arc lamps, the hydraulic mining firm was able to double production by working through the night. By the end of the decade, several mines had installed arc and even incandescent lighting, including the Spring Valley Hydraulic Gold Mine in Butte County and the Wildman Gold Mining Company in Amador County. In 1887, the Eocene Placer Mining Company and the Big Bend Tunnel and Mining Company, which had organized some years earlier to drive a tunnel through the neck of the Big Bend oxbow on the Feather River, opened a hydroelectric plant installed by the Sprague Electric Company to operate mine pumps and hoists. More significantly, Sprague sent power over an eighteen-mile line to thirteen nearby mining operations and a sawmill, with an estimated loss of no more than 30 percent of the falling water's kinetic energy after being converted into electricity and delivered for use. This meant that miners everywhere could convert their hydraulic waterpower to electricity without a significant energy loss. The 1890s saw several other gold mining companies put in hydroelectric plants, among them the Standard Consolidated Mining Company, which opened a plant, in 1893, with a twelve-and-one-half-mile-long transmission line to its mill in Bodie.[6]

By 1900, mines throughout the gold regions had electric lighting, and several mines used electrically operated stamps, compressors, crushers, locomotives, hoists, pumps, and ventilating fans. A.M. Hunt and Wynn Meredith explained electricity's success in mining operations: "It offers higher flexibility for bringing the power to the point of required use, than any other means. There is no limitation imposed on the location of the machinery, as is so often the case where water power is used direct. . . ." Hydroelectricity offered as much as 60 percent savings over steam power by eliminating fuel costs. Because the mining industry and mining communities were located in the Sierra Nevada, often quite close to waterpower sites, they were among the first beneficiaries of hydroelectricity. As the state mineralogist observed in 1893, "the relatively low cost at which electric power can be delivered in many localities, as compared with the cost of steam power, where the rapidly diminishing wood supply is the only source of fuel, reduces the cost of operation, and enables mines to be worked at a profit which would otherwise be worked at a loss." Old mines reopened and new ones began operation where steam costs were prohibitive and direct waterpower could not be employed efficiently.[7]

Since the high cost of coal and wood plagued all California, the appeal of cheap power reached well beyond the mines. By the mid-1880s, every major California city possessed steam-generated electricity, so a ready market seemed to exist for hydroelectric power. But, to deliver the electricity, one had to overcome the environmental condition posed by hydropower sites located far from urban markets, and the prevailing direct current technology did not permit power to be sent much over ten miles. "The most important problem now presented to electricians," observed San Francisco engineer, E.J. Molera, before an 1885 meeting of other California engineers, "is the question of the transmission of power at great distances." Direct current resistance caused 40 to 60 percent transmission line power losses. An increase in the copper wire size could overcome this problem, but that solution faced both economic and physical limits, and insulation problems arose with the use of increased voltage. Molera had no answer, but when one was found, he argued, hydroelectricity from "out of the way places . . . would be sufficient . . . to run cars, machines for lighting and warming houses, and could be used for cooking purposes." As late as 1895 in the East, however, many engineers agreed with well-known engineer Herman Haupt that "electricity could not be sent over 5 miles and delivered in successful competition with steam."[8]

Alternating current, the solution to long-distance transmission of electric power, was tested first in Europe. Like Californians, Europeans in some regions suffered from the scarcity and high cost of fuel and had abundant but often adversely located waterpower. Some Europeans estimated that developing their waterpower would cost from a fifth to a tenth of the overall costs of steam power. This provided a tremendous incentive for the cultivation of hydroelectric power and its long-distance transmission, and various experiments and demonstrations were undertaken beginning in the 1870s. A culminating test of the feasibility of long-distance transmission occurred at the 1891 Frankfurt Exposition, where over thirty-three thousand volts of three-phase AC current was transmitted 112 miles between Lauffen and Frankfurt. This demonstration, followed by another in 1892 between Tivoli and Rome, Italy, gained wide attention in the electrical industry, especially in the United States, where it attracted George Westinghouse and engineers in the Far West. Because of their state's geography and energy resource situation, Californians particularly heeded the European successes, and it was in their state that hydroelectricity and long-distance transmission of power found "its true habitat."[9]

Westinghouse had become interested in AC technology during the mid-1880s, acquiring certain patent rights and developing, with William Stanley, a single-phase current AC system. After an initial installation in Pittsburgh, his firm came west in 1890 to install at Oregon City a hydroelectric plant that provided single-phase power for Portland, thirteen miles away. The next year the firm built a second installation near Telluride, Colorado, sending power some three miles to a mine. But single-phase current was suitable only for lighting and failed to power motors efficiently. In part, this problem led many eastern electric power engineers to remain more interested in short-range DC power distribution from steam-powered central stations than in long-distance transmission, but they soon faced formidable competition. Westinghouse continued to build on Nikolas Tesla's

1888 patents in polyphase AC technology to develop a two-phase current system, winning contracts to provide power for both the 1893 Chicago Exposition and a hydroelectric project at Niagara Falls. In the meantime, the General Electric Company was formed through a merger of the Edison and Thompson-Houston firms, Elihu Thompson bringing to the new company the three-phase AC technology that held the promise of efficient, economic multipurpose power. The new firm settled on it.[10]

In 1891, self-taught engineer Almerian Decker left the Brush Electric Company in Cleveland, Ohio, which also had started work on AC technology. He immigrated to California for health reasons and settled in Pomona, where he joined with southern California entrepreneur Henry Harbison Sinclair and Pomona College President Cyrus G. Baldwin in a local hydroelectric scheme. He realized their biggest problem would be transmitting power fourteen miles to Pomona from the waterpower site they had chosen, but the Lauffen-Frankfurt effort convinced him that this could be solved through AC transmission. Decker and Baldwin tried, unsuccessfully, to convince Westinghouse to design a three-phase system for their project, but the firm would agree only to a ten-thousand-volt, single-phase system. In response, they solicited the help of William Stanley, who then headed his own electrical manufacturing firm in Pittsfield, Massachusetts, to persuade Westinghouse to provide the needed equipment. In 1892, their San Antonio Light and Power Company opened a fourteen-mile, ten-thousand-volt, single-phase transmission line to Pomona, followed by an additional twenty-eight-mile line to San Bernardino later in the year. Still convinced of the superiority of three-phase technology, however, Decker signed a contract in 1892 to build one for Henry Sinclair's new Redlands Electric Light and Power Company. In February 1893, the new General Electric Company, now eager and able to compete with Westinghouse in AC technology, agreed to supply the project. Eight months later the Mill Creek power plant opened, the first commercial American three-phase system with generators turned by Pelton waterwheels under a head of 377 feet. Unfortunately, Decker did not live to see the system opened.[11]

The Mill Creek plant proved practical the commercial transmission of three-phase AC power, and other California hydroelectric power entrepreneurs contracted with General Electric. The company placed James A. Lighthipe, who had worked with Edison in Menlo Park, New Jersey, as chief engineer in its Pacific Coast office. His first task was to design and supervise installation of a three-phase system on the American River at Folsom, where Horatio Livermore and his sons had been working on a hydraulic power project since the 1860s. Lighthipe oversaw placement of four 750-kilowatt generators for the Livermores and construction of a twenty-two-mile-long transmission line to Sacramento. Beginning eleven-thousand-volt power transmission in July 1895, the new system represented a significant advance over Mill Creek and preceded by a year the acclaimed Westinghouse line from Niagara Falls to Buffalo, New York. By the time Buffalo received its first power, Folsom already supplied Sacramento with power for 12,662 incandescent lamps, 582 public and private arc lamps, up to 35 streetcars using 41 miles of track, and motors giving 537 horsepower in breweries, print shops, flour mills, and Southern Pacific Railroad shops. On 9 September 1895, California's Admission Day, a great electric carnival drew some thirty thousand people to the state capital.[12]

A few months before the completion of the Folsom plant in the Sierra Nevada foothills, Carl E. Grunsky, retiring president of the Technical Society of the Pacific Coast, applauded the tapping of waterpower: "Hand in hand with the development of the mining industry . . . [comes] the utilization of power to be generated by the flow of our mountain streams. By electric transmission it can be delivered at those points where it will be of greatest service in reducing cost of manufactures, it will enable and encourage the establishment of industries, which, without it, would have been impossible." Despite the excitement of a few Californians, the success at Folsom was obscured for their eastern contemporaries by the Niagara Falls project. Louis Hunter reflects the general tenor about Niagara:

> It demonstrated to the world the possibilities of large-scale production and long-distance transmission of electrical energy and the special value of alternating current in bringing into service remote waterfalls hitherto running to waste. . . . All the early developments in long-distance transmission, abroad and in this country, and the technical advances on which they rested were overshadowed . . . by what was by far the most impressive hydroelectric venture of the time and one of the leading engineering events of the nineteenth century. This was the pioneer installation at Niagara Falls. . . .[13]

While publicity made Niagara important in the public mind, its greatest technical significance was in convincing hesitant eastern engineers and entrepreneurs of the importance of hydroelectricity and polyphase (in this case, two-phase) AC technology. For Californians, this impact was much less important. Almerian Decker, persuaded by the more efficient three-phase technology in 1892, contracted with General Electric for the Mill Creek system two months before Niagara engineers had made their decision to go with Westinghouse's two-phase system. Niagara's influence on Folsom seems only to have been in General Electric's decision to employ the same type of governors as those used at Niagara and, possibly, to use turbines rather than Pelton-type wheels; neither found much favor in subsequent California installations. In the end, AC technology came to California electric power engineers in the same way as mechanization came to its farmers, "among the very first needs, instead of being a late fruit of long discussion and experience."[14] Whether Niagara had been undertaken or not, Californians, through Decker, already had committed to three-phase long-distance transmission of electricity generated by water-power.

A statewide effort was already underway by California engineers to bring mountain-generated hydroelectricity down to valley communities and farther. Folsom and Mill Creek launched a hydroelectricity craze. Relief from expensive coal, the possibility of electric lights further replacing gas and kerosene, the expansion of electric railways—these and other energy hopes spurred on regional engineers and entrepreneurs. On 1 June 1895, a year before the Niagara hoopla, the *San Francisco Call* reported the mania: "A new kind of 'hustler' has arisen, and within the past three or four months he has been rapidly multiplying and filling the earth. He is the promoter of new electrical enterprises, and especially the promoter of schemes for the long-distance transmission of electric power. The air of the whole Pacific Coast has all at once become filled with talk about setting up water wheels in lonely mountain places and making them give light and cheaply turn other wheels in towns miles away."[15]

Before Folsom transmitted its first power, General Electric already had agreed to supply three-phase equipment for another more ambitious system in the Sierra Nevada foothills near the San Joaquin Valley town of

Fresno. Designed by civil engineer John S. Eastwood, Fresno's city engineer, the San Joaquin Electric Company undertaking tapped water from two branches of the North Fork of the San Joaquin River. It typified what would become California's hydroelectric style: seven miles of ditch conducted the water to a small reservoir above the power plant located east of Fresno; there it plunged down a 4,020-foot penstock (high-pressure pipe) to three tangential waterwheels under a head of 1,410 feet; generators fed eleven thousand volts of power into a transmission line that stretched thirty-five miles to Fresno. The system comprised the greatest hydraulic pressure and longest commercial transmission line in the world. Eastwood accomplished what others had only theorized, actually reaching beyond the Sierra Nevada foothills and into the mountain range itself to harness its waterpower potential and deliver electricity to a municipality.[16]

Development of Hydroelectric Power

By 1900, twenty-five hydroelectric plants stretched from Eureka to Pomona (map 9.1). Following the style established by Eastwood, companies organized to supply power to valley communities dissatisfied with expensive gas lighting or steam electric plants, or to nearby mining operations that lacked cheap fuelwood and coal. In Chico, for example, a dispute over cheap energy became so intense, in 1898, that the city ordered its electric street lights turned off. Because the Chico Gas Company took "only a desultory interest . . . in the development of the electrical end of the business," local entrepreneurs organized the Butte County Electric Power and Light Company. Acquiring two abandoned mining ditches, the firm hired consulting engineer E.W. Sutcliffe to build a powerhouse. The large number of similar projects around the state caused delays in acquiring plant equipment, but in May 1900 the Centerville powerhouse finally opened. Centerville fed water through a medium-high head of 590 feet to two fifty-eight-inch Pelton water wheels that drove two 400-kW Westinghouse generators. Its 800-kW capacity delivered 60 cycle (Hz) current at 2,400 volts to three transformers that stepped it up to 16,500 volts for transmission over two lines, one fourteen miles to Chico and the other thirty-two miles to gold dredges operating on the Feather River. The *Chico SemiWeekly Record* crowed: "At last the town has been freed from the [gas

Map 9.1. Electric Power Transmission Systems, 1900. Source: W.G. Vincent, Jr., "The Interconnected Transmission System of California," *Journal of Electricity*, 54 (June 15, 1925): 556.

company] that for years maintained . . . and dictated to the City Trustees top-notch prices for inferior lamps."[17]

These successful early hydropower projects inspired some hydroelectric entrepreneurs to deliver their electricity to California's largest urban communities. Among them was San Francisco entrepreneur Eugene J. de Sabla, Jr. Born in 1865, he came with his father to San Francisco at the age of five. After finishing primary and secondary school, he worked for a time in an Arizona copper mine and, in 1886, was made a full partner in his father's import and export company. He managed the business for several years, dabbling also in other enterprises. In the early 1890s, de Sabla obtained an interest in the Peabody gold mine in Grass Valley because of his continued contact with one of the mine's partowners, Alfonso Adolphus Tregidgo, previously the copper mine superintendent for whom de Sabla had worked in Arizona. Tregidgo was interested in hydroelectricity for running mining equipment and, in 1892, decided to build a hydroelectric plant on the South Yuba River to supply Grass Valley and Nevada City area mines with electricity. He secured water rights and incorporated the Nevada County Electric Power Company in September, inviting his friend de Sabla to join the venture as the company's vice-president. Work hardly had begun when the Panic of 1893 put the project in jeopardy. The depression closed down area mines and also forced de Sabla to liquidate his import and export business and to devote his attention to settling business debts.[18]

After de Sabla resolved his business problems, he returned his energies to the hydroelectric scheme. In 1894, he persuaded San Francisco business acquaintances and Grass Valley and Nevada City area mine owners to refinance the power company by selling bonds. One of the men he approached introduced him to John Martin, a businessman and San Francisco agent for the U.S. Cast Iron Pipe Company. Following a luncheon discussion with de Sabla, Martin sought out the Stanley Electrical Manufacturing Company and became its California representative; within months, he entered a partnership with de Sabla and arranged for installation of Stanley equipment for the Nevada County power plant's electrical system, teaching himself electrical engineering along the way. The new powerhouse was completed in 1896 and began serving the nearby mining communities.

Convinced that the power market was large, de Sabla and Martin decided to add another generator to their plant. In San Francisco, while planning a trip to New York to seek financing from an uncle, de Sabla met Romulus Riggs Colgate, grandson of the founder of the Colgate Soap and Perfume Company and an investor in California gold mines. Colgate took an immediate interest in de Sabla's plans, visited the Nevada County powerhouse, and agreed to provide financial support. He also agreed to support Martin in forming the Yuba Power Company, for which Martin completed a second plant in Brown's Valley in March 1898. Meanwhile, the impact of a dry winter, in 1898, had greatly impaired output of the Folsom plant on which the capital city depended for power. Always alert for opportunities, de Sabla and Martin again enlisted Colgate's financial support to build a third plant. Named after Colgate, the new plant opened in 1899 and delivered forty-thousand-volt power from the middle fork of the Yuba River to Sacramento, a distance of seventy-six miles. This attracted widespread attention among electric power engineers, and the accomplishment placed the two entrepreneurs squarely among the leaders of California's hydroelectric energy movement.[19]

In 1900, de Sabla and Martin felt ready to send power to the Bay Area. Seizing on the opportunity to provide electricity to the Oakland Transit Company, they gained additional eastern backing, consolidated their Yuba Electric Power and Nevada County Electric Power companies into the Bay Counties Power Company, issued $3 million in bonds, and began to enlarge their Colgate plant. Neither Westinghouse nor General Electric were willing to guarantee transmission above thirty thousand volts, but the Stanley Company accepted the challenge to supply the record-breaking 142-mile, forty-thousand-volt transmission line. To reach its final destination in Oakland, the line crossed the Carquinez Strait over a world-record 4,427-foot aerial span of four 6,900-foot-long steel cables, each bearing twelve tons on their anchorages. Sierra Nevada hydroelectricity reached the Bay Area for the first time in April 1901, and the current was raised to an unprecedented sixty thousand volts two years later.[20]

As the Colgate-Oakland line neared completion, de Sabla and Martin recognized they needed both retail distribution capabilities and further hydroelectric capacity to meet the potentially vast urban power market.

Martin took on the first challenge, organizing the California Central Gas and Electric Company, in March 1901, and acquiring twelve community gas and electric utilities from Butte to Marin counties; in December, he consolidated the several utilities into the California Gas and Electric Corporation. Meanwhile, de Sabla expanded their Bay Counties Power Company's hydroelectric resources, explored the Sierra Nevada north of the Colgate plant for new power sites, and reached Butte County near Chico just as Martin acquired the Chico Gas and Electric Company. De Sabla created the Valley Counties Power Company to purchase the Centerville powerhouse and the Cherokee Mining Company's water system, which drew on both Butte Creek and a branch of the Feather River. Eight miles upstream from the Centerville plant, he took Cherokee ditch water to supply his new De Sabla powerhouse, beginning with a 4,000-kW capacity in 1903 and expanding to 14,000-kW by 1906. The plant set a new distance record with a static high head of 1,528 feet, and current from it traveled 242 miles at sixty-six thousand volts to Marin County, in 1904, and 378 miles to Calaveras County, in 1905, setting another record.[21]

Meanwhile, de Sabla and Martin continued to purchase other firms and build more powerhouses, inspiring the evolution of California's hydroelectric culture. They acquired the Standard Electric Company in 1904, which provided hydropower for Stockton and San Francisco via San José, and the United Gas and Electric, which served the peninsula south of San Francisco. With control of virtually the entire Bay Area, they merged their principal California Gas and Electric Corporation with San Francisco Gas and Electric Company to form, in 1905, the Pacific Gas and Electric Company (PG&E). The new firm eliminated competition and garnered additional powerhouses by acquiring Northern California Power (1919), Great Western Power (1930), and San Joaquin Light and Power (1930). During the 1910s and 1920s, as part of its own hydroelectric expansion program, it acquired and developed, with five plants, the holdings of the South Yuba Water Company. By 1940, its interconnected, synchronized hydroelectric electric power network blanketed the Sierra Nevada and served virtually the entire northern California market. PG&E was one of the largest public utilities in the nation.[22]

Wherever PG&E and its predecessor companies tapped waterpower, they naturally left their mark on the environment. Their access roads, transmission lines, and generating facilities all modified the landscape, but their water storage and conveyance systems perhaps made the greatest impact. Initially, most of PG&E's hydroelectricity came from run-of-the-river plants dependent on the water flow from the gradual Sierra Nevada snow melt. Installations of this type involved small diversion dams, ditches, flumes, tunnels, and generally small forebay reservoirs. But several years of drought led the firm's engineers to reevaluate this technique and construct more and larger dams to reservoir sufficient stores of water to allow plants to operate during the driest of years. In 1931, PG&E completed the Salt Springs Dam, one of the largest rock-filled dams in the world, as part of its $40-million development of the Mokelumne River. Its last run-of-the-river system, in the far north during the 1920s, harnessed the Pit River and its tributaries to four powerhouses, but storage became part of the system when it was expanded through 1966 to nine plants. By the 1960s, PG&E owned ten of the forty-one major dams in California that were part of hydroelectric facilities.[23]

In southern California, the same process of growth and consolidation in the industry resulted in construction of impressive hydroelectric projects. Development in the industry was urged by enormous growth in and around Los Angeles that accelerated with mid-1890s oil discoveries, an antiunion environment that kept labor costs down, and city boosterism. It continued through the 1920s. With a fast-growing market, a race occurred between competing firms to get hydroelectricity to Los Angeles. Southern California Power Company organized in 1896 to tap the Santa Ana River eighty-three miles east of Los Angeles, and, a year later, the San Gabriel Electric Company formed and constructed the Azusa powerhouse twenty-three miles from Los Angeles. Competing with each other, as well as with companies generating electric power at steam plants, a number of rate wars occurred. Out of the ferment, Edison Electric Company in Los Angeles merged with Southern California Power in 1898, went on to absorb a number of other companies, and reorganized as Southern California Edison in 1909. Meanwhile, the City of Los Angeles, determined to gain municipal control over its electric as well as gas systems, formed its own Bureau of Power and Light in 1911.[24]

De Sabla and Martin's counterparts in southern California were William G. Kerckhoff, Allan C. Balch, and Henry E. Huntington. Kerckhoff, born in Indiana in 1856, visited southern California as a young man and convinced his family to migrate there in 1878. He and his father started a lumber company, which impelled him on an entrepreneurial path. He was drawn into shipbuilding, acquired an interest in the San Gabriel Valley Rapid Transit Railroad, and, in the 1890s, formed the waterpowered Azusa Ice and Cold Storage Company to chill oranges for shipment east via the Santa Fe Railroad. This last venture steered him and various partners to file on San Gabriel River water rights, and, in 1896, he joined with Allan C. Balch to form the San Gabriel Electric Company. Balch had graduated in electrical and mechanical engineering from Cornell University, and moved to Portland, Oregon, to manage a steam-powered electric company. Early power transmission efforts in California attracted him to Los Angeles, where he met Kerckhoff. Their San Gabriel Electric Company opened the Azusa powerhouse in 1898, supplying power for Los Angeles's street railways and manufacturing, and they began plans to develop the Kern River, one hundred miles north of Los Angeles.[25]

In 1901, Kerckhoff and Balch entered into an association with Henry E. Huntington. Born in 1850 in New York, Huntington had gotten a start in the railroad business from his uncle, Collis P. Huntington, one of the original "Big Four" founders of the Southern Pacific Railroad. The elder Huntington, who had no children, recognized his nephew's abilities and hoped that Henry would succeed him at the helm of Southern Pacific, a plan that he sought to strengthen by arranging Henry's marriage to his adopted daughter's sister. Henry rose to vice-president of the railroad in the 1890s, but after his uncle died in 1900, his climb to the presidency was thwarted by the Union Pacific Railroad's Edward H. Harriman, who gained control of Southern Pacific. Huntington sold $50 million in Southern Pacific stock, which he had inherited from his uncle, and moved from San Francisco to Los Angeles, in 1901, with plans to invest in real estate and develop an extensive electric railway system. Earlier acquisition of the San Gabriel Valley Rapid Transit Railway by Southern Pacific had introduced Huntington to William Kerckhoff; now, with Kerckhoff and Balch supplying electric power to Los Angeles's street railways, the three men became natural allies.[26]

Working with Huntington, Kerckhoff and Balch acquired the Los Angeles Electric Railroad Company in 1902. The three men then incorporated the Pacific Light and Power Company, which absorbed the street railway as well as the San Gabriel Electric Company. Although they had the overall aim of acquiring all the existing power companies in the Los Angeles area, they turned first to hydroelectric power development. Following up on Kerckhoff and Balch's interest in the Kern River, they began construction of a 10,000-kW powerhouse in 1902, and soon after 1904 began commercial power transmission at sixty-six thousand volts over their 125-mile line to Los Angeles. Concurrently, Kerckhoff and Balch built a hydroelectric plant on the Santa Ana River and purchased the San Joaquin Electric Company in Fresno, while Huntington began building the interurban Pacific Electric Railway Company. By 1909, they had five hydroelectric and three steam-powered generating plants supplying Huntington's fast-growing railway network, as well as providing power for irrigation pumping, manufacturing, and lighting in several cities.[27]

When they acquired San Joaquin Electric, they became acquainted with its vice-president and chief engineer John S. Eastwood. He had been exploring the southern Sierra Nevada for waterpower sites since 1890, and his eye had fallen on Big Creek, a principal tributary of the San Joaquin River. He began acquiring water rights in the area in 1900, but he was unable to secure financial backing for a major hydroelectric development. Eastwood presented his ideas to Kerckhoff and Balch, who hired him, in 1902, to finish obtaining water rights and to do preliminary surveys for the project. During the next year, Eastwood completed his work and submitted a detailed report of his plans for a huge hydroelectric undertaking. Pacific Light and Power was not yet ready for such an endeavor, but they kept Eastwood working on its development under an agreement with Balch that provided Eastwood with reimbursement of living expenses and a promise of 10 percent interest in the company, if and when the project came to fruition. By the end of 1910, Huntington and Kerckhoff had checked out Eastwood's cost and electric power estimates, consolidated their holdings in the Pacific Light and Power Corporation, and acquired all of Eastwood's plans and water rights in exchange for the promised 10 percent interest in stock.[28]

The Big Creek hydroelectric complex was started in 1911. After building a fifty-six-mile-long railroad to reach the area, within two years the company had constructed two electrically powered incline railways to remote construction sites and erected a dozen work camps in which to house as many as thirty-five hundred workers at the project's peak. The headquarters camp alone was a small town, with warehouses, offices, barns, a hotel, a movie house, railroad depot, sawmill, concrete plants, bunk houses, commissary, guest house, and tents. The most formidable of California's hydroelectric systems, Big Creek's initial phase was completed in 1913. Three dams fed water into a 3,880-foot tunnel connected to a eighty-four-inch diameter, 6,480-foot-long pipe that led to two 4,500-foot-long penstocks that dropped the water 2,131 vertical feet to powerhouse Number 1. Once used, the water collected in a fourth dam below the powerhouse, whence it passed through a tunnel 21,300 feet long into penstocks that dropped 1,858 vertical feet to powerhouse Number 2. With a combined generating capacity of 28,000 kW, the powerhouses sent current at 150,000 volts over a 243-mile line to Los Angeles. The engineering world looked with great respect on this accomplishment.[29]

To complete Big Creek, Huntington had turned all his efforts to the project, even buying out Kerckhoff and Balch. But, in 1916, he was nearing the end of his career and had been frustrated in his attempts to achieve a monopoly over electric power in Los Angeles by Southern California Edison's president, John B. Miller. Consequently, Huntington exchanged his Pacific Light and Power stock for shares in SCE and stepped down from day-to-day work in the industry. SCE, of course, had been doing its own expanding during Big Creek's construction, developing the Kern River with three powerhouses near Bakersfield and extending its service lines. The merger made it the fourth largest electric utility in the nation, but, more importantly, it became dominant in the Los Angeles area, the fastest growing region of California. Hence, SCE undertook to continue development of Big Creek, generally following Eastwood's basic plan. Between 1919 and 1929, SCE expanded Big Creek's capacity to 398,000 kW by enlarging the first two powerhouses, adding three new ones with related water conduits, building two major dams and several smaller ones, and drilling several tunnels, including one stretching over thirteen miles from Florence Lake to Huntington Lake. SCE's only regional competitors by

1930 were the San Diego Consolidated Gas and Electric Company, serving the far southern end of the state; Southern Sierras Power Company which supplied power from Bishop, in the Owens Valley, to the Imperial Valley; and the City of Los Angeles Department of Water and Power.[30]

Theoretical and Technical Education in Electric Power Transmission

Evolution of California's hydroelectric industry occurred in relative isolation of the East Coast, heartland of American electric research and development during the 1880s and 1890s. Entrepreneurs, from Sinclair to de Sabla and Huntington, were guided in their decisions more by the dictates of regional geography and resource characteristics than by eastern electric power practices. Similarly, the engineers who worked with and for them, from Decker and Lighthipe to Balch and Martin, had to work out on their own the technology needed to tackle western challenges. They also drew from the existing international pool of electrical theory through technical literature, correspondence, and occasional personal contact. And, General Electric, Westinghouse, the Stanley Electrical Manufacturing Company, and some other eastern electrical equipment companies sent sales representatives and engineers to assist with Pacific Coast hydroelectric installations. But California engineers were the ones who came to understand and adapt theory in resolving many of the practical problems inherent in long-distance, high-voltage power transmission. By 1914, their success resulted in California having more long-distance, high-tension transmission systems than any other region of the world, seven of thirty-one systems of 100,000 volts or more (map 9.2).

In 1884, California's civil, mechanical, and hydraulic engineers already had responded to their geographic isolation from the discourse of America's professional engineering organizations by forming the Technical Society of the Pacific Coast. In 1886 and 1887, two papers concerning electricity were read before the Society, but the first formal discussion of AC power and long-distance transmission came, in May 1893, when W.F.C. Hasson read "Electric Transmission of Power Long Distances." Although many of the region's engineers had training in mathematics, Hasson's presentation suggested their lack of acquaintance with electric theory. He carefully de-

Map 9.2. Electric Power Transmission Systems, 1910–1915. Source: W. G. Vincent, Jr., "The Interconnected Transmission System of California," *Journal of Electricity*, 54 (June 15, 1925): 572.

fined "volts," "amperes," and "watts" and compared this terminology to that of waterpower:

> The *size of pipe* necessary to conduct a given water power *increases* with the *quantity* of water, the *distance* and *efficiency* of transmission, and *decreases* with the *pressure.*
>
> The *size of wire* necessary to transmit a given electrical power *increases* with the *quantity* of current, the *distance* and *efficiency* of transmission, and *decreases* with the *electro magnetic force.*

So began California engineers' formal theoretical and technical education in electric power transmission.[31]

But the Technical Society was a general engineering organization. While electrical engineers felt welcome, they soon sought their own more focused forum for discussion. The American Institute of Electrical Engineers enhanced formation of a community of discourse focused on critical electric power problems through its *Transactions,* but headquartered in New York, it was far removed from practical realities on the Pacific Coast. Therefore, California electric power interests formed the Pacific Coast Electric Transmission Association in spring 1897. Composed of corporations, firms, and individuals engaged in general and long-distance transmission, it met quarterly during 1897 and 1898 and settled into annual June meetings through 1905. Never a large organization, with only thirty-two regular members through 1900 and fifty-four after that year's meeting, devotees nevertheless presented and discussed papers focused on long-distance transmission problems, as well as enjoyed social outings and tours of electric power sites.[32]

Pacific Coast Electric Transmission Association (PCETA) proceedings were published in *The Journal of Electricity, Power and Gas,* begun two years earlier in San Francisco by George P. Low, a journalist strongly interested in electricity. The *Journal*'s comprehensive, illustrated articles detailed the technological characteristics of each new electrical system on the Pacific Coast. Low supplemented regional pieces by reprinting eastern journal articles and professional papers he felt would be of interest to Californians. The work of Westinghouse engineer Charles F. Scott on transmission line corona and insulation received particular attention. Low also published regular book reviews, announcements of new electrical equipment, personal notices, and advertising for consultants and equipment manufacturers. Finally, his editorials urged California's electric power community "to continue to lead the world in electric power transmission."[33]

During this same period, the region's educational institutions responded to the need for formal studies in electrical engineering. In the early 1890s, A. Van der Naillen's School of Practical Engineering in San Francisco began offering a program in electrical engineering, while the city's Heald's College offered a practical training program. On a higher level, theoretical training centers developed. In 1891, Throop Polytechnic Institute opened in Pasadena, and Stanford University admitted its first class in Palo Alto. Each began small programs in electrical engineering, and the University of California at Berkeley launched a full program in 1892. Professors came from outside the region, Berkeley organizing its department around Professor C.L. Cory, who had received his electrical engineering degree from Cornell University, and Stanford hiring Frederic A.C. Perrine in 1897. As young engineers graduated from these programs, they took the place of eastern consultants in designing and implementing hydroelectric plants and transmission systems.[34]

Through the PCETA and its publishing organ, *The Journal of Electricity, Power and Gas,* California's entrepreneurs, engineers, and university professors established a regional community of professional discourse. PCETA members identified and discussed critical regional problems, always highlighted by the field experiences of the region's practicing engineers. Transmission issues, which generally received less attention in eastern literature, received regular consideration: insulation, line regulation, switching, line construction, and power losses on lines. Those issues prevailing in East Coast literature and practice, such as lightning protection, were found to be "very trifling," and rarely were raised more than once. In 1898, PCETA President C.P. Gilbert, stated the situation well: "Situated as the membership of this Association is, off the line of travel of eastern electrical interests, we are more dependent upon each other than they, and at the same time being scattered over a coast line of over twelve hundred miles, association with each other is more difficult than among the electrical companies of the more thickly settled east."[35] Thus, Californians went their own way, inevitably engendering reaction from the East. In 1912, *Electrical World* observed that Californians had "no respect for authority . . . [on] hydraulic, mechanical and electrical matters. If an impossible dam has to be erected . . . they build it . . . if an altogether unheard-of bit of tunnel has to be made to connect with a quite impracticable flume . . . they bore the tunnel and build the flume . . . if three or four stations must be operated together in defiance of all precedents, in go the switches and the plants operate."[36]

Through the PCETA, a pragmatic working relationship between science and technology emerged in California's electric power industry. Field experience led to invention and innovation, university laboratory experiments, and cooperative experiments and testing. For example, when de Sabla and Martin contemplated transmitting power longer distances at higher voltages in 1898, Martin tackled the problem of line regulation by developing his own high potential oil switch. They then sought out a fellow member of the PCETA, Stanford University's Frederic Perrine, to help them. In May 1898, the *Journal* reported the results of Perrine's work with some of his senior students, Frank Vanatta of de Sabla and Martin's Yuba Power Company, and Theodore E. Theberath, visiting the West Coast from the Stanley Company. They had conducted a thirty-thousand-volt field test on a 16,700-volt, fourteen-mile line. Observers, armed with telephones and stationed along the line to observe arcing and shorting, reported no corona, and Martin's own high potential oil switch successfully broke thirty thousand volts. Perrine concluded that slight modification of insulators, transformers, and other equipment would allow higher voltages than the sixteen to seventeen thousand volts then being installed in the region. De Sabla and Martin, armed with the results before they were published, were already working on their seventy-six-mile, forty-thousand-volt Colgate-to-Sacramento transmission line. It began service in early 1899.[37]

In April 1898, Professor Perrine took a two-year leave from Stanford to consult with the new Standard Electric Company of California. Prompted by rising copper prices, he ran practical experiments with aluminum wire for the company, enlisting the University of California's mechanical laboratory to perform tensile strength tests and Professor C.L. Cory and its electrical laboratory to perform conductivity tests. A year later, Perrine presented his "Tests and Calculations for a Forty-Mile Aluminum Wire Transmission Line" to the Transmission Association, a work that helped to introduce aluminum wire for common usage in California systems. During the same period, power company and consulting engineers also designed and tested eucalyptus wood and iron insulator pins, porcelain and glass insulators, pole designs, and oil and open-air switches. In early 1901, Perrine's colleague at Stanford, Professor E.E. Farmer, employed an oscillograph, forerunner of the oscilloscope, to study line capacity and self-induction.[38]

The interchange between professors and practicing engineers, working both in the laboratory and in the field, sustained high-tension, long-distance transmission advances. Perrine more than once sought the opinion of "these practical men" about various issues, and Professor Cory was praised for his valuable, plain-spoken presentations to the Association:

> Professor Cory's simple and clear-cut style of presenting electro-technical matters tears the mask of ambiguity from his subject. . . . He does not lead one through an interminable labyrinth of mathematics—we were going to say through catacombs of theory—but instead transports the reader over these dreary wastes by at once opening up a panorama of results. . . . This is just what the busy commercial man of today wants. . . ."[39]

In 1902, Charles Van Norden, in his presidential address, clearly stated the relationship of science and technology as evidenced through the PCETA:

> We are not, strictly speaking, a scientific body, and yet we are engaged in the application of the newest and abstruse science. We follow after the scientists and yet we also go before them. We are their heirs, and also their ancestors, furnishing them with queries of great amount, making hints of much value and applying their consequent generalizations.[40]

Although Californians had made the best of their isolation from other centers of electric power activities, they were not altogether happy with their situation. They received little outside recognition for their accomplishments and complained that the American Institute of Electrical Engineers, which was unwilling to permit a somewhat autonomous West Coast chapter, failed to understand "the local conditions that must govern the gatherings of a body of men located some three thousand miles away."[41]

While they applauded that "the Institute Council, practically for the first time, has recognized the West" in its presidential nomination of Chicago's Bion J. Arnold, they wanted a local chapter only if it could be similar in character to the PCETA, mixing theory and practice. They did not want a forum for purely theoretical talk, which the AIEE seemed to represent, yet they were pleased when their association was invited to join the AIEE and other societies at the 1904 International Electrical Congress held in conjunction with the St. Louis World's Fair. Finally, in January 1905, a San Francisco Chapter of the AIEE was formed. The arrival of this national body in California marked the end of the Pacific Coast Electrical Transmission Association's geographic isolation. Following the fate of the Technical Society of the Pacific Coast and other regional groups born of earlier conditions, the association ceased to exist within a few years, the professional interests of its members now served by the specialized national organization.[42]

Formation of the San Francisco chapter of the AIEE also coincided with increased theoretical and laboratory work in California, plus national recognition of the region's leadership in high-tension, long-distance electric power transmission. The *Journal of Electricity, Power and Gas* denoted the turn toward greater theoretical concerns by publishing a seven-part scientific laboratory report on AC current testing by Stanford Professor George H. Rowe. More significantly Professor Harris J. Ryan, research pioneer in transmission line loss at Cornell University, left the East in 1905 to head Stanford's electrical engineering department. Over the next fifteen years, he published a score of articles dealing with high-voltage and transmission problems, and he installed the first high-voltage laboratory in the West at Stanford in 1913. As a result of his contributions to understanding the phenomena of corona and his work with high-tension insulators, "a closer cooperation between Stanford and the power companies" developed and "gifts from various electrical concerns" made it possible for him to erect the first two-million-volt university laboratory in America at Stanford in 1926.[43]

Meanwhile, California engineers increasingly gained national recognition. Frederic Perrine had been the first, leaving Stanford in 1900 to accept an invitation to join the Stanley Electrical Manufacturing Company in

New Jersey, a company he would eventually head. Five years later, F.W. Peek, Jr., graduated from Stanford and joined the General Electric Company in Schenectady, New York. There he carried on high-voltage research, achieving permanent recognition through twenty scholarly papers published in the AIEE *Transactions* between 1911 and 1931. Rudolph W. Van Norden, an 1896 Stanford graduate, served as chief engineer for the Central California Electric Company through its 1905 merger with Pacific Gas and Electric. A year later, he opened his own consulting practice in San Francisco. In his lifetime, he designed thirty hydroelectric plants and fifty dams, and during the 1930s he served as technical advisor to the U.S. Secretary of the Interior on the construction of Boulder Dam. Frank G. Baum, another Stanford graduate who had worked with Perrine, stayed in California, serving as a professor at Stanford, as a private consultant, and as chief engineer for PG&E's hydroelectric department. He became a leading expert in high-tension, long-distance power transmission, and one of a handful of early California engineers to work, speak, and publish widely outside the region. He consulted with the federal government about Muscle Shoals power potential as well as with the German government, and in the 1920s he became a strong advocate of superpower and turned to inventing.[44]

Increased laboratory and scientific work in California, however, did not undermine the practical nature of the science/technology relationship that had been established through the PCETA. Its cooperative style reached into southern California, where Professor Royal W. Sorenson at Throop Institute of Technology worked in cooperation with Southern California Edison during the late 1910s to improve long-distance transmission. In 1919, he joined with company engineers H.H. Cox and G.E. Armstrong to present a proposal for a 220,000-volt, 1,100-mile interconnected California transmission bus before the Pacific Coast AIEE convention in Los Angeles, having worked out theoretical data in the Throop Institute laboratory and carried out practical tests on SCE lines. Based on this work, Sorenson began to guide his institution, soon renamed California Institute of Technology, toward stronger work in electric power transmission. His efforts culminated, in 1923, with the opening of the first million-volt laboratory outside those maintained by eastern equipment manufacturers. Thus university laboratories expanded their efforts to investigate the critical problems that challenged consulting and power company engineers, and the latter continued to innovate based on laboratory experiments and tests. Furthermore, material costs and other commercial considerations, plus social concerns, such as fuel conservation during and after World War I, continued to motivate and influence research and development.[45]

Summary

By the 1920s, California engineers and entrepreneurs had successfully confronted the principal environmental conditions affecting waterpower development within their state, bringing California into a position of national leadership in both hydropower and long-distance power transmission. Seeking efficiency through economies of scale, they reorganized and consolidated small companies into greater ones, PG&E and SCE emerging as two of the largest utilities in the nation. They pushed high into the Sierra Nevada and achieved stunning hydraulic and mechanical feats to milk energy from mountain rivers and transmit it to the San Francisco Bay Area and the Los Angeles basin. They reinvigorated and enlarged mining ditch and storage reservoir systems, drove tunnels through granite peaks and ridges, erected monumental dams that created expansive reservoirs, and contrived new flumes and ditches to feed water at higher head through longer penstocks to larger waterwheels turning bigger generators.

To be sure, hydroelectric power people greatly impacted the environment. They took water from natural stream courses, cut ditches and flumes across hillsides, inundated canyons and valleys with water held back by dams, and slashed transmission lines across the landscape.[46] But, until the last half of the twentieth century, this concerned only a few conservationists. Until then, providing California with inexpensive energy remained paramount, and the power industry created an exceptional hydroelectric system that helped to overcome the debilitating high cost and scarcity of traditional fuels. An interconnected power grid served customers from Oregon to the Mexican border (map 9.3). By 1930, California, with 3.5 percent of the nation's population, produced 9.5 percent of the nation's hydroelectricity. Californians consumed electricity at two-and-a-half times the national rate, and falling water generated 83 percent of the

Map 9.3. Electric Power Transmission Systems, 1920–1925. Source: W. G. Vincent, Jr., "The Interconnected Transmission System of California," *Journal of Electricity*, 54 (June 15, 1925): 574.

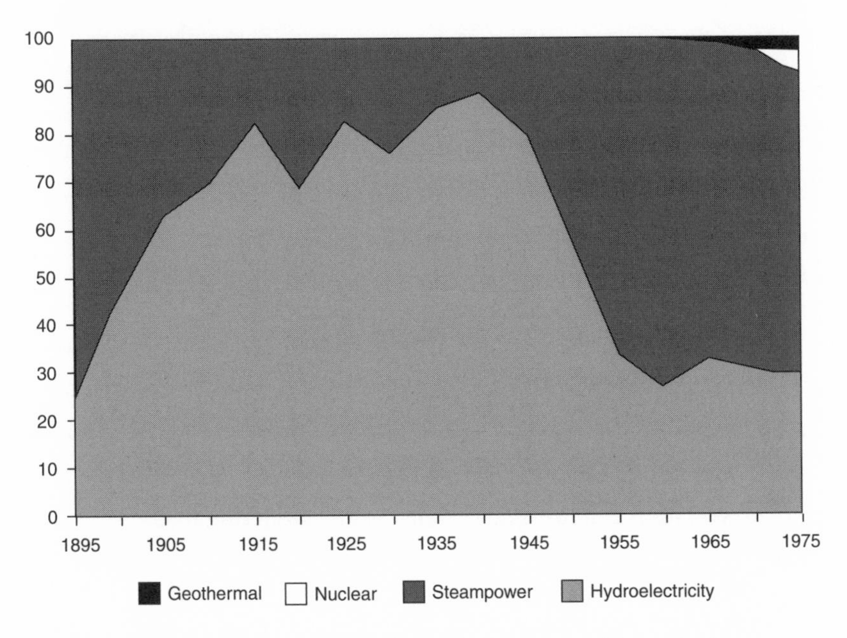

Figure 9.1. California Electric Power Production by Percentage of Generating Sources, 1895–1975. Source: *Supra,* Table C-14.

state's electricity. Not until after World War Two would fossil fuels generate more electricity than Sierra Nevada waterpower (figure 9.1).

Give-and-take between the environment and technology conditioned, in a number of ways, the path Californians followed in harnessing the force of falling water for electrical generation. The character of mountain streams, for example, guided technological development in hydraulics, causing miners to devise an entire waterpower system. The technological success of miners and the usefulness of portions of their system for early electric power companies led hydropower entrepreneurs and engineers to adopt high-head waterpower, ignoring the low-head waterpower model that eastern hydropower development followed. The snowpack in the Cascade Mountains and Sierra Nevada range, and the fact that its gradual annual melt provided sufficient year-round flow in major streams to allow run-of-the-river hydropower, initially prompted hydroelectric companies to avoid constructing large artificial storage facilities. But this approach changed after a string of dry years, another characteristic of California's climate, began in 1917. Thereafter, power companies built more reservoirs and, in fact, developed dams specifically to fit the high mountain conditions that they faced.[47]

Finally, the geographic location of waterpower sites, scores of miles from coastal and valley locations attractive to and settled by people, required transmission of electric power enormous distances if it was to reach sufficient urban customers to make investment in it worthwhile. This might have blocked major hydroelectric development altogether, but other environmental factors counteracted it. The absence of regional coal deposits and great physical distances between California and coal regions led to persistently high fuel prices. Substituting oil for coal was experimental, and oil supplies, in any case, were unstable. Consequently, entrepreneurs worked with local engineers and scientists in conquering the power transmission problem.

At the nexus of this technology-environment interaction, of course, were people, those who created the hydroelectric system and eventually those who became its users. Engineers and entrepreneurs conceived, designed, and built the hydroelectric power systems, and they also helped shepherd acceptance of electricity by its users. Joined by marketing specialists, publicists, and boosters of all things progressive and modern, they worked hard to bring hydroelectricity to the businesses, farms, and homes of Californians. Largely because of their efforts, people imbued hydroelectricity with the idea of progress, and it became an indelible part of California's society as well as its landscape.

CHAPTER X

Electrifying the City and the Home

PEOPLE EXPERIENCED ELECTRICITY from many different perspectives before it became a natural part of the fabric of twentieth century life. As historian David Nye points out, "the electrical landscape sprang up in patches." In most of the United States, as well as in other nations, electric power was overwhelmingly directed first toward transportation, industrial, and commercial applications. Its use in homes, even in urban households, was slow in coming, delayed by lack of wiring in housing stock and the cost of replacing existing tools, appliances, and lighting fixtures with electric ones. Moreover, electricity spilled over to the countryside only after urban areas were fully electrified.[1] But this was not the experience in California. The state's poor coal endowment and generally high energy costs, development of hydroelectricity and its transmission long distances to valley and coastal communities, and the hyperbole of hydropower developers and marketing efforts of others in the electric power industry united to overcome skepticism and inertia among potential domestic and farm users. California assumed national leadership in domestic electrical use and, in the process, crafted a model marketing framework for long-lasting cooperation within the industry. Westerners, bragged the *Journal of Electricity, Power, and Gas* in 1914, knew how to "do it electrically" (map 10.1).

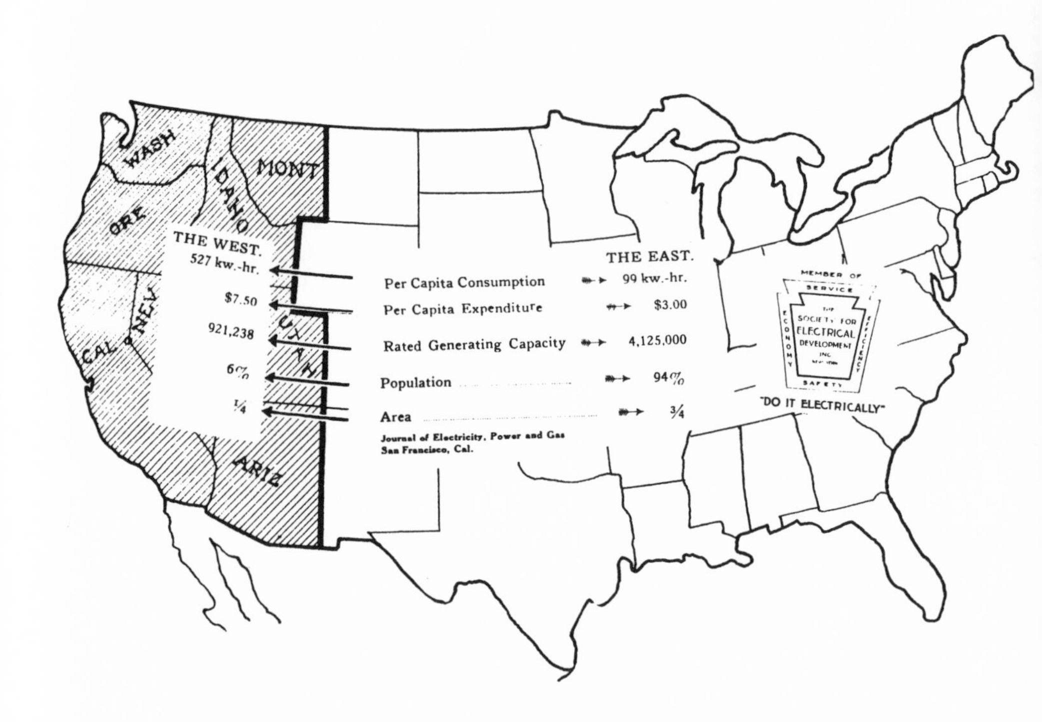

Map 10.1. The West Knows How to "Do It Electrically." Source: *Journal of Electricity, Power, and Gas*, 32 (March 28, 1914): 277.

The Electrical Mythos

The cultivation of hydroelectricity captured the public's imagination. Electricity's magical qualities made it the quintessential energy technology, fully exemplifying the technological essence of modernity in the early twentieth century. Clean, magical, an unseen force for better living, the electrical mythos was absent in older, more familiar ways of harnessing energy. Invisible, silent, powerful, and efficient, it was clearly the child of science, the world's "great comforter, civilizer, and enlightener," according to Cornell University's Robert H. Thurston, dean of America's steam engineering professionals. All other energy applications seemed to pale before electricity. Boosters suggested even the electric fan and the flatiron were vital elements in the advance of civilization, and others were convinced electric power would remake society. In 1915, California engineer Frank

Baum spoke for many members of the industry when he predicted before an International Engineering Congress in San Francisco that "some time in the future a nation's civilization will be measured largely in terms of kilowatt-hours consumed per capita."[2]

As elsewhere in America, Californians first experienced the thrill and convenience of electricity's magical power in a public forum. Street lighting, easily contracted for between power companies and local government, symbolized local success and progress and was the first sign in California of what writer E.H. Mullin called "great electric waterfall cities." Coincident with the arrival of hydroelectric power to Pacific Coast cities and towns during the 1890s, early arc lighting devices gave way to improved arc and incandescent lamps. Communities sought to outdo one another. They installed ornamental lamp posts, curb-to-curb incandescent arches, and, by 1906, San Franciscans had illuminated Fillmore Street with intersecting arches at each street corner. In the same year, Western Electric Company's house journal, *Energy*, suggested that "if the City of Los Angeles continues to extend its system of lighting as planned, it will undoubtedly be justified in its claim to the title 'The Electric City.'"[3]

Whenever a community reached a new lighting threshold, a great electric light carnival, like the one in Sacramento in 1895, might draw thousands of people to christen the dawn of the new age. Spectacular lighting efforts became an essential part of community celebrations. The 1909 San Francisco festival commemorating the discovery of San Francisco Bay by Gaspar de Portola featured "electrical illuminations . . . on a larger scale than had ever before been projected on the Pacific Coast." Over thirty thousand incandescent lamps outlined the Ferry Building and other principal buildings, and festooned downtown streets. At the intersection of Market and Third streets, two thousand lamps fabricated an immense bell, and a parade of electrified floats capped the festival's final evening.[4]

Planning the illumination of the 1915 Panama Pacific International Exposition in San Francisco began four years ahead of time. Event officials retained Walter D'Arcy Ryan, General Electric Company's lighting laboratory director, who had illuminated Niagara Falls and eastern festivals. Outdoor illumination, exhibition building architecture, and landscaping of the exposition ground's 625 acres were planned together. Inasmuch as the opening of the Panama Canal would symbolize commerce between eastern and western hemispheres, decoration and illumination was to "typify in a measure the traditions of both the Orient and the Occident." The exposition's colors of vermilion, burnt orange, and Oriental blue would be especially highlighted, "at night the play of the illumination . . . directed so that colors will hold their absolute values." When the exposition opened, it was lit with the most advanced lighting technology, from high-current luminous arc lamps and colored incandescent bulbs to flame arc lamps and searchlights. Although only one of a number of major national fairs to highlight electricity after 1890, fifteen years later the Exposition's "Architecture of the Night" continued to be applauded.[5]

The exposition's success led San Francisco's Downtown Merchants Association to bring high-current luminous incandescent lamps to the city center. General Electric had marketed its earlier arc lamp street lighting systems as the "Great White Way," and now Walter Ryan created downtown San Francisco's new lighting plan as the first of General Electric's improved "Intensive White Way." Willis Polk, well-known bay region architect and chair of the exposition's architectural commission, designed a tri-unit lighting fixture to cap existing downtown trolley poles that he had designed earlier. Pacific Gas and Electric Company assumed the entire cost of installation. On 4 October, the "largest crowds ever assembled on the streets of the city at night," witnessed an electric pageant. A parade of floats mounted on trucks and streetcars depicted the history of lighting from "The Cave Man" to "The Exposition by Night." When San Francisco's "Path of Gold" finally was switched on, the *San Francisco Chronicle* described the effect: "A warm white light, the most brilliant that ever shone through a city thoroughfare after the sun had gone down, flooded the canyon of Market street. . . . the stars disappeared. . . . light flowed everywhere, touched and enveloped everything. . . . Such was the wizardry of this marvelous light."[6]

The electric streetcar also captured electricity's mythos for Californians, as it did elsewhere. It *was* the "machine in the garden," permitting escape from urban squalor. Linked to the imagination of science, not the reality of industrialism, the trolley operated with great efficiency. Cheaper than cable railways to install, it was fast and did not emit sparks, smoke, or other noxious byproducts. Its noiselessness, cleanliness, and efficiency promised better city sanitation, more leisure time and pleasure, and an end to tenement living by providing a healthier suburban life for Americans. As the railroad locomotive before it had symbolized the energy and strength of steam power and the age of industry, the trolley symbolized the positive moral influence, the civilizing thrust of electricity and the scientific age.[7] It rapidly became a part of California's urban landscape, and streetcar systems lingered, even after crowded conditions undermined their initial magic and people turned the automobile.

Electric power also served Pacific Coast industry, but in a manner that diverged from the initial eastern experience. Partly because high fuel costs and the state's colonial status with the East had discouraged industrialization, California power companies did not have a large, established industrial market to serve. After the mid-1890s, most of the state's established urban-based manufacturers switched from imported coal to oil for their principal fuel. And, as oil fuel passed its experimental stage and supplies become stable, some new industrial development was fostered. The quality of newness about electricity was matched to some degree by oil as fuel, and this worked against utilities' efforts to convert over to electricity established companies using steam power or industrial heat processes. Thus, California's electric power industry tended to serve new industries rather than convert the energy configurations of old ones. Companies adopting electricity built facilities from scratch to take advantage of the flexibility offered in operating independent electric motors for tools. Low, generally single-story, factory buildings extending horizontally across spacious acreage came to characterize the California industrial landscape, as opposed to multistoried plants with central ground-floor steam engines powering overhead belting systems in workspaces stacked one on top of another.[8]

Hydroelectric companies did seek both to supply the energy needs of existing industries and to help them expand. The mines dotting the Sierra Nevada offered an obvious market and hardly needed to be sold on electric power. Although many mining firms had waterpower available for their own use, those that did not generally were located sufficiently near early commercial hydroelectric plants to become ready customers. Some

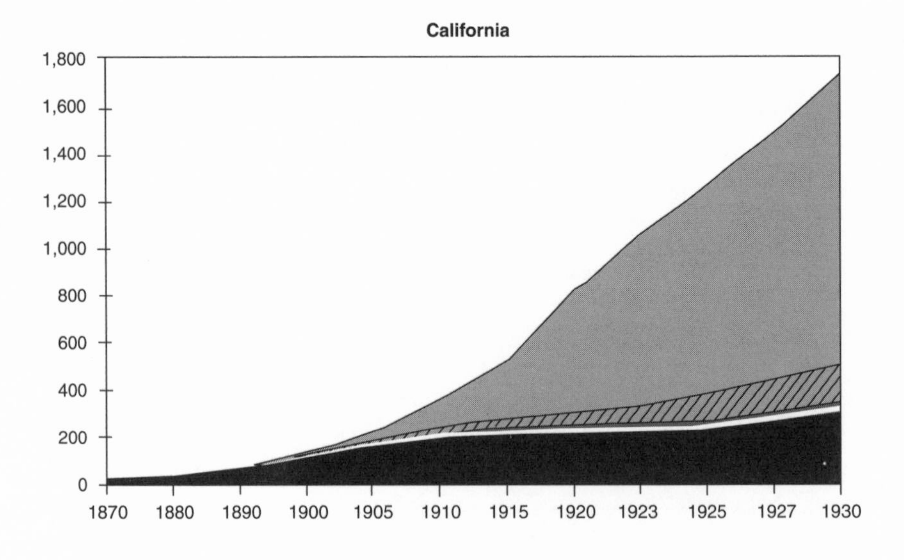

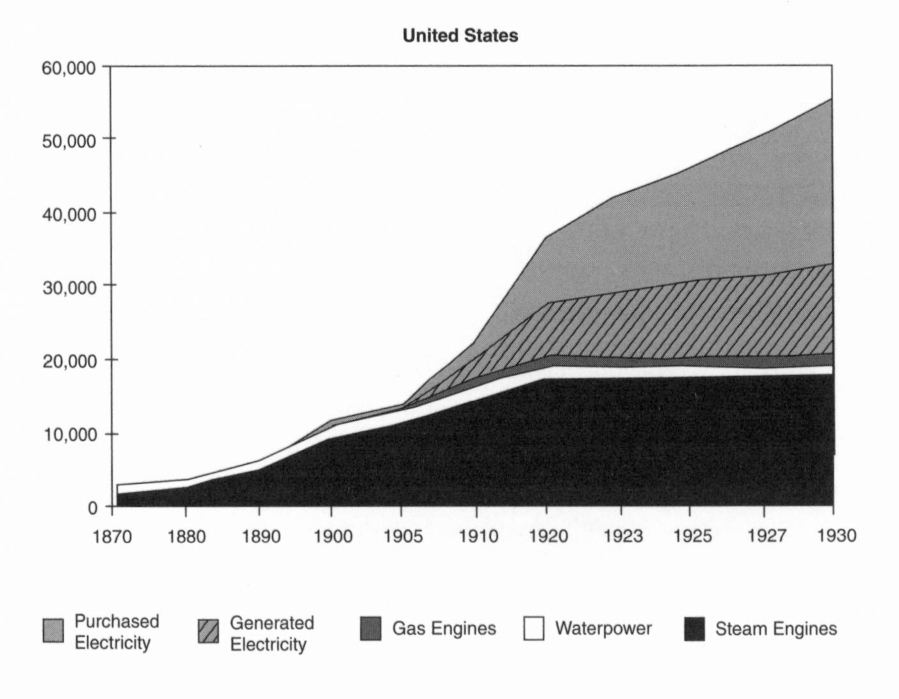

Figure 10.1. California and United States Manufacturing Horsepower, Selected Years, 1870–1930. Source: *Supra,* Table C-15.

hydropower companies based their own expansion on such a market. The Northern California Power Company in Shasta County, for example, installed three new hydroelectric plants, in the early 1900s, to serve the nearby copper mining and smelting industry. Northern California Power also tried to develop a commercially successful electric iron smelting company, thus establishing a new industry in the state. Although it failed, the rest of the power industry showed great interest in the project, and they encouraged the state's iron and steel fabrication industry to adopt electric power between 1910 and 1920.[9]

After 1910, most industrial firms expanding or establishing themselves in California adopted electric power (figure 10.1), and some represented industries born from the new age of electricity and petroleum itself. Following San Francisco's great earthquake, rebuilding with seismic safety in mind stimulated rapid expansion of the Portland cement industry. Robert Sibley, editor of the *Journal of Electricity, Power and Gas,* reported that this demand "could have never been met had not the electric motor with its flexible independent drive been a possibility." Long-established as well as new machine shops also adopted electricity, and electric utilities provided power to operate well pumps in the new oil fields opening up through the 1920s. The power industry also tried, with some success, to entice electrochemical firms to establish in the state.[10]

In 1920, a survey of industrial electric power users in California by *Electrical World* concluded that "there is probably no other section of the country which can show such a mastery of the industrial situation by central stations." Electric power delivered by the state's power companies was essential to almost every industry, from food processing (the state's biggest industry in value of products), through paper, glass, shipbuilding, machine shops, lumber, and textiles. But the greatest harbingers of electricity's future, users that clearly promoted its magical qualities, came in the form of entirely new industries: the forty-six studios of the motion picture industry in and around Los Angeles, in 1919, all depended on electricity, and in northern California, Stanford University electrical engineering graduate Cyril F. Elwell founded Federal Telegraph Company in 1910, the first of many companies that eventually paved the way for a midcentury microwave electronics revolution and the evolution of Silicon Valley.[11]

Stimulating Domestic Power Sales

Street lighting, trolleys, and industry provided good markets for electric power companies and, in general, set people toward harmonizing electricity into their daily experience. But utilities had to put domestic power sales high on their marketing agendas, if widespread use of electricity was soon to be brought into the home.[12] California's utilities were cognizant of their state's history of high energy costs. Furthermore, they operated under regional environmental conditions that urged development of hydroelectric over fossil-fuel power generation. Both these things led them quickly to find the domestic market a potentially lucrative one. In electrifying the home, the interplay between technology and the environment proved to be a basic factor in the California experience.

No matter how electricity was generated, it could not be easily stored. This was particularly true with alternating-current transmission systems, for batteries worked with direct-current. Consequently, a power company needed a generating capacity at least equal to its total customer demand or maximum load. Moreover, to prevent idling large numbers of generating plants and to attain the most profitable and technologically efficient operating economies, it needed customers whose power use offset each other. It was no good to have the power consumption of all one's customers peak and plunge inconsistently or simultaneously, so, to avoid this, companies expanded their geographic service areas and sought to develop a diverse customer base. One way of measuring their success was by their load factor, that ratio representing average power demand and maximum power demand on a company's system during a given period of time, such as a day, week, or even season.[13]

Operating conditions in California differed from those elsewhere. In the East and Midwest, utilities relied heavily on fossil-fuel generated electricity, as opposed to hydroelectricity. They also developed predictable industrial and transportation power loads that were generally much heavier than anywhere on the Pacific Coast. Consequently, although some eastern and midwestern companies expanded and diversified in pursuit of higher load factors, notably Chicago's Commonwealth Edison, they also found it economical simply to shut down generating plants during predictable low-demand periods. This practice saved enough in the cost of coal that they did not feel greatly impelled to invest in opening up other markets that might balance load and raise low load factors. In California, however, utilities generated the vast majority of their electricity by waterpower, a "fuel" that particularly paid great benefits if plants did not have to be shut down. Therefore, the state's utilities found it extremely desirable to develop diverse classes of customers whose power consumption offset each other across various time periods.

Southern California Edison was particularly successful in marketing small household appliances. Its own hydroelectric development occurred so rapidly that surplus power and load problems plagued the firm soon after the turn of the century. SCE's general agent, S. M. Kennedy, described the plight his firm faced in 1904: "The Edison company had water going to waste during the daylight hours; it [also] had big steam turbines standing idle, waiting for a day load. . . ." Despite the fact that the Los Angeles metropolitan area was the fastest growing in California—the population increased fivefold there between 1900 and 1920—and that most houses built there after 1900 were wired for electric lighting, SCE's load dissolved during certain time periods. In domestic appliances that could be plugged into lighting fixtures recalled Kennedy, "a large and satisfactory load was ready to be taken, only wanting an invitation. . . . the customers needed the appliances, although they did not know it, and the company needed their business and knew it."[14]

Because SCE managers thought the pace of lamp-socket appliance sales at local hardware stores was too slow, the firm undertook aggressive, direct marketing. It opened showrooms at its various offices and advertised the advantages of the appliances in newspapers, circulars, and letters. It loaned flatirons and percolators, sold them on an installment plan, and took old irons and coffee pots in trade. It focused on the sale of appliances that, said Kennedy, were "popular, efficient, durable, and likely to be regularly in use." By 1907, SCE managers determined that "the only way to sell large quantities [of appliances] and do it quickly was to send expert salesmen to call on customers in their homes."[15] A sales department was organized with fifteen sales representatives, all male. Through regular house-to-house canvassing, they recorded each household's appliances, gauged the market, and pushed only those devices which they believed received regular use. In 1911, the company reported it had sold 15,438 flatirons, 4,634 coffee percolators, 3,440 toasters, and 2,000 other miscellaneous devices. The salesmen soon accounted for 80 percent of the company's total appliance sales, which, by 1915, reached 141,705 in a service population of 500,000, and SCE achieved a national reputation for its marketing of domestic lamp-socket appliances.[16]

Edgar A. Wilcox summarized many California utility sales techniques in his 1916 manual for selling appliances to householders and businesses. A Great Western Power Company electric heating specialist, Wilcox offered salespeople basic information about electricity and its benefits over other fuels. He reviewed appliance types and models from southern California's Hotpoint "El Grillo" table top grill and "El Bako" lamp-socket oven to full-size ranges, hot water heaters, space heaters, and commercial stoves and ovens. He emphasized the modern, progressive qualities of electricity and advised salespeople how to "educate" housewives in proper appliance use. He underlined the perceived freedom which electricity gave housewives, "making housework an enjoyable pastime." And he urged salespeople to try to install electric appliances in school domestic education departments, emphasizing the important inculcation of female children with a photograph of the Hughes Jr. Electric Range "for early training of housewife." Armed with his book, Wilcox thought, the salesperson could approach any customer.[17]

The electrical industry particularly tapped the newly emergent domestic science field to help with marketing. They hired young female college graduates to represent them to the principal consumer of home electricity, the housewife. The *Journal of Electricity, Power and Gas,* for example, employed as an associate editor, Clotilde Grunsky, a distinguished 1914 graduate of the University of California and daughter of a noted Pacific Coast civil engineer. She became "recognized nationally as perhaps the best woman writer on technical subjects pertinent to the electrical industry" and was one of three women members of the American Institute of Electrical Engineers in 1922.[18] She kept direct contact with women in the utility company's domestic science programs, and she championed both the housewife's electrical needs and women working within public utility organizations. In 1921, when the National Electric Light Association established a Women's Committee on Public Utility Information within its Public Relations Section, Grunsky was a member, and she led the women in California's electric industry as chairperson of the Woman's Public Information Committee of the Pacific Coast Electrical Association.[19]

The Woman's Public Information Committee represented the interests of women within the industry and at home, insofar as it upheld the pecuniary interests of the utilities. It sought out the best ways to get housewives the electric message, urging use of the radio, organization of an industry information service, and development of utility-run home economics laboratories. In 1925, the Committee surveyed twelve California and three Arizona power companies about women's roles in the industry. Most firms employed from one to seven women demonstrators, five maintained demonstration and exhibition kitchens, and six or seven ran cooking schools, alone or with local newspapers or manufacturers. Women helped develop newspaper advertising and company-sponsored newsletters, such as *PG&E Progress* which included one page in each issue devoted to "Helps for the Homemaker." They reached housewives informally by joining local women's clubs, with dues paid by their companies, and they represented home electricity at county fairs and other community events.[20]

Displays at public events played an important role in getting out the electrical message. Electrical exhibitions had been held by the industry since the turn of the century, the largest on the Pacific Coast at the Panama Pacific International Exhibition. But local, regular displays meant constant exposure. One knew immediately that women had conceived the San Joaquin Light and Power Company's 1920s display at the Fresno District Fair. Integrated as a central and essential part of the exhibit was a Free Day Nursery, rest room, and emergency hospital. Female attendants watched the children, nurses cared for infants, and a physician was present. Electricity ran a refrigerator to keep milk fresh and heated milk in a warmer. Regular meals also were provided. On a year-round basis, the firm also maintained a regular exhibit at the Fresno free market: a "model wash room and sewing room" staffed by women demonstrators.[21]

Between 1910 and 1940, utilities carried out aggressive sales campaigns. Most maintained display booths and sales centers, both temporary and permanent, at county fairs, shopping areas, and their own headquarters.

Figure 10.2. PG&E operated a special car demonstrating electricity in cooperation with the Northern Electric Railway between 1912 and the early 1920s. From *The Journal of Electricity*, 45 (November 15, 1920): 460.

Each new electrical appliance prompted a crusade, from coffee percolators and vacuum cleaners to ranges and refrigerators. Utilities intermixed direct mail and personal contact sales techniques, and worked with local jobbers, contractors, and electrical equipment supply firms. They offered free electricity during promotional periods, and their advertising campaigns filled California's newspapers and were reported faithfully in their industry's trade and technical journals. PG&E ran a farm and home electricity display and demonstration car over the lines of the Northern Electric Railway from 1912 into the early 1920s.[22] Meanwhile, with an eye toward persuading recalcitrant householders through their children, the Mount Whitney Power and Electric Company installed electric ovens, stoves, and water heaters in a local high school classroom, making Domestic Science there "so attractive the girls would be anxious to take the course."[23]

Promoting Electrical Products in Homes and Industries

Unfortunately, utilities, appliance manufacturers, jobbers, contractors, and dealers often found themselves competing with each other. This was particularly problematic where dealers and contractors perceived that the central station power company undercut their appliance sales. Some marketing cooperation occurred, with various attempts at coordinated sales efforts during the 1910s. The Mount Whitney Company worked with local hardware stores to sell ranges and heaters, the Federal Sign System, based in Chicago, developed an easy payment plan program with nineteen power companies, and PG&E conducted a lamp-socket appliance campaign in cooperation with seven San Francisco jobbers. In 1915, the National Society for Electrical Development proclaimed 29 November to 4 December to be "Electrical Prosperity Week." Coinciding with a general "shop early for Christmas" campaign, the society developed advertising supplements for newspapers with the slogan "Do it Electrically," and urged the California industry to tie in. Nevertheless, friction persisted.[24]

In April 1917, this discord was faced at the first convention of the Pacific Coast Section of the National Electric Light Association. Riverside convention delegates agreed that "cooperation and educational development are the means by which to ensure . . . improved and enlarged service." They adopted a resolution "that the member companies of the Pacific Coast Section lend active cooperation and support to the California Association of Electrical Contractors and Dealers. . . ." Their commercial committee met to draft a plan with representatives of the contractor-dealers and of the Electrical Supply Jobbers of California. Although contractor-dealers may have felt some loss of pride that power companies thought they needed education to become "more aggressive, practical and progressive" businessmen, they nevertheless agreed to participate. Thus began, in "a spirit of genuine cooperation" among all divisions of the industry, the California Electrical Cooperative campaign.[25]

The campaign sought to coordinate activities between central stations, dealers, and jobbers. Power companies wanted to help dealers get business and improve advertising and displays, and all participants wanted to halt debilitating competition between branches of the industry. To do this they formed a campaign advisory committee including representatives of three contractor-dealers, three central stations, one jobber, and one manufacturer. The committee then solicited $12,000 to finance a 1918 program and employ two field men, one in San Francisco, the other in Los Angeles. They received 50 percent of the required funds from central power stations, 25 percent from jobbers, 16.6 percent from contractor-dealers, and 8.3 percent from manufacturers.[26]

Campaign field representatives began 1918 by conducting meetings to foster acquaintances and harmonious relations between central station managers and contractor-dealers. They got the two groups to agree to sell appliances only at manufacturer's list prices, as well as to adopt identical deferred sales terms and repair charge scales. Central stations also agreed to withdraw field sales personnel wherever contractor-dealers could take over. Southern California Edison was among the first to do so. At this point, concern that the campaign might undercut national energy conservation efforts in support of America's involvement in World War I, plus resistance to the cooperative campaign from the weak contractor-dealer association, led to suggestions in July and August that the whole cooperative idea should be abandoned. The advisory committee decided otherwise, concluding that "every effort should be made to hold the contractor-dealers' association together." They supported energy conservation and encouraged dealers "to maintain themselves in business by directing their efforts toward work which would help win the war." They agreed to defer appliance sales until the war's end.[27]

Committee members met with campaign participants at an October get-together, where they gave pep talks and gained renewed support from contractor-dealers and jobbers. They launched a newspaper advertising campaign to educate the public and amalgamate "the separate interests of the industry in merchandising," preparing copy for ten newspaper display advertisements, which central stations could sponsor between 21 October and the week before Christmas, and providing fourteen coordinated window display cards for dealers. Recognizing World War I constrictions on their industry's expansion, the campaign softened direct selling appeals. "The consumer was urged as a patriotic duty to have all electrical appliances repaired and brought up to the utmost efficiency as a means of economizing fuel. . . . The equally patriotic duty of using high-efficiency lamps was likewise woven into the educational fabric of the campaign." The armistice on 11 November resulted in recasting the final advertisements, adding to each the phrase: "The war is over. Now for a happy, practical Christmas."[28]

The war's end and success of the 1918 program assured the campaign's continuation. Industry pledges were renewed after the holidays, a third staff member was assigned to work in northern California, a demonstration assistance program developed, and a salesmen's auxiliary was organized. Field representatives solicited new members for the California Association of Electrical Contractors and Dealers and worked with dealers to relocate or remodel 133 stores so as to facilitate better appliance displays and sales. Dealers responded to the campaign's effort by putting 167 salespeople into the field, while contractors urged new home builders to add more electrical sockets to wiring plans. The campaign coordinated a 400 percent increase in newspaper advertising, which, in turn, generated more general news coverage about the industry. As 1920 began, the campaign adopted the slogan "better service to the public" and, with greater manufacturer support, increased its budget to $27,000. It began the new year with a staff of seven: an executive secretary, three field representatives, architectural and advertising representatives, and an office assistant.[29]

The invigorated campaign launched a major assault against one of the major obstacles to household use of electricity: inadequate house wiring. Staff member Walter Price approached builders and architects about putting more than the standard two base plug electrical outlets in new houses, but he faced stiff resistance. Architects were slow to accept the use of electricity in homes, and builders saw only that the cost of electric wiring and modern kitchens added substantially to their construction costs. Price therefore decided that in order to win over builders and architects, the electrical industry had to first sell the house-buying public on the need for more outlets. He launched a model electrical home program, beginning in San Francisco. Builder Duncan McDuffie agreed to equip one of his new homes with thirty-six outlets, tripling the normal wiring. PG&E and the San Francisco Electrical Development League pitched in to help, and jobbers and furniture companies furnished the house. Widely advertised and open for seventeen days in June with appliance demonstrators, it attracted twenty thousand visitors. The campaign immediately planned other homes, opening five more by January 1921. The fifth, Los Angeles's "Adobe Electrical Home," attracted fifty-seven thousand visitors in twenty-five days.[30]

Successful electric homes spawned related efforts to attain more thoroughly electrified homes. The campaign distributed twenty thousand model wiring diagrams and sponsored informational gatherings for architects, builders, realtors, and high school trades classes. It distributed an "Adobe Electrical Home" motion picture to California theaters and adopted the slogan "An Outlet for Every Appliance." In addition, a "convenience outlet" campaign was started, sending a series of illustrated stories to rural and urban newspapers, producing a slide show, getting the Associated Manufacturers of Electrical Appliances to stop illustrating advertisements with lamp-socket appliances, and featuring convenience outlets at the Los Angeles Industry and Trade Fair and San Francisco's California Industries Exposition. It worked with cities to adopt local building codes that required concealed contacts for convenience outlets, and it had representatives go incognito as potential home buyers to new home tracts, where they demanded more outlets from sales people. Finally, the campaign convinced eastern manufacturers to drop the term "flush receptacle" in favor of "convenience outlet" on their packaging. By 1922, the campaign had succeeded in largely clearing the market "of the various non-interchangeable [outlet] types which had in the past confused the mind of the layman."[31]

The campaign's wide-ranging advertising and publicity efforts included a statewide newspaper appliance campaign for June Bride Week and awarding appliances to winners of an essay contest, "Why is Electricity the Modern Servant in the American Home," which ran in the *San Francisco Bulletin* and *Los Angeles Express*. The campaign provided the newspaper editors with materials for their home feature sections, achieving particular success through the *Los Angeles Examiner*'s "Home Economy" writer and personality, Prudence Penny; she became a national spokesperson in electric appliance advertisements during the 1930s. It also worked with both private and public schools to release a series of films about electric power production, electrical equipment manufacturing, and electricity's use in

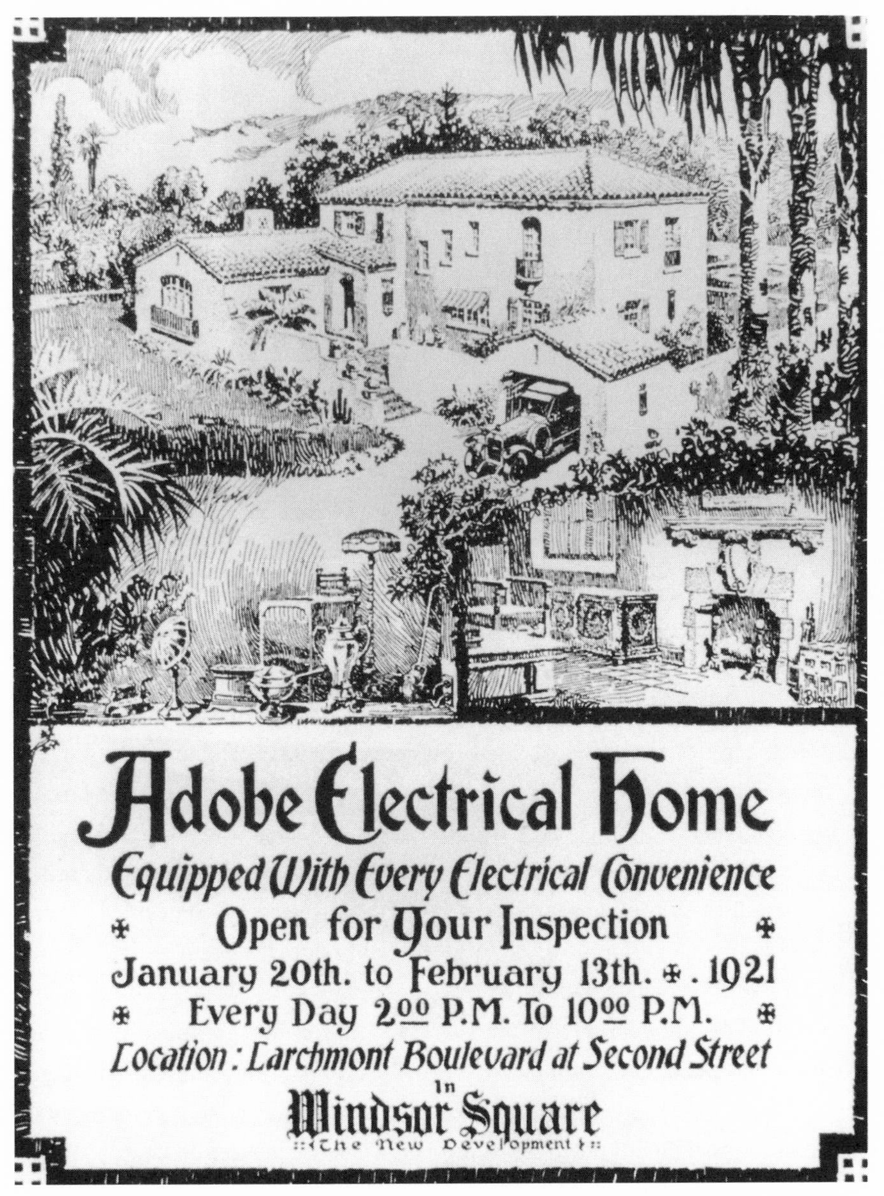

Figure 10.3 The California Electrical Cooperative Campaign's second all-electric demonstration home, the Adobe Electrical Home, opened in Los Angeles in 1921. From *The Journal of Electricity and Western Industry*, 50 (March 15, 1923): 208.

the home and in industry. Finally, the campaign worked with the producers of Buster Keaton's feature film, *The Electric House,* to set up displays promoting home electrification wherever it was shown in the state.[32]

By the mid-1920s, the campaign's staff doubled, its annual budget approached $65,000, and its message reached every corner of the state. A twelve-foot-long truck carried an "Educational Lighting Exhibit" to small town merchants in northern California, a field representative worked among architects and builders in the expansive Sacramento and San Joaquin valleys, and two campaign-owned railroad cars carrying demonstration exhibits "spread the electrical message into the small communities and rural districts."[33] The campaign crusaded for convenience outlets in apartment buildings, developed a curriculum for high school electricity classes, commissioned Santa Claus outfits emblazoned with "Give Something Electrical," and pushed for better commercial window display lighting. In 1925, expanding beyond its own successful work with model homes, it adopted the National Society for Electrical Development's "Red Seal Electric Home" program.[34]

The successful California Electrical Cooperative campaign garnered national recognition almost from its inception. In February 1920, the Engineers Club of New York heard a full report about the program from a campaign representative, and, in early summer, the National Electric Light Association (NELA) met in Pasadena and got the story first-hand from participants. Within a month after the NELA meeting, Ohio decided to try "the California cooperative plan," and at the 1921 campaign get-together, "L.H. Newbert, chairman of the advisory committee, . . . pointed out that the outstanding evidence of the success of the California movement was that it has been adopted in several other sections of the country." In 1925, the campaign changed its name to the California Electrical Bureau, becoming a permanent fixture in California's electric power industry.[35]

Summary

As early as 1920, 79 percent of California's population lived in houses wired for electricity, albeit inadequately wired in the eyes of the power industry. Only Utah, with its population clustered around Salt Lake City, joined California in exceeding 70 percent, and no other states reached the

60 percent mark. By the end of the 1920s, California's total electric power consumption had surpassed that of Pennsylvania, a state less than one-third its size but almost twice as large in population. By 1930, only New York, with less than one-third the land area of California and 12.6 million versus 5.7 million people, consumed more total electricity. Californians, comprising 4.6 percent of the national population, accounted for 7 percent of the nation's domestic electricity consumers.[36] Engineers and entrepreneurs had harnessed Sierra Nevada waterpower, and the power industry's purveyors of electricity had folded it into the everyday lives of Californians with remarkable alacrity and success.

In 1915, Southern California Edison's S.M. Kennedy argued that selling electricity on the Pacific Coast was probably easier than elsewhere. Because California was a relative young state and many of its people were newly arrived, he felt Californians had fewer "deep-rooted prejudices to overcome." Inasmuch as many of them were "looking for the easy way to get through the day's work," they quickly opted for electricity.[37] On one level, Kennedy probably was right, yet regional environmental and technological conditions clearly played a fundamental role in the rapid electrification of the state's households. The high cost of wood and fossil fuel had led Californians to exploit Sierra Nevada waterpower during the nineteenth century. Hydroelectric developers, taking advantage of the foundation laid by miners, then built the region's electric industry around waterpower rather than fossil fuel. To gain the full benefit of hydropower over fossil fuel, however, the industry had to develop better than tolerable operating efficiencies, and that meant cultivating diverse markets. Therefore, it went after the home electric market with a vengeance, taking the whole state with it and gaining national recognition for its efforts. It also achieved rural electrification far ahead of the rest of the country.

CHAPTER XI Electrifying the Farm

RURAL ELECTRIFICATION reached a highly developed stage in California by the mid-1920s. With hydropower generated in the Sierra Nevada, extending down the eastern spine of the region, and most urban consumers located along the western seacoast, power lines crisscrossed the great Central Valley and smaller agricultural valleys on their way to San Francisco, Los Angeles, and other urban areas. Once the first trunk lines were in place, uneven urban loads and unmet rural irrigation needs combined with technical possibilities in power transmission to persuade power companies to distribute electricity to consumers en route to the urban centers. They did and the results gave them high load factors, precipitated continued rural power marketing, and spurred both private and public research and development in the application of electricity to rural needs. Elsewhere in the nation, electricity remained an urban phenomenon well into the 1940s. Cities offered concentrated domestic, industrial, and commercial markets; farming regions promised only high service costs and little or no profit. Most of the nation's power companies did not consider rural service feasible, and farmers remained untended until after Franklin D. Roosevelt created the Rural Electrification Administration in 1935.

Figure 11.1. Cover of a California power company promotional booklet. From *The Journal of Electricity*, 45 (November 15, 1920): 459.

Early Efforts To Harness Electricity

A few farmers, fortunate to have running streams on their land, began experimenting with hydroelectricity as early as 1889, when San Francisco's Pelton Water Wheel Company introduced small, light, and inexpensive impact water wheels. Streams harnessed by either a six- or twelve-inch motor could turn a dynamo producing sufficient energy to light a house and outbuildings or power small machines. H.J. Lewelling of St. Helena, north of San Francisco Bay, used the flow of several small springs on his property to achieve 140 pounds pressure. With his twelve-inch water motor, he received ample lighting at night plus daytime machine power. In another case, a dairy operator, at the foot of the coast range north of Santa Cruz, dammed Wilder Creek with a one-million-gallon reservoir, built a 4,000-foot redwood flume to a penstock that gave a 216-foot drop to a twenty-four-inch Pelton wheel and 13.5 kW dynamo. He lit inside and outside lamps and powered a hay cutter and pumpkin grinder, while two more Pelton wheels operated hydraulic cream separators and powered a machine shop. Meanwhile, J.A. Haines of Clipper Gap created a unique system around 1898. He constructed a small flume to feed a five-and-one-half-foot overshot wheel from the outlet of the South Yuba Ditch Company's Lake Theodore. An employee of the ditch company, Haines belted the twenty-three-RPM wheel to a 115-volt, one-horsepower Westinghouse dynamo, providing light for his home, the dam gates, and a nearby school building. He built the wheel and flume from odds and ends for $15, purchased the generator and wire for $100, and did his own wiring. In ten years of service, he claimed, he had spent $.25 on two generator brushes and bought "a little oil for the bearings."[1]

In 1908, P.J. O'Gara, a U.S. Department of Agriculture employee, encouraged farmers in the Sierra foothills to harness waterpower. By way of example, he described Ellis Franklin's success in Colfax, forty-eight miles northeast of Sacramento. Franklin had enlarged an old miner's ditch into a one-hundred-thousand-cubic-foot reservoir, built spillways, headgates, and a 445-foot steel penstock to drop water seventy-five feet to a twenty-four-inch Pelton wheel connected to a 4.5 kW generator. A quarter-mile transmission line carried current that fed a storage battery and could light seventy-five lamps, as well as power a pump, circular saw, cream separator, churn, washer and wringer, and cooking and heating devices. Cables attached to winding drums and strung between the headgate and his house allowed Franklin to regulate his system. Based on an average $.10 per kWh residential commercial power rate, Franklin's plant paid for itself in two years.[2]

For general farming applications, interest in electricity began in the late-nineteenth century. In 1890, San Francisco's Electrical Engineering Company introduced an interesting but impractical fifteen-horsepower motor "for propelling or drawing gang plows in the Sacramento Valley, the circuit wire winding and unwinding by means of a reel on the traction engine. . . ." Two years later, the *Pacific Rural Press* speculated that arc lamps, with intensity-cutting opal globes, could serve as a night substitute for sunlight, increasing the growth and richness of plants. A magazine contributor suggested that electric current in the soil would stimulate plant growth by provoking the action of soil atoms. In 1899, San Francisco's *Journal of Electricity, Power and Gas* noted that it might be time to "devote some attention to the development of . . . electric farming."[3]

By 1907, the *Journal* rooted for farm hydroelectric possibilities. After describing the replacement of kerosene lights and a windmill by a farmer's waterpowered 12.5 kW dynamo, its editor, George Low, assessed a bright future for California farmers:

> Not content with his several crops a year and other manifold advantages, the western farmer is preparing to harness a multitude of miniature Niagaras that leap downward from their sources in the mountain tops, and make them generate enough electricity to render the arduous task of agricultural endeavor mere child's play. . . . It is a practical reality on a small number of farms and its scope is being so rapidly extended that another decade may prove the electrical farm the rule rather than the exception in the Golden State.[4]

Unfortunately, farming was not confined to the Sierra foothills or along coastal mountain streams. Therefore, some rural residents without access to their own waterpower adopted gasoline engine power plants. The Western Electric Company advertised systems, "particularly selected for farms, estates and country homes located away from the lines of the public service electric companies," which cost between $278 and $536. Karl A. Hedbert's smaller San Francisco firm offered the Uni-Lectric Gasoline-Electric Unit Lighting System, a 110-volt DC plant that could light from one to fifty lamps and power motors, irons, vacuum cleaners, toasters, churns, and washing machines. And, perhaps sensing that farm families envied access by city dwellers to electricity, some catalogs played on the civilized character of electricity:

> Is Your Residence a Real Home?
>
> After all just what *is* home? Is it the type of house? No! Is it the atmosphere of comfort and happiness and love? Yes!
>
> Can a home be cheerful and happy if the only light is the sickly, yellow

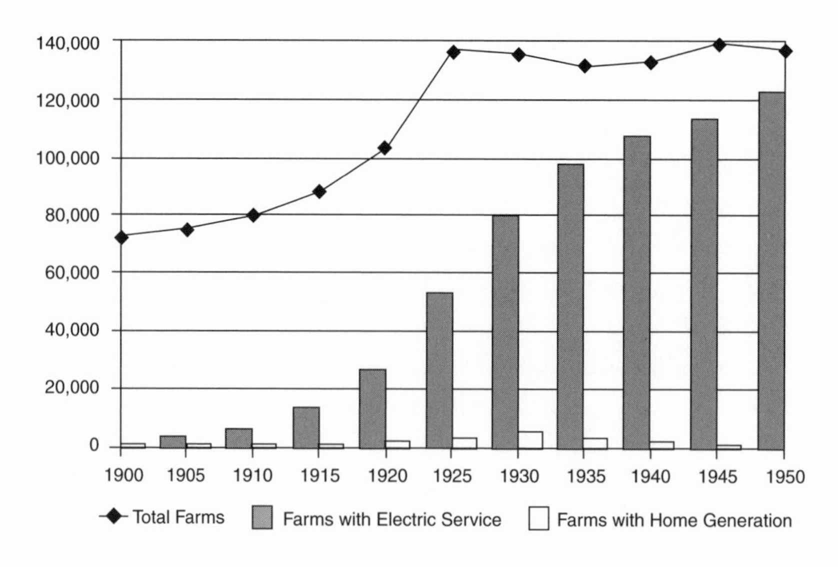

Figure 11.2. California Farms With Electricity, 1900–1950. Source: *Supra,* Table C–16.

> flame of a kerosene lamp? . . . Can the home be a place of health and good cheer when all the work must be done by hand? . . . No!
>
> And isn't it true? The finest stock and the best farm land in the world are worth but little if a man's home is cheerless, the women dead-tired and the children discontented.
>
> A real home is what every man wants and works for. And one big step towards this happiness is the installation of the Fairbanks-Morse Home Light Plant.[5]

Home electric plants remained a popular State Fair attraction into the 1920s, but the number of California farmers employing them probably did not exceed 10 percent of the number of farms with central station service (figure 11.2).

The Electrical Farm

The rapid expansion of long-distance electric transmission by central station power companies fulfilled George Low's prediction for the electrical farm in California. Approximately 18 percent of California's farms had electric service at the outset of the World War I. By 1924, California led all other states with 23.5 percent of its farms receiving electricity. Only Washington, Utah, and Idaho exceeded 10 percent, while the national percentage stood at just 2.6 percent. A decade later, the national figure had risen to 11.3 percent, but California still led all other states with 60.1 percent of its farms getting power. By 1940, several states had surpassed the 60 percent mark, and Massachusetts, New Jersey, and Rhode Island all had reached 80 percent. But, among the large agricultural states, only California (81.3 percent) had passed the 70 percent mark. The nation would not achieve 80 percent rural electrification until after 1950.[6]

Various factors prompted power companies to provide farm service. In southern California, expensive fuel alone seemed enough to instigate rural electrification, but technical choice and capability influenced its development north of the Tehachapi Mountains. During the 1890s, engineers debated the merits of polyphase electricity transmission frequencies ranging from 25 to 133 cycles (Hz). Most manufacturers and power companies felt 25 Hz provided the best efficiency for transmission and running large motors, but in California only a few mining companies adopted it. Since large manufacturing motors were few and many communities already had 60 Hz general purpose steam-generated systems, most state power companies chose 60 Hz to facilitate power sales. Frank Baum, a San Francisco-based electrical engineer, defended this decision in 1903, responding to a colleague who supported the better efficiency of 25 Hz: "these towns being already equipped with 60 cycle apparatus, practically force the power companies to supply that frequency."[7]

Along with the adoption of 60 Hz, the early choice of transmission voltage became important for rural Californians. Power companies used sixty-thousand-volt transmission for the first northern California power plant interconnections, and it became a standard because it facilitated economical delivery of power to users along main transmission lines. According to A.H. Markwart, vice-president of Pacific Gas and Electric Company during the 1920s, long-distance hydroelectric transmission lines "could be tapped en route because this voltage permitted substation costs that bore a proper economic relation to the size of loads to be carried." Markwart also observed that "this economically suitable voltage made it possible for industry to locate in rural communities. . . ." So, with the general purpose frequency, sixty-thousand-volt transmission permitted profitable delivery of power to agricultural consumers and small towns along transmission lines and also encouraged community growth.[8]

Finally, World War I fuel needs led electric power companies to expand their Sierra Nevada hydroelectric projects, drove up petroleum fuel prices, and created scarcities. Fuel oil conservation became the watchword and waterpower the panacea; the crisis urged all Californians to convert from petroleum to hydroelectricity. In 1918, the *Journal of Electricity* proclaimed "electricity on the farm is a patriotic necessity." It would take the place of the hired hand where labor was scarce, it would lighten chores of boys and girls and so pinch off their desire to flee to the city, and it would relieve family drudgery as well as conserve other fuels for the war effort. The civilizing characteristics of electricity merged with the practical crisis: "It is the difference between losing and winning, between conquering and being conquered." Rural electrification, which despite other encouragements still might have taken a gradual path in California, was pushed forward rapidly by wartime fuel conservation.[9]

Delivering electricity to rural areas provided power companies with a significant benefit: they achieved load factors that were higher than those in almost any other part of the nation. In the 1890s, a load factor of 20 to 25 percent was common throughout the electric power industry, and companies that achieved higher load factors generally cultivated electric street railway and industrial motor loads. By 1915, eastern power companies had reached average annual load factors ranging from the Boston Edison Company's 32.5 percent to New York Edison's 34.7 percent. The Niagara Falls Power Company gained an unusual 81 percent load by providing power for an immense, round-the-clock electrochemical industry. But California's companies, covering much wider territories, achieved much higher load factors. Only Los Angeles's Pacific Light and Power, which obtained its principal load from the Pacific Electric Railway, followed typical eastern load-building practices. Elsewhere, traction was less important. The San Joaquin Light and Power Company reached a 64 percent load factor through heavy irrigation pumping, and the Great Western Power Company achieved a 65 percent load through diversity.[10]

The importance of customer diversity, including agricultural use, was evident in the character of Pacific Gas and Electric Company's average 60 percent load factor: "Most of [its] hydroelectric plants give local service to the mining districts along the foothills. In crossing the valley, the trunk transmission lines furnish power for agricultural needs during the summer and carry reclamation pumping loads in the lowlands of the Sacramento and San Joaquin deltas during the spring overflows. In the Bay region and the cities of Stockton and Sacramento, the lines supply power for all the uses common in urban and manufacturing districts and, at many points on the system, for special industries, such as cement mills, rock crushers, and miscellaneous manufacturing establishments."[11]

Irrigation and reclamation pumping particularly evened out the load factor for some companies. Pumping provided high spring and summer demand opposite high urban lighting demand in fall and winter. In the San Joaquin Valley, the irrigation load was so attractive that the Mount Whitney Power Company directly participated in opening new lands to farming, and at least one coastal valley farming community, Atascadero, was founded on the expectation of electric power. Albert Wishon, Mount Whitney's electrical irrigation pioneer, joined the San Joaquin Light and Power Corporation in 1903, where he continued transforming arid land with electric pumping. He also encouraged the state's first commercial cotton farming, introduced electric power for oil pumping, and reduced rural power costs by developing unattended, outdoor high-voltage transformer stations. In 1912, electrical engineer Putnam Bates observed that it was hard to tell whether power companies made the irrigated farms or farms guaranteed the success of power companies. "Both interests seem to have worked together."[12]

Following Wishon's lead, Southern California Edison, Pacific Gas and Electric, Great Western Power, and other companies had developed substantial irrigation pumping loads by the 1920s. One firm helped land developers create new 40- to 140-acre electrified farms. Others reclaimed marsh and delta lands for new crops and undertook controlled flooding of arid plains to begin the state's rice industry. A 1927 electrical energy use survey indicated that irrigation comprised 12.27 percent of the state's total electrical consumption. Of eleven western states, California accounted for 80.2 percent of the electricity used in irrigation pumping (figure 11.4).[13]

But pumping demand could become so heavy that it hurt load factors. Mount Whitney and other utilities found they had to develop domestic farm electricity markets to balance the irrigation load. Electric power, in addition to operating pumps and transforming cultivation patterns, also

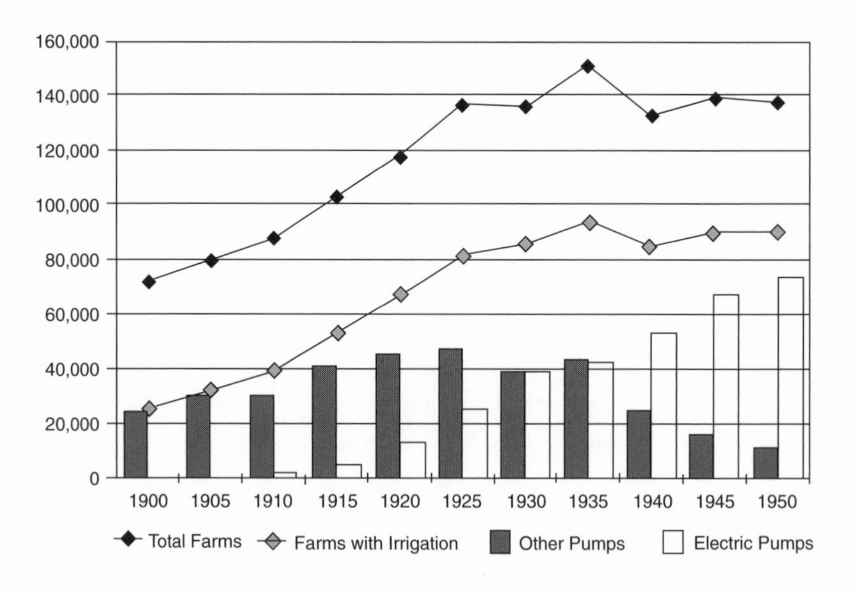

Figure 11.3. California Irrigation Pumping, 1900–1950. Source: *Supra,* Table C–17.

affected rural lifestyles. R.B. Mateer, regular contributor to the *Journal of Electricity, Power and Gas,* predicted it would make farm life much easier. Mundane chores would be eliminated and family ties strengthened by providing warmth on cold winter evenings. On a practical level, it would do these things more cheaply than oil, coal, or wood, plus reduce insurance rates. Since prosperity was rooted in the soil, and the farmer was the "indispensable man," it would emancipate the farm family from previous servitude.[14]

As in city households, electric appliances performed a growing variety of tasks in newly wired farm homes. The *U.C. Journal of Agriculture* observed, in 1916, that, "electrical manufacturers have been busily at work building electrical household appliances that will give to the farmer's wife every comfort and convenience that her city sister enjoys." By 1925, according to a *Journal of Electricity* survey, about one million wired homes in California used electric appliances, sixty thousand of them farm homes. For the most part, farm householders used fewer appliances but the same sort as those adopted in city homes. Flatirons, vacuum cleaners, toasters, and percolators were the most popular. But farmers adopted far more

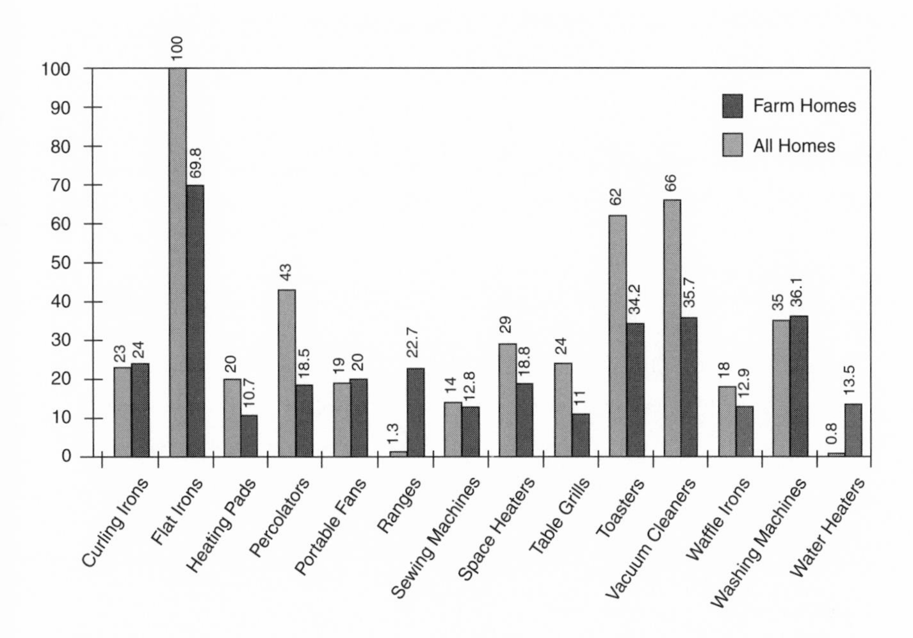

Figure 11.4. Percentage of Selected Electric Appliances in California Electrified Homes, 1925. Source: *Supra*, Table C–18.

electric ranges and hot water heaters than the number of city residents served by gas utilities (figure 11.5).[15]

By the 1920s, load development by power companies had resulted in dramatic farm electrical expansion. A 1925 University of California Agricultural Experiment Station farm survey showed an average of 10.6 electric lamps inside main farm houses and an average of four in outbuildings and yard areas. Motors operated irrigation pumps on 77.6 percent of California's wired farms, accounting for 91.1 percent of the motor horsepower used, but smaller motors served farmers in several other ways. The survey revealed 23,625 other motors: 12,500 domestic water supply motors, 1,570 cream separators, 1,400 shop motors, 1,370 milking machines, 1,280 feed grinders, 1,090 refrigerators, 600 dehydrator fans, 400 wood saws, and 3,415 miscellaneous motors. Other special appliances included 5,130 battery chargers, 5,700 incubators, 4,400 poultry brooders, plus sheep shears, horse clippers, and some 20 electric trucks. Electric plowing, although it had been tried in Europe, experimented with at the University of California, and encouraged by the *Journal of Electricity, Power and Gas,* was found nowhere.[16]

Dairy farmers were early electricity users, in part because they were located near urban areas and early long-distance transmission termination points. Around Dixon, near San Francisco Bay, dairies received power soon after 1900, and consistent irrigation pumping permitted five to six superior alfalfa crops per year. Dixon farmers also used milkers and cream separators. By the mid-1920s, 92 percent of the state's dairy farms employed electricity. Over nine hundred ran feed grinders, milkers, cream separators, homogenizers, pasteurizers, churns, bottle and can washers, bottling machines, refrigerators, and freezers. Whether they agreed with electric boosters that "no inspector could condemn milk from an electrically clipped cow, housed in an electrically lighted barn, fed with forage chopped by an electric motor, and milked by an electric milking machine," dairy farmers certainly took advantage of electric power.[17]

Poultry farmers also employed electricity, the *Journal of Electricity, Power and Gas* suggesting in 1908 "the pregnant possibilities of electric incubation." Eight years later, two electric poultry equipment manufacturers sold their wares from northern California's chicken capital, Petaluma, where some ten million chicks were hatched each year. There, the Great Western Power Company provided power for the poultry industry, while interested parties studied various heating methods. Before accepting electric heating, farmers used kerosene, gas, steam, and solar heat devices to hatch and brood chicks. But each had drawbacks. Coal, oil, and gas-fired systems brought noxious fumes to brooders and incubators, created fire risk and ventilation problems, and often failed at even heat distribution. Solar heating systems required a fuel-fed back-up system, and all methods demanded regular attention to temperature control.[18]

Edgar Wilcox, Great Western Power's heating specialist, devoted a full chapter in his *Electric Heating* manual to the theoretical and practical advantages of electric incubation and brooding. It saved labor, created a healthy environment, and easily achieved perfect heat distribution and ventilation. "Electrically hatched chicks always begin to pip about twelve hours quicker than those hatched by other artificial means . . . [and] a much higher percentage is hatched." In fuel-type brooders, said Wilcox, only 50 percent of the chicks survived, while 85 percent survived in electric brooders. "An electrically brooded chicken is usually ready for the roost about two weeks sooner than one brooded by fuel heat, and is universally stronger and more vigorous." Although electric power was not as cheap as fuel-fed equipment, "the advantages to the user of electrically heated apparatus will more than offset this added expense."[19]

In the 1920s, practically all large California poultry hatcheries used the "electric hen." Southern California Edison supplied power to a hatchery in Artesia, south of Los Angeles, which hatched thirty thousand chicks a week. But Petaluma remained "the World's Egg Basket," producing 35 million dozen eggs in 1925. The Must Hatch Incubator Company hatched twenty-five thousand chicks daily between February and June, reaching three million a season. Adjoining its 560,000-egg capacity incubator rooms, Pacific Gas and Electric Company set up a substation with three 150 kilowatt transformers. In 1927, the firm tripled its capacity to 1.8 million eggs, using 3,780 incubators. Ben D. Moses, Agricultural Engineering Professor at the University of California-Davis campus, estimated that about half the state's chick brooders also were electric.[20]

Electric poultry operations were attractive to both poultry farmers and power companies. Besides incubation and brooding, electrically lighted hen houses lengthened winter laying time and increased egg production. Automatic, timer-controlled dimmers simulated dusk and sunrise at any time of the night, in the end giving hens as little as three hours of "nighttime." Power companies gained greater balance in electric loads because most poultry power needs came during the winter season when irrigation was at a low point. Unfortunately for the chicken, the application of electricity, coupled with other changes in poultry farming, proved devastating. In order to make artificial light effective in egg farming, hens were not just kept in hen houses but confined five together in close caging, their beaks clipped so they could not peck each other to death, hybridized to get the best layers, subjected to forced molts, and fed food with chemical additives to stimulate their appetites. Confinement led to increased susceptibility to disease, so hens were inoculated with various vaccines created just to solve this new problem. In the end, historian Page Smith argues convincingly, the chicken was debauched and enslaved by human beings and their technology.[21]

In addition to dairy and poultry farmers, citrus growers also developed an early interest in electricity, primarily because they used large quantities of fuel to generate heat for frost protection. Although growers principally burned heating oil and occasionally used coal, carbon briquettes, and wood, electrical boosters saw an important market in frost protection. In 1912, a year when fuel oil prices were rising, the *Journal of Electricity, Power and Gas* urged citrus farmers to adopt electric heating devices. Experiments had mixed results. During the winter of 1912, the *Journal* said a Santa Barbara lemon grower effectively protected 225 acres by electric heating, but a study several years later by the University of California suggested that 1913 efforts at electric frost protection in Riverside stopped tree damage but not fruit loss.[22]

Nevertheless, experiments continued as farmers sought to overcome the limits of climate and geography with technology. In 1916, a San Fernando Valley farmer ran wire through underground tile conduits between tree rows, installing at each tree an upright tile outlet supporting a resistance coil. During cold weather he heated the coils, regulating electrical current by a thermometer. Other orchardists near Covina tried the process during the 1917 winter, but the system died. A successful electrically powered protection system was not developed until the late 1930s, when fans or blowers were mounted on towers reaching above the trees. With aeroplane type propellers, they used 50- to 250-horsepower electric motors and cost about $2,500, but a large one protected as much as twenty acres. They were used near Lindsay and some two dozen other locations around the state in 1938, freeing citrus communities from oil heater pollution. After World War II, they became the principal frost protection device for many crops, including citrus groves and vineyards.[23]

Electricity also served orchardists in dehydrating fruits and nuts. In 1919, the *American Fruit Grower* applauded the comfort and convenience of electricity on the farm, noting that the electric fan had particular value in drying fruit. During the next decade, electric fans in individual farm dehydrators blew air heated by gas and oil-fired furnaces, the first large-scale application of dehydrator fans occurring in 1923 at an Anderson-Barngrover Company plant near Stockton. Electricity also displaced cheaper gas and oil for generating heat: 15 percent of California's 435 walnut dehydrators used electricity. Edgar Wilcox argued that "the better quality of the product dried electrically and other advantages of electric operation offset the higher heating cost" of about $2.00 more per ton. By the 1930s, electric blowers served wherever nuts, fruits, and vegetables were dehydrated, and all electric operations were not uncommon.[24]

The related packing and canning industry also adopted electric power wherever and whenever it became available. In 1912, electricity operated practically all the canneries and fruit-packing plants in areas served by the Mount Whitney and San Joaquin Light and Power companies. A year later, the Imperial Valley cotton industry employed gins run by electric motors. Heavy purchasers of electric power, agricultural processors used fruit elevators, conveyor belts, sorting tables, and lights. The steam engine, on which the canning and food-preserving industry had relied entirely in 1890, gradually disappeared. Electricity provided 23 percent of the industry's power, in 1910, and ten years later it furnished 78 percent. By 1930, the industry's total horsepower increased fivefold, and electricity provided 90 percent of it.[25]

Public Sector Support for Rural and Farm Electrification

Special support for rural and farm electrification and other technological development also came from the public sector. The University of California's College of Agriculture took a leading role. Its Davis Experiment Station, operating under authority of the 1887 Hatch Act, had grown substantially by the 1910s, carrying out farm technology research and operating a state-funded Farmers' Institute. Passage of the Smith-Lever Act in 1914, prompted in part by the national interest in rural America sparked by President Theodore Roosevelt's Country Life Commission, led to a further expansion and federal support of the Experiment Station's extension program. Thus authorized and funded by state and federal programs, the University of California ran an aggressive publication and outreach program. It published hundreds of bulletins and a *Journal of Agriculture*, provided lecturers and displays for events, such as the annual State Fair "Power on the Farm" display, and experimented with all sorts of farm technology.[26]

In 1908, Professor Warren T. Clarke suggested that the College of Agriculture, the State Horticultural Commission, and Southern Pacific Railroad put together a four-car farm demonstration train for a sixty-four-day trip through rural northern California. It drew 37,270 visitors at 197 stops, and the University added two more cars with displays on field crops, dairies, veterinary science, soils, fertilizers, insecticides, viticulture, horticulture, poultry, public health, and home economics. Between 1909 and 1912, the train reached 254,511 people at 683 stops, and, in 1913, a dairy special toured the San Joaquin Valley and a "frost special" traveled southern California. Meanwhile, Pacific Gas and Electric added displays to the "Agriculture Special," increasing the train to fifteen cars, and it continued running into the 1920s.[27]

Because much of the University's work followed the interests of researchers and extension personnel, support for energy industry activities relied on the predilections of university employees. But a national movement for rural electrification did provide the impetus for a coordinated effort to apply electricity to agriculture. In May 1924, Dr. E.A. White, director of the newly established national Committee on the Relation of Electricity to Agriculture, came to California to see if a state committee might be organized to correlate various state agency "investigations in many phases of rural utilization of electric energy." White and Professor L.J. Fletcher of the University's Agricultural Engineering Division at Davis invited thirty-four representatives of power companies, farmers' organizations, electric appliance companies, and the University to meet in San Francisco. Attendees discussed the concept at length and agreed to elect a state committee of seven, chaired by Fletcher, to plan and direct a California Committee on the Relation of Electricity to Agriculture (CREA).[28]

A few days later, the committee met in Fresno. The Executive Committee expanded to ten members drawn from major power companies, the University, and the California Farm Bureau Federation. Four subcommittees were formed to study existing farm electricity use, seek new applications, study utility regulation and rates affecting agriculture, and gather comparative data with other fuel resources. In November, Professor Ben D. Moses was elected CREA Executive Secretary, the university paying his salary and power companies providing $3,000 to maintain the committee's work for a year. Fletcher observed that California had much to show the rest of the nation: 24 percent of California's farms received central station power versus under 5 percent of farms in the rest of country, and over 60 percent had pressurized home water systems compared to 10 percent nationally. "Electricity, the willing servant which, with its four hands furnishes us with heat, light, power, and chemical energy, has revolutionized our manufacturing and transportation. Who dares to prophesy what electricity has in store for agriculture?"[29]

Over the next decade, CREA carried out a coordinated research, development, education, and information program. A 1925 postcard survey of farms provided statistical data about farm electrification. San Joaquin Light and Power and CREA constructed a scale-model diversified electric farm to illustrate gross farm earnings versus electrical costs and urge further electrification, and a truck-mounted kitchen lighting display toured the Sacramento Valley. The university offered farmers seminars on electricity and provided electric power sales and service people with a rural electrification short course. Numerous articles appeared in industry journals, and CREA progress reports disseminated research work about electric poultry brooders, walnut dehydrators, automatic water heaters, refrigeration, utility motors, household appliances, insect electrocutors, the effects of electric light on codling moth behavior, electrically heated soil and plant growth, and the lethal effect of alternating current on yeast cells.[30]

California's CREA prospered, helping expand farm electrification. By the 1930s, few farms in other major agricultural states had electricity, while California served as a national model for rural electrification. In 1934, for example, 87.6 percent of California's farmers used electricity, while 76.3 percent of Iowa's farmers relied on kerosene, gasoline, acetylene, and piped gas for lighting (figure 11.5). Additionally, under 1 percent of California's electric farms used home power plants compared to 43 percent of Iowa's, and the distance from power lines for the average California farm was one mile compared to 2.4 in Iowa. Twelve and six-tenths percent of California farmers had refrigerators and 20 percent cooked with electricity, while Iowans had few refrigerators and used even fewer electric stoves (figure 11.6). On the other hand, the influence of California's mild climate led many of its farmers to continue using inefficient fireplaces and stoves for space heating, while 29.4 percent of Iowans had adopted furnaces.[31]

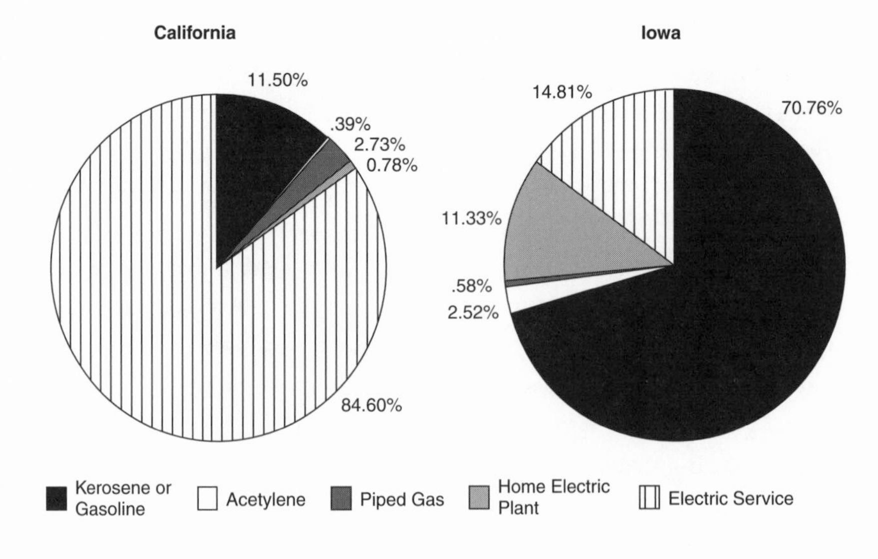

Figure 11.5. California and Iowa Farm Lighting, 1934. Source: *Supra,* Table C–19.

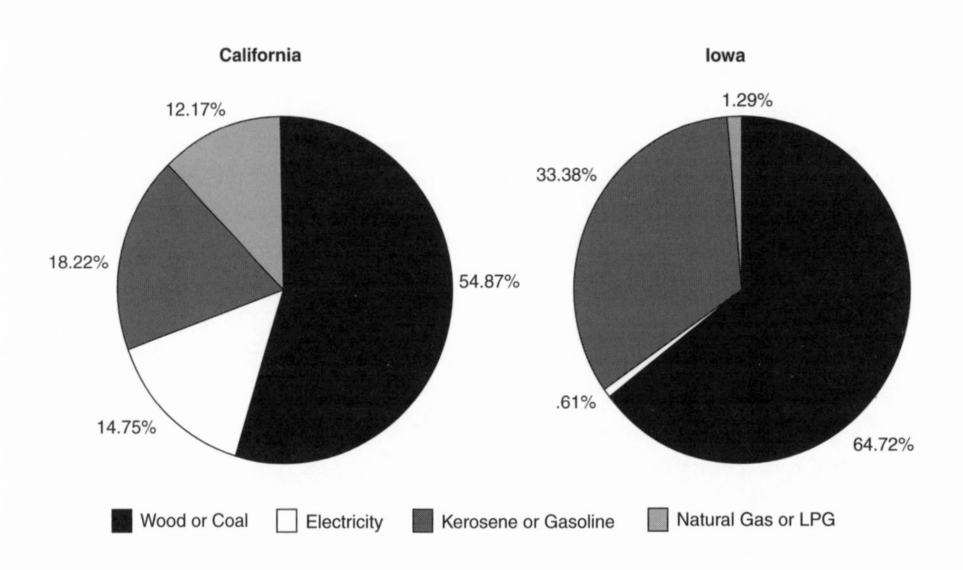

Figure 11.6. California and Iowa Farm Cooking, 1934. Source: *Supra,* Table C–19.

Summary

Whereas midwestern and eastern power companies had concentrated on electrifying industry and street railways, California's central stations served both city and country. During the 1930s, when Franklin Roosevelt's New Deal created the Tennessee Valley Authority as an immense display of the beneficent bond between electricity and democracy, decentralization, and landscape reclamation, California's irrigated and reclaimed agricultural valleys already illustrated the power of electrical redemption. And, when the Rural Electrification Administration set out to light the rest of the nation's farms, despite some fears that farmers would not use enough electricity to make it worthwhile, California provided a continuing example of success and profit. Although the Rural Electrification Administration served but four tiny projects in isolated corners of California, it raised states like Iowa to levels of electrification in 1950 that rural Californians had enjoyed a decade earlier.[32]

The environment joined with regional choices in electric power technology to play a significant role in making possible California rural electrification. During the hydroelectric industry's early years, the population centers possessing clearly profitable markets in public and commercial lighting, street railway and manufacturing power, and, eventually, domestic light and power were located in the San Francisco Bay Area and around the Los Angeles basin. Central Valley communities, such as Sacramento and Fresno, also offered gainful markets. For power companies to be successful, they had to tap these markets. To do so, they had to transmit their electricity great distances, across farmland and by smaller towns. In choosing to deliver 60 Hz at sixty thousand volts, hydroelectric companies made it economical for themselves to provide power to farms and small towns along their trunk transmission lines. Not only did power companies find diverse rural market opportunities, they discovered that wherever irrigation was an agricultural necessity, particularly in the semiarid environment of southern California, the amount of power it demanded and its seasonal character gave them more than favorable load factors.

The interplay between electrical power technology and the environment thrust California agriculture into the modern world ahead of almost any other region in the nation. For farmers, a 1908 editorial in the *Journal of Electricity, Power and Gas* said it all: "There is probably no single factor that has done, and is doing, more to change [people's] environment than electricity. This is particularly noticeable with the intelligent farmer of the present compared to the agrarian of the past. He is surrounded with books and papers that tell him today what the world did yesterday. He can go to the city quickly, safely and cheaply in an electric car. He need never be lonely, for the telephone puts him in instant communication with his neighbors. These lift out of the narrow rut of provincialism, and energize into a reasoning, thinking being, what might otherwise be a mere clod."[33]

CHAPTER XII

Resolving the Power Network

AT THE BEGINNING OF THE twentieth century, Americans lived through a restructuring of every aspect of society, one characterized by the rise of centralization, systems, and experts in the face of traditional localism, decentralization, and self-reliance.[1] The electrification of California and the growth of its hydroelectric power companies into regional public utility monopolies, occurred during this period of change. As elsewhere in the nation, the industry's development embodied the tensions inherent in America's evolving political, economic, and social life. Therefore, despite remarkable achievements by public utilities, the public hardly could remain uncritical of them. The great reform movement that swept the nation in response to societal changes ensnared the industry.

Important parts of the progressive reform era included the conservation of natural resources and the regulation of public utilities, conflict between advocates of private and public ownership of electric and other public utilities, and the evolution of immense multiple-purpose water projects. Quite simply, the rise of an urban-industrial society posed compelling questions: Who should control natural resources, such as water? Whose interests should be foremost, those of natural resource preservationists, conservationists, hydroelectric developers, or consumers? Should electric power resources be regulated private monopolies or publicly owned? Just what should be the structure of the electric power industry?

Regulation of Water Rights

California's first hydroelectric entrepreneurs had been aided by the open character of western lands. The situation in the Sierra Nevada during the early years of the industry had been much the same as when tens of thousands of argonauts were drawn to California during the gold rush fifty years earlier. Just as mineral rights were unregulated in 1849, waterpower rights on the streams flowing out of the Sierra Nevada were open. Indeed, miners had established the legal doctrine of prior appropriation in determining water rights, and it subsequently was adopted by the California State Legislature and the U.S. Congress and used by hydroelectric promoters. All a person had to do was post a notice where he intended to appropriate the water, indicate how much he would use, and record it with the local county government. If recorded and actually put to use, the claim gave the owner a perpetual water right. Furthermore, the water rights could be sold. Virtually no state control existed. Hydroelectric developers quickly purchased any rights that seemed useful and claimed new ones.[2]

As waterpower entrepreneurs swarmed through the Sierra Nevada, however, other Americans were becoming concerned with commercial exploitation of natural resources in the vast public lands of the West, particularly forested lands. In 1891, Congress enacted the General Land Law Revision Act that authorized the establishment of forest reserves. Six years later, the Forest Management Act granted the Secretary of the Interior the power "to regulate the occupancy and use" of national forest reserves. In 1898, Gifford Pinchot, a Yale graduate who had studied scientific forestry in Europe, became Chief of the Division of Forestry, located within the Department of Interior's General Land Office. Unable to move forward as he wished with rational scientific management of the forests within the General Land Office, he campaigned to transfer the division to the Department of Agriculture. In 1905, with support from western congressmen and others, the transfer finally was accomplished. Pinchot then set out to establish rational use of the nation's forest reserves, broadly interpreting the powers of the 1897 Forest Management Act. Meanwhile, President

Theodore Roosevelt expanded the reserves from 46.4 million acres in 41 national forests to 150.8 million acres in 159 national forests, California accounting for 20.2 million acres.

The national forest expansion program, coupled with Pinchot's plan to carefully regulate commercial use of forest lands, including rights-of-way permits, changed the playing field for hydroelectricity entrepreneurs. Until then, hydroelectric promoters virtually had a free hand in claiming and developing Sierra Nevada water rights. Even the 1901 Act of Congress granting the Secretary of the Interior the power to issue temporary, revocable rights-of-way for water development had made little impact. Permits subsequently had been granted for dams, reservoirs, and some other hydroelectric facilities, but further controls or regulations were not tied to them. Pinchot changed this. In 1906, he got specific conditions included, for the first time, in a special law granting a waterpower permit to the Los Angeles Edison Electric Company's Kern River project: a time limit for the easement, a requirement that the project be finished in a reasonable and definite time period, and an annual forest conservation fee to the Department of Agriculture. Western power companies bristled at the new restrictions, and conflict quickly emerged between them and the Forest Service.[3]

Charles H. Shinn, supervisor of the Sierra National Forest near Fresno, felt caught in the middle. Within his forest, most of the permits under which Pacific Light and Power and San Joaquin Light and Power operated had been issued before 1906, and he had not seen all of them. Shinn harbored serious concerns about the power companies indiscriminately removing young trees to clear the way for power lines, cementing ditches that made it impossible for horses and men to cross them, and interfering with public use of the forest by fencing ditches and pole lines. Urging issuance of comprehensive project permits, he wrote Pinchot an official letter in 1907: "I do believe in great enterprises, but I want them to pay for what they get: . . . a yearly rental for each transmission pole, and . . . a yearly payment per horsepower or power-house or otherwise according to the privilege and the territory they control." Yet, Shinn was appalled with the general attitude and character of the power company people:

> In brief, they have surveyed and estimated all the power in this Forest, and have filed on most of it. They expect to reservoir and use the whole water-

> shed of the Sierra Nevadas, with as little payment as possible, and with no attention to the broader demands of higher civilization for outdoor life. My personal relations with the managers are excellent, but we deal with men of entirely primitive capitalistic instincts and training. It is one chain: Wishon and Eastwood, . . . Huntington and Harriman (agents, attorneys, principals, etc.)—to one and all of them the entire modern Rooseveltian theory of public utilities is lunacy, ignorance and diabolism.

In an appended personal note, he added: "The whole power situation is a hard one—As usual, I suppose it will be wise to charge them about half of what is *really right.* But I think the American people are with us in struggling for a fair deal."[4]

The power companies hardly saw the permit restrictions and privilege payments suggested by Shinn and the Forest Service as a "fair deal." Company officials and engineers alike lined up against the Forest Service and gained substantial support in Congress. In 1907, they tried unsuccessfully to thwart the Forest Service permit system in the courts, and their success in getting a congressional measure passed that authorized perpetual leases with almost no payment was vetoed by Roosevelt. Unable to circumvent the new policies, they met in Washington, D.C. with Pinchot and other federal officials in the spring of 1908, but, while Pinchot agreed to fifty-year leases that could only be revoked if lease conditions were violated, power company representatives held fast to perpetual leases with nominal payment. Shortly after the failed Washington conference, the position of California's hydroelectricity community, localistic in its nature, was made clear in a paper presented by Frank G. Baum at the annual AIEE conference in Atlantic City, New Jersey. Baum, chief electrical engineer of Pacific Gas and Electric, asserted that rights-of-way and land needed for waterpower development in the public domain should be sold to power companies. Annual forest conservation charges could be accepted only if the fees were based on the actual cost for conservation of the land involved and only if the proceeds were used for conservation within the watershed from which they were derived. Finally, since the streams were under state control, "the right to charge a privilege or franchise tax [such as Shinn recommended for the power generated] should remain with the state and subject to local laws."[5]

Frank Baum's position reflected both the autonomous atmosphere in which California's hydroelectric industry had thus far developed and the colonial character of the state's history. The federal government, three thousand miles away, for over fifty years had left Sierra Nevada water use largely up to California. In the 1850s, a strong democratic bias had flourished in the state, one embodying distrust of central authority in favor of localism. Consequently, the state government had invested in the counties' authority for management of natural resources, which included flood control and reclamation in the Sacramento and other valleys and the appropriation for use of Sierra Nevada water. During and after the Civil War, California followed the national political shift embodied in the rise of the Republican Party and its assumption of the old Whig Party mantle of paternalistic "public spiritedness." In terms of natural resources, Republicans' concern for community well-being increasingly displayed itself in "corporate, team-spirited, community-centered attitudes" and an interventionist state role. During the 1880s, this particularly played itself out in faith in engineering expertise and in bold efforts by the state engineer's office to control and reclaim flood and swamp lands statewide. But this Republican interventionism did not affect Sierra Nevada water use beyond the bitter fight by Central Valley farmers to stop hydraulic miners from washing silt into the valley's rivers. Thus, when hydroelectricity development began in the 1890s, engineers and entrepreneurs harnessed Sierra Nevada waterpower in a purely individualistic and laissez-faire atmosphere. When Gifford Pinchot, described by historian Robert Kelley as the quintessential Whig, put centralized, federal waterpower regulation in place, California hydroelectric engineers and entrepreneurs, many actually Whiggish by nature, responded with prickly opposition born of the democratic individualism and laissez-faire action that characterized their exploitation of mountain waterpower.[6]

California's hydroelectricity developers found support for their position against federal regulation from western, populist Democrats who strongly opposed Roosevelt's overall natural resource management program, but they and other western states' hydroelectric people stood alone within the Whiggish national electric power industry. All but one of the discussants of Baum's paper at the AIEE meeting opposed him. Baum's contention that waterpower development in California's national forests was only a local matter prompted immediate antagonism. J.H. Finney, arguing the government's position, said Baum ignored the fact that the federal government was trustee of natural resources for all the people. The right to exploit waterpower, he contended, should be based on "a policy that is embraced in what is termed 'a square deal' to all concerned; to the investor, to the corporation, to the power user and to the nation, and, incidentally, to posterity." Waterpower development should be based on a plan "that will do equal justice to all and contain special privileges to none." Forest Service employee P.P. Wells echoed Finney and sketched out once again Pinchot's position as presented at the Washington meeting. More significantly, electric power people stood against Baum, among them General Electric's Charles P. Steinmetz, one of the country's best known and most influential electrical engineers: "If the problem were merely to develop water powers, then indeed it would be local; but the problem is much broader, it is the problem of preserving and protecting our national capital."[7]

Even though they stood alone and watched hydroelectric development slow, the Californians held their ground against the Forest Service. They soon discovered, however, that their argument for state control might well be undermined in their own backyard. Inadequate state water law long had been recognized as an important issue by Californians, but debate about it had revolved largely around the conflicting riparian and appropriative water doctrines as they related to irrigation. By 1900, most of the normal flow of California's largest rivers had been appropriated, and what remained for irrigation involved snowpack "flood water" that required storage. Drought, in the winter of 1899, bore witness to the need for such storage, and within a year the California Water and Forest Association formed to push for water legislation. A 1902 bill died, and Edward F. Adams, who was convinced the Water and Forest Association was too narrowly focused, formed the Commonwealth Club in San Francisco to act as a new forum to discuss state matters. The club began by looking at a wide range of public issues, but it devoted itself, in 1904, to water laws and drafted a measure to take to the state legislature, only to have it set aside in the turmoil caused by the 1906 earthquake.[8]

Four years later, Progressive Republican Hiram Johnson's victory in the gubernatorial race brought forth in Sacramento the same regulatory-prone centralism that California hydroelectric people were fighting in Washington. A water bill was in the works immediately. Seeking to influence its writing, in January 1911, hydroelectric engineer John D. Galloway wrote former Governor George C. Pardee, who had prevailed on Hiram Johnson to establish a fact-finding State Conservation Commission to study California water issues. While Galloway conceded that ownership of state waters should remain with the people, he urged wide prerogatives for power companies: forty-year licenses with a forty-year renewal privilege and purchase of properties by the state only at appraised value; no rights to irrigation companies or districts to appropriate potential waterpower; and fees based on actual kilowatt hours produced, not on plant capacity. Reflecting the progressive era's emphasis on efficiency and expertise, he also urged that the envisioned state water commission be chaired by an experienced civil engineer, with membership comprised of two good assistant engineers plus the governor and state controller. The engineers should each be paid $10,000 per year, he argued, for "three men at $4,000 per year will lead not to good men but to political prize seekers."[9]

A bill, not to Galloway's liking, was introduced in December 1911 and defeated largely through pressure from the power companies, but the State Conservation Commission formed by Johnson held hearings on a new bill in 1912. In January 1913, the commission reported strongly on the dangers of a state waterpower monopoly, and Governor Johnson put a water bill at the top of his agenda for the year. After a bitter fight, it passed and established a state water commission empowered to grant licenses which, after twenty years, gave any local government the right to purchase the works and property. The commission also could discard water appropriations filed before 1913 and not put to use. The power companies launched a referendum campaign against the measure, arguing that the law would require elaborate proofs and surveys; necessitate burdensome state administrative costs; undermine the rights of miners, irrigation districts, and municipalities; and prove a boon to speculators. The referendum was defeated in November 1914.[10]

After losing within their own state the campaign to stop waterpower regulation, many hydroelectric people must have wondered if a victory against Pinchot would be a hollow one. Nevertheless, they had felt heartened when Woodrow Wilson won the White House in 1912, for they hoped he would come down on their side at the federal level. Six years later, as World War I focused the nation's attention on petroleum shortages and people saw hydroelectricity as a way to save oil, the federal permit issue remained at an impasse, but the California hydroelectric community saw an opportunity to push forward its position. PG&E's vice president, John A. Britton, and chief hydroelectric engineer, Paul M. Downing, led the way. In early 1918, Britton argued in *Electrical World* that the Forest Service had held up hydroelectric development by "placing such restrictive conditions upon occupancy of government lands" that investors were not willing to back projects. In July, he made clear that "such a holding up is directly traceable to the hysteria of ultra conservationists, and to an absolute and complete misunderstanding of the rights of the several public land States to the control of their natural resources." Downing fanned the flames ignited by Britton: "The radical ideas and rulings of the conservationists have for years been a stumbling block in the way of hydroelectric development." Moreover, he added, the federal government's "short-sighted policy" over the years had forced California's electric power companies to "take care of their increased load with [oil-fueled] steam generated energy." Thus, the Forest Service had exacerbated the petroleum problem.[11]

In effect, Californians tried to turn the Forest Service's whole argument of natural resource conservation on its head. Britton had said in his *Electrical World* article that he knew "of no stronger argument in favor of further hydroelectric development than the conservation of the natural resources which are exhaustible against other resources which are practically inexhaustible." The members of the Pacific Coast Section of the National Electric Light Association, at their May 1918 convention, made clear California's stance that the Forest Service was a prophet of false conservation. With a bill to deal with the permit issue once again before Congress, the NELA conferees adopted a resolution urging immediate enactment of remedial legislation to "encourage development of the millions of horsepower now wasting to the seas, that true and not false conservation may prevail, and that the waste of fuels in generation of electricity may be minimized." Hopeful now that congressional action finally would resolve the issue in favor of the California position, Britton took a final shot at Pinchot in the conclusion of his July article: "When the day of the ultra conservationist is passed and a sane view of the necessities of development of the resources of each state is taken by those responsible, then will the era of prosperity that we have been waiting for so long come upon us. . . ." It could not have been an accident that the *Journal of Electricity*'s editor inserted two patriotic poems, "Your Flag" and "The Flag," immediately following Britton's article. But President Wilson disappointed his western supporters by continuing to advocate a national resource program, and when the Water Power Act finally passed in 1920, it codified essentially those permit conditions offered by Pinchot in 1908.[12]

Expansion of Alternative Energy Resources

As hydroelectric people struggled to win a favorable permit process, petroleum's indispensability to the war effort, plus ongoing fear that oil reserves might be exhausted soon anyway, prompted an organized effort to expand alternative energy resources. In California, this meant hydroelectricity. However, new waterpower plant construction snagged not only on the permit wrangle and tight financing, but also on labor shortages occasioned by conscription and material scarcities necessitated by essential war industry needs. Any new projects had to overcome these obstacles, which required additional permits and approvals from the national wartime Fuel Administration, War Industries Board, Railroad Administration, and various Selective Service boards in addition to the Forest Service. Then, even after surmounting these hurdles, no one expected new power plants to come on line in time to do any good during the wartime emergency. Consequently, the California State Railroad Commission, which had been empowered in 1911 to regulate electric utility rates, joined with the Committee on Petroleum of the State Council on Defense, in 1917, to suggest that power companies interconnect their major transmission lines. The commission anticipated that at least an additional 21,500 horsepower in statewide electric power distribution, the equivalent of not less than 1.1 million barrels of oil, could be gained by transmitting power from localities where hydroelectric plants produced below capacity to regions where demand outstripped power production. Larger power companies took a keen interest in the plan.[13]

Interconnection work moved quickly, the Railroad Commission observing that "at the present time we are faced by an emergency sufficiently grave to require unusual measures. . . . Without pooling of power and a central control in distributing it the present shortage threatens to seriously affect the various war industries and food production so vitally necessary to the Nation." Acknowledging increased labor and material costs, as well as the need to leverage capital to make interconnections, the commission approved several rate surcharges, and H.G. Butler, the commission's power administrator, worked with the companies to hurry along interconnections. Within a year, almost every company in northern California was connected into a power pool, and similar interconnections were completed in southern California. Yet, when the task was completed, the interconnected power grid did little to save oil. The winter of 1917 began a period of drought that lasted eight years and cut hydroelectricity production by at least 20 percent statewide, while state population growth accelerated demand for electric power. Interconnection, it seemed, simply had staved off a greater catastrophe.[14]

Wartime interconnections, however, had another far-reaching impact. They accelerated a prophesy made in 1915 by San Francisco engineer Frank Baum that "transmission and distribution systems . . . [would] ultimately form an electrical, metallic screen over the entire country"; and, as interconnections from Oregon to the Mexican border began, a contributor to *Electrical World* in January 1918, noted that "in the emergency of war and facing fuel famine the economies of monopoly in public utility management" not only had become apparent but were "urged on the companies themselves as patriotic duties." Californian Franklin K. Lane, Woodrow Wilson's Secretary of the Interior, transformed California's interconnection idea into the concept of superpower and organized an Atlantic states superpower survey. Between 1918 and 1921, William S. Murray, a consulting engineer in New York City, carried out the survey for the Department of the Interior and designed a plan to tie together huge privately operated base-load generating stations with very high-voltage trunk transmission lines in a zone from Maine to Maryland. Two years later Gifford Pinchot, then governor of Pennsylvania, advanced a scheme for publicly operated "Giant Power" in the mid-Atlantic region.[15]

Meanwhile, superpower was a reality in California. The introduction by PG&E and SCE, in 1924, of 220-kilovolt-ampere (kVA) transmission lines, observed Westinghouse engineer Stephen Hayes, was "the real start of the Super-Power System in the United States." Pacific Coast interconnections were so successful that a West Coast Superpower Survey Committee studying additional efforts in Oregon, Washington, and Idaho, felt comfortable in recommending against a superpower effort there, since "existing networks of power systems will, through natural expansion, become more interconnected." Frank Baum, important in bringing 220-kVA transmission to PG&E and always looking toward the stars, summed up his life's work with a proposal for a national superpower system. With twelve regional power districts, it would be a public utility system, and, though Baum had bristled against the idea of regulation when opposing Pinchot's plan to control waterpower on public lands, he implied it would be overseen by federal and state regulatory commissions.[16]

Interconnection and superpower meant more to many people than superior engineering. People began seeing electricity in terms of dollars and politics. Critics attacked William Murray's plan as a private utility strategy to effect a gigantic power combination, and Pinchot's plan was attacked as socialism. In California, the middle ground of public utility regulation, which had been born as part of the progressive era's solution to the rising power of corporations, found itself being worked out during the war years in an unanticipated atmosphere of close cooperation between power companies and the Railroad Commission. Protracted drought and increasing demand for electricity perpetuated into the postwar years this collaborative spirit between companies and the commission. The commission continued to facilitate new hydroelectric plant construction and interconnections, and power administrator Butler actually began to argue on behalf of greater consolidation in the industry. The logical way to conserve resources and supply electricity economically, he contended, was to determine the natural zones in which power can best be handled by a single system and "unify generation within these zones." California, he concluded, was just such a natural zone. Consumers became unhappy. The Railroad Commission, wrestling with drought-induced power shortages and growing power demand, did not order rollbacks of wartime rate surcharges. To many people, it appeared that the commission had been captured by the very companies it was supposed to regulate. Frustrated with high prices and perceptions of poor service, a clamor arose in favor of public ownership against state-regulated public utilities.[17]

Municipal vs. Private Ownership

The idea of public ownership of electric utilities stretched back into the nineteenth century. Early utilities—gas and water, as well as electric—got their start by negotiating franchises with local officials. Community perceptions that promised cheap and efficient service had not been fulfilled resulted in disputes across the state, much like the one the City of Chico had with its gas and electric company during the 1890s. Communities responded in various ways, such as granting competing franchises or seeking regulatory assistance from the state. Particularly where these steps did not bring relief or where corruption appeared to flourish, communities turned to forming municipal utilities, usually by issuing local bonds and purchasing the offending company's system. In 1884, the City of Santa Cruz was the first community to choose this path, followed by Alameda in 1887. As Californians read more and more during the 1890s about the fraud, bribery, and chicanery often perpetrated locally by utility corporations, municipal ownership became part and parcel of the larger progressive reform movement. Anaheim formed a municipal system in 1894, followed by Colton, Riverside, and Santa Clara in 1896. Palo Alto, Healdsburg, and Ukiah joined the movement before 1900. Healdsburg had its own hydroelectric plant, whereas all others but Colton produced their electric power by steam generation. Between 1910 and 1930, these early municipal systems abandoned their generation equipment in favor of purchasing power from private companies. Another sixteen communities and two irrigation districts ventured into municipal ownership after 1900, seven of them purchasing as well as producing some of their power by steam generation or hydroelectricity, and the rest only distributing purchased power.[18]

Because the bulk of California's electric power came from hydroelectricity, whenever water became an issue, so too did electric power. The greatest impetus in California toward municipal ownership of electric utilities came from Los Angeles, which experienced such population growth at the turn of the century that it sorely needed a new municipal water supply. In 1898, the city council, long frustrated with the high water rates and poor service received under its contract with the private Los Angeles City Water Company, acquired the company. Fred Eaton, who moved from the water company to become the city's engineer, conceived of bringing water by aqueduct from the Owens River, over two hundred miles northeast of Los Angeles, and he started acquiring land for the purpose. In 1904, amidst community fear of a water famine, William Mulholland, Eaton's successor and protégé, became interested in the idea and pushed it forward. Eaton agreed to sell his land to the city, quietly purchased more on its behalf, and in a special election the next year, Los Angelenos passed a bond issue to acquire rights to the water. Although the U.S. Reclamation Service had formed a plan to irrigate the Owens Valley and residents launched a bitter campaign to stop the project, Theodore Roosevelt sided with Los Angeles, and the project moved forward.[19]

The same year that voters passed the aqueduct bond issue, they also became discontented, as we have seen, with the gas service being provided by the Los Angeles Gas and Electric Company. After a manufactured gas famine during the 1906—1907 winter, broad citizen support emerged for municipal ownership of the gas system, and they formed the Municipal Ownership Party to clean up political corruption. John Randolph Haynes stood out among the movement's supporters. A successful medical doctor and real estate investor, Haynes had earlier created the Direct Legislation League, which succeeded, in 1903, in getting the city to adopt a new charter that included the recall, initiative, and referendum. In 1906, Haynes and other progressives prevailed in electing reform candidates to almost every city office. During the same year, William Mulholland became aware that hydroelectric plants planned as a part of the Owens Valley Aqueduct would produce more than sufficient electricity for aqueduct pumping and could be a significant source of revenue for the city. This fit right into the hands of the municipal ownership movement with which Mulholland agreed, and a bond issue to complete the aqueduct was scheduled for June 1907.[20]

Southern California electric power companies tried to prevent the aqueduct project from the start. The Edison company unsuccessfully mounted formal opposition to the 1905 bond issue, and Edison and other companies tried to block federal right-of-way approval in Washington. In the 1907 bond election, the power issue became more significant, but it also was a difficult one for the power companies to fight. Henry Huntington and William Kerckhoff, chief stockholders in Pacific Light and Power, probably sensing they would be defeated and seeking to profit from it, had acquired significant landholdings in the San Fernando Valley, which stood to become much more valuable with aqueduct water. Beyond an ineffectual effort to convince Mulholland to lease aqueduct power to Pacific Light and Power, they largely backed away from the battle. Power companies mounted opposition through the *Los Angeles Evening News*, but Mulholland successfully played down the power issue and the opposition campaign was stifled. Voters approved the $23 million bond issue by over a ten to one margin.[21]

By 1910, the city was committed to producing electric power as a part of the aqueduct project, and Los Angelenos were presented an additional bond issue to finish power plant construction. Ezra F. Scattergood, who had been appointed the city's chief electrical engineer, reportedly promised voters that electric power generation would provide "a good net profit, guaranteeing against any possible taxation for aqueduct or power bonds from the beginning of operation." The mayor was quoted as saying the project would do more than pay for itself: it would "wipe out the entire city indebtedness."[22] The bond passed, and the next year the city formed a Bureau of Power and Light (renamed the Department of Water and Power in 1925). In 1913, the city sought another bond issue to begin building an electric power distribution system. Defeated in a tough campaign mounted by private utilities, it passed the next year, and the city started stringing distribution wires and seeking to purchase private utility properties. The city and Southern California Edison came to an agreement in 1919, and three years later SCE turned over its properties in the City of Los Angeles for $13.5 million. Los Angeles Gas and Electric fought the city for two decades, finally selling its properties for $46 million in 1937.[23]

Los Angeles's venture into municipal ownership, despite its many setbacks and unanticipated costs, inspired advocates of municipal ownership throughout California. In San Francisco, a struggle with the Spring Valley

Water Company over its water-service franchise made the water supply issue part of a general reform movement. In 1900, a new city charter authorized a municipally owned system, and the city engineer recommended a water source that had been looked at as early as 1879, the Tuolumne River running through the Sierra Nevada's Hetch Hetchy Valley. The city, led by reform mayor James D. Phelan, filed water appropriation notices, in 1901, and applied to the Secretary of the Interior for dam construction permits. The acquisition of Hetch Hetchy by San Francisco became embroiled in one of the most striking conservation battles of the twentieth century. It pitted the city, notes historian Norris Hundley, against the Spring Valley Water Company, farmers in the Central Valley Turlock and Modesto irrigation districts that claimed prior rights to the Tuolumne's water, and wilderness advocates "dedicated to preserving forever in the natural state unique and beautiful wild places."[24]

In 1905, John Muir's Sierra Club won a long struggle to get the federal government to establish Yosemite National Park, which included the Hetch Hetchy Valley. Then Muir, the Sierra Club, ardent wilderness preservation groups and individuals nationwide, and the General Federation of Women's Clubs rallied to fight the project. All were committed to strict preservation of natural resources. At the Department of Interior and, finally in 1913, in hearings before Congress on the Raker Bill to approve the project, Muir and his allies faced Gifford Pinchot and others who believed in scientific conservation for rational use of natural resources. A bitter contest ensued. The preservationists could not offer sufficient hard data to convince Congress that San Francisco had viable alternatives, such as the water from the Mokelumne River. Meanwhile, amendments to the Raker Bill were adopted that recognized the two irrigation districts' prior water rights and prohibited San Francisco from making project water available for resale by others. With farmers in Turlock and Modesto and the Spring Valley Water Company placated, Congress finally approved the Raker Act in December 1913. It shattered Muir, and he died a few months later.[25]

San Francisco turned to Hetch Hetchy for a water supply, but its plans also identified three hydroelectric sites for future development. Not surprisingly, PG&E became an early opponent of the plan, and the company's concerns proved accurate. When San Francisco's city engineer M.M.

O'Shaughnessy started work on the project, he made hydroelectricity a central part of the project. Knowing the water supply aqueduct would take years to complete, he wanted to build a power plant as quickly as possible. He explained his position in 1922:

> Electric power development is of vital importance to the industries of California, and for many years the market will absorb new power as fast as it can be made available. It is to the advantage of San Francisco, therefore, to begin as soon as possible to generate the power which is a by-product of the water development, so that the revenue from power sales can be used to pay interest and redemption charges on bonds. . . .

To dispose of the generated power, the city could either wholesale it to a public utility "with a guaranty of power for municipal needs at minimum prices inside the limits of San Francisco," or build a municipal transmission line, take over the municipal railway load, and gradually acquire a full distribution system.[26]

Because the Raker Act included language that seemed to prevent San Francisco from selling its hydroelectric power to a private corporation, the potential of 70,000 kW of power aroused advocates of public ownership. Progressives, such as former mayor James Phelan, used the Los Angeles example to counter opponents who claimed that municipalities were ill-suited to operate utilities. Because corporations would never provide cheap power, Phelan argued, "San Francisco, under private ownership of utilities such as water and power, is and always will be under a competitive disadvantage." But consulting engineer Louis F. Leurey asserted that more than adequate power came from northern California's interconnected transmission network—"the admiration of the nation"—and with the Railroad Commission preventing "onerous rates," there were no grounds for city ownership. In response, Henry F. Boyen challenged all public utilities, pointing out William Murray's superpower plan as an undemocratic "attempt of the power kings to build an electric ring from Washington to Boston and tie it under one control." Civil engineer John D. Galloway could not be held back at such a comment and retorted that municipal ownership was simply socialistic: "the state in business means the building up of a political machine that will in time become a Frankenstein and destroy its maker." Ironically, O'Shaughnessy, who began as a municipal ownership advocate, shifted from advocating public power and noted that "with the boast of prowess of Los Angeles, they have not power enough . . . to serve the needs of their customers, and they are buying from private companies thirty per cent of their energy. . . ."[27]

In the end, San Francisco followed O'Shaughnessy's skepticism and contracted with PG&E as agent for Hetch Hetchy power, but public power advocates did not give up. In 1923, former congressman William Kent argued before the Commonwealth Club that, while the Raker Act's language might be clumsy, "the meaning is clear, as is the intent, and there is no possible excuse for . . . cheap pettifogging and subterfuge as to its meaning." The city was not to sell or contract power with privately owned utilities. During the 1930s, persistence by public power supporters convinced Secretary of the Interior Harold Ickes that the city's contract with PG&E violated the law, and the federal government brought suit and won. The victory was to no avail. San Francisco had no funds either to buy PG&E facilities or to construct its own municipal distribution system, and city voters defeated needed bond issues eight times between 1927 and 1941. The bitter battle involving local, state, and federal officials finally ended, in 1945, with a settlement in which the city leased a transmission line from PG&E to deliver its power to San Francisco for street railway, lighting, and other municipal uses; only power that would be "wasted" might be sold. Public power was defeated.[28]

League of California Municipalities' Water and Power Initiative

While the Los Angeles and San Francisco struggles were being fought, the advocates of public ownership found a juicier opportunity at the state level. River navigation, flood control, reclamation, and irrigation in the Central Valley had challenged Californians since the gold rush. Problems were approached at both the local and state levels through the nineteenth century, and, in the 1870s, an investigation by Colonel B.S. Alexander for the Army Corps of Engineers provided what was the first comprehensive water plan for the valley. Too ambitious for the times, however, Alexander's plan languished. In 1919, Robert Bradford Marshall, chief hydrographer for the U.S. Geological Survey, drew on this previous work, developed another plan, and offered it to the state. The so-called Marshall Plan recommended building a number of reservoirs and encircling the entire Central Valley with a system of canals, thus providing flood control and transferring water from the Sacramento Valley south to the San Joaquin Valley. In 1921, the state legislature considered the plan and appropriated the funds to study it.[29]

The Marshall Plan found a ready reception among members of the League of California Municipalities. Wartime power shortages and reduced service from public utilities, drought-reduced hydroelectric production, increasing rates, dissatisfaction with Railroad Commission utility regulation, and a perception that public utilities had taken the place of the Southern Pacific Railroad as the corporate political power brokers had transformed the league's members into champions of public ownership. Spurred forward by the successful public ownership experiences of Los Angeles and the province of Ontario, Canada, they formed a committee to look into the waterpower question, in October 1920, and then took their concerns to the state. They asked the Railroad Commission to order private utilities to make public specific employee salaries, they lobbied successfully for a bill to permit cities to apply jointly to the Federal Power Commission for hydroelectric projects, they requested the legislature to investigate aggressively the Marshall Plan, and they pressed hard for a bill to create a state hydroelectric power commission. When the last measure failed in early 1921, the league decided to redraft it as an initiative measure to amend the state constitution.[30]

In crafting their initiative, the League of California Municipalities allied themselves with a number of long-time progressive reformers. All were known for their strong support of public ownership of utilities or for their opposition to corporate "malefactors of great wealth." William Kent, chair of the initiative committee, had been instrumental as a key member of Congress in getting the Raker Act passed; John Randolph Haynes, vice-chair, had led Los Angelenos in municipal reform; and Rudolph Spreckels, executive director and treasurer, was a well-known reform-minded San Francisco millionaire. Other eminent members of the league's initiative committee included Franklin Hichborn, a journalist known statewide for uncovering corruption in the state capital; Mrs. Herbert A. Cable, a long-time California leader in the General Federation of Women's Clubs; and

Photo 12.1. Map of the Marshall Plan for the Sacramento and San Joaquin valleys issued by the California State Irrigation Assocation, November 1920. (Courtesy of the University of California Library.)

Francis J. Heney, famous still as the federal prosecutor in San Francisco's well-known post-1906 earthquake graft trials. Finally, these luminaries were joined by city engineers, labor leaders, attorneys, city managers, mayors, and commissioners representing municipally owned utilities.[31]

This formidable, largely Whiggish-minded group, grounded the Water and Power Act on the Marshall Plan, but it plainly concentrated on public ownership and development of waterpower. It called for the creation of a water and power board of five appointed members who were granted broad authority to develop a statewide, publicly owned, water and power system in cooperation with local government units. They would have the power to condemn and acquire power company properties as necessary, and the authority to finance their program through the sale of up to $500 million in bonds. The Canadian Ontario Power Commission served as its model; it had, according to league members, successfully purchased and taken over the province's private electric power properties at fair prices and subsequently operated with rates reduced from one-third to two-thirds for consumers in 217 cities and farming districts.[32]

In September 1921, the league met in Santa Monica to ratify and build support for the measure. C.W. Joiner, Pasadena's city manager, opened the conference by presenting their position on the Water and Power Act as one of stimulating California's economic growth and independence from eastern manufacturers, something private companies were not doing as quickly nor as efficiently as necessary. As other conferees agreed, the high cost of electricity from public utilities stifled industry, and "under regulation and recognized monopoly," the necessary competitive incentive for sufficient power development could never exist. But it was a greater moral opposition to private ownership of utilities that propelled and titillated league members. The paramount issue was: "How shall the complete monopolization of the water power of the State be prevented, and how shall that which has already been monopolized be reclaimed for the benefit of all concerned?" While some members intimated that the initiative carried on the heroic struggles of Bunker Hill and Valley Forge, Jay A. Hinman, secretary, San Joaquin Valley Association of Municipalities, cut to the quick of it: "God gave us this water and the consequent power, and he never intended it to belong to a privately owned public utility."[33]

Appropriately, Gifford Pinchot delivered the conference's keynote address. After being introduced by solemnly invoking Theodore Roosevelt, scientific conservation of natural resources, and reclamation in the West, he too went to the moral heart of the matter:

> There isn't any reason why the people of the State of California should turn over to the big corporations the job of developing their water and their power. . . . I know how they work. Their motto is . . . "All the traffic will bear." They take what they can, and what they can't take, they don't take. . . .
>
> The grip of the power corporations has extended, little by little, over the State of California. . . . [But] if you pass this law it will be a declaration of industrial independence. . . .
>
> There is no other such opportunity anywhere in the United States. . . . Water and power at cost to the people, at cost for the people; or water and power at a profit for the corporations.[34]

This moral high ground simply could not tolerate opposition, and after A.H. Redington presented the City of Hillsborough's resolution against the initiative, the response was unmerciful: "How many [council members] own stock in or are directors of the Pacific Gas & Electric Company?" "Is it not true that it is necessary to hold stock in a public utility corporation in order to exercise the franchise in Hillsborough?" Hillsborough, said the next speaker, must have an awfully good tax rate "if they have money to spare to send out such 'bunk.'"[35]

The League of California Municipalities voted eighty-eight in favor and one abstention to go forward with the Water and Power Act Initiative, and the debate over the measure became heated across the state. At the Commonwealth Club in San Francisco, for example, strong opposition prevailed. The club assigned its sections on irrigation and waterpower, both comprised largely of engineers, to study and report on the initiative. They presented to the club four reports. The first, by civil engineer John D. Galloway, concluded that the Water and Power Act was "superfluous and wholly unnecessary" since waterpower was being developed as quickly as possible; the second, by former head of the California Water Commission Charles H. Lee, concluded that the success of Los Angeles and other municipal public ownership experiences did not necessarily mean success at the state level; the third, by H. F. Jackson and Lewis A. Hicks, looked at Ontario's experience and concluded that California would do better with state regulation of privately-owned public utilities; finally, Louis Bartlett, Mayor of Berkeley and President of the League of California Municipalities, submitted a minority report in support of the measure. In March and June, the general membership then devoted two full evenings to its discussion, at which opposition prevailed in a vote against the measure, 101 to 7.[36]

Public utilities naturally mounted substantial opposition to the measure. They spent as much as a half million dollars to defeat it. SCE and other southern California power companies funded a front organization called the People's Economy League and northern power companies, led by PG&E, underwrote the dummy Greater California League. Perhaps feeling the pressure, the League of California Municipalities decided to discuss their initiative again at their October meeting; they reaffirmed their position, this time with a vote of forty-eight to one. San Francisco had come out against it. When the people finally had their say in November, the measure lost 597,453 to 243,604 votes. Although the power companies' campaign, which emphasized the socialistic aspects of the proposal and its potential impact on the taxpayers, certainly had much to do with its defeat, the Whiggish morality of the Water and Power Act's supporters may well have predetermined its failure at the polls. In 1924, and again in 1926, the League placed the initiative on the ballot, only to have it defeated two more times.[37]

Central Valley Project Act

The question of public ownership and development of electric power that focused debates during the 1920s over the Water and Power Act was really but a part of California's overall evolving water policies. Robert Bradford Marshall's plan for water in the Central Valley stemmed from much larger issues than that of state power development, indeed receiving its inspiration from the report Alexander had written before hydroelectricity became a factor in water planning. Similarly, the largest single new hydroelectric undertaking in California toward the end of the decade, construction of Boulder Dam on the Colorado River, was conceived not as a hydroelectric project at all. Rather, it was the U.S. Bureau of Reclamation's second multiple-purpose water project in the West. It focused on flood control, irrigation, and municipal water supply. Plans for hydroelectric generation came only as a by-product of the larger enterprise, as well as a way to convince Congress the project would pay for itself.[38]

The first development of the Colorado River came from entrepreneurs who considered irrigating part of the Colorado Desert, a low-lying region spanning the California/Mexico border between Yuma and San Diego. Although a bleak wasteland, the extremely rich soil made the agricultural potential of the region extraordinary. All it needed was water. In 1896, the California Development Company organized and negotiated with Mexico to draw off the Colorado River and to bring water by canal across the border. After some false starts, the company hired George Chaffee, in 1900, to complete the task. Chaffee, an immigrant from Canada who had developed irrigation systems successfully for the southern California communities of Ontario and Etiwanda, diverted water into a canal that ran fifty miles northward through Mexico along ancient dry watercourses. It crossed the border at two townsites, which he named Mexicali and Calexico, and pushed twenty miles further into California. Chaffee renamed the desert area the Imperial Valley, a change that helped to attract a substantial number of settlers, and then he left the firm in 1905. Soon after Chaffee moved on, a bypass was added around the main diversion headgate for the canal, and it washed out during severe rain storms. The entire flow of the Colorado River diverted into the canal system for almost two years, transforming the Salton Sink into the Salton Sea. The Southern Pacific Railroad, which saw its route to Yuma flooded, took over the California Development Company and, with prodigious effort, finally turned the river back into its course in 1907.

The Imperial Valley disaster convinced its residents that more had to be done to control the Colorado River. During the flood, the federal government had refused to take action because the problem was in Mexico. But the Bureau of Reclamation was looking at an overall plan for the Colorado that one of its senior engineers, Arthur Powell Davis, developed in 1904. Imperial Valley residents were among the first people to openly advocate Bureau of Reclamation action. Not only did the potential of future floods worry them, but they were dissatisfied with the agreement that they had with Mexico. In 1911, they formed the Imperial Valley Irrigation District, approved bonds to purchase the canal, and then called for the building of a new canal that would be located only in the United States—an All-American Canal. A decade later, Secretary of the Interior Albert B. Fall joined with Davis, then Director of the Reclamation Service, in publicly recommending an extensive plan for the Colorado River. Phil Swing, who had been chief counsel for the Imperial Valley Irrigation District for several years and then represented the region in Congress, was joined by California's former governor, influential Senator Hiram Johnson, in introducing the Fall-Davis proposal in Congress in early 1922.

The Swing-Johnson bill opened up the issue to immediate and prolonged controversy. Congress already had authorized states through which the Colorado or its tributaries flowed—Arizona, California, Colorado, New Mexico, Nevada, Utah, and Wyoming—to negotiate an agreement over the waters. Secretary of Commerce Herbert Hoover facilitated talks among the states during 1922 and succeeded in getting their representatives to agree to a Colorado River Compact in November. Six states ratified the agreement quickly, but Arizona refused and began prolonged opposition to the whole scheme. American landowners in Mexico who relied on the existing Imperial Canal, also opposed the plan, calling it socialistic. Most important among them was Harry Chandler. Part of a syndicate that owned over 850,000 acres in Mexico and publisher of the *Los Angeles Times,* which he had inherited from father-in-law Harrison Gray Otis, Chandler employed his influence and the newspaper to condemn the plan, particularly the All-American Canal, arguing it would deprive Mexico of its rightful share of Colorado River water.[39]

Response of the California electric power industry to the Colorado River plan, in fact, was divided. Utility companies, particularly SCE, which had filed in 1920 to begin power development on the river, were irked when the Federal Power Commission withdrew the Colorado and its tributaries from power permit consideration. They also resolutely opposed federal involvement in electric power production, and while they fought the state Water and Power Act, they also lobbied successfully enough to bottle up the Swing-Johnson bill in congressional committees. After 1924, President Calvin Coolidge and his Secretary of the Interior Hubert Work agreed that construction of the dam itself was enough federal involvement and that electric power production should be a private matter. This heartened the utilities, yet they realized the government could most efficiently carry out the power operation in "that the physical conditions where the dam would be constructed necessitated the construction of power facilities as an integral part of the dam itself."[40] Consequently, some private power people continued passionately to object to the project. As John D. Galloway insisted, it was socialistic, would overturn existing waterpower development policy as managed by the Federal Power Commission, remove control of water from the states, violate the rights of Arizona, and smirch the nation's honor by robbing Mexico of water. Nevertheless, Coolidge's support of private sector interests led SCE to temper its opposition. In 1926, SEC's vice-president and general manager, R.H. Ballard, publicly expressed support of the project's overall aims.[41]

Municipal power systems, of course, supported the project wholeheartedly. The City of Los Angeles, which also had started investigating the Colorado in 1920, was pleased particularly with the Bureau of Reclamation's plan. With its population approaching 600,000, its Bureau of Power and Light predicting shortages of electric power, and its reliance on purchasing a substantial amount of its electricity from SCE, Los Angeles planners plainly saw the river as a potential source of municipally controlled hydroelectricity. Therefore, it immediately came out in support of the Bureau plan and was soon joined by Pasadena, Glendale, and Burbank, each with its own municipal system. Los Angeles also took the lead in organizing ten other communities into the Metropolitan Water District (MWD) of Southern California to plan an aqueduct to bring water from the project for regional domestic use. This was added to the Swing-Johnson bill in 1926, and the MWD became an important lobbying agent on behalf of the overall project.

Southern Californians in favor of the project found a staunch supporter in Secretary of Commerce Herbert Hoover. He had developed a strong attachment to California during and after his collegiate years at Stanford in the 1890s, coming to view the state as his home. Although Hoover believed strongly in private enterprise and the continued strength of regulated public utilities, he also saw municipal power as a local matter. Los Angeles's system was well established, so he trod a middle ground on the power issue and worked for the overall project. In the end, his ongoing support became a crucial factor. A month after he won the 1928 presidential election, Congress bowed to his position, ended its deadlock, and approved the Boulder Canyon Project Act. Likewise, President Coolidge signed it, knowing Hoover ultimately would approve it if he did not. When power finally was allocated in 1930, public power systems gained the largest share, 55 percent of the total (3.185 billion kWh), but private power interests were not ignored. SCE and two other private power companies shared 9 percent of the total allocation (1.145 billion kWh); SCE, along with the City of Los Angeles, received the right to lease and operate the power plants; and in recognition that Los Angeles would cancel its purchase of power from SCE when Boulder Dam went on line, the private firm received a three-year grace period before it would be required to purchase its allocations of power.[42]

Boulder Dam proved to be the federal government's first major step into the power business. When Franklin D. Roosevelt replaced Hoover as president in the midst of the Great Depression, resource planning and other broadly aimed efforts at government planning became an important part of his administration. Public power development particularly took on importance under Harold L. Ickes, the new Secretary of the Interior. Ickes had a commitment to public power dating back to his efforts to stop Samuel Insull from building a power monopoly in Chicago. As head of Interior and also of Roosevelt's new Public Works Administration (PWA), he found himself in a good position to advance public power. In 1934, he funneled $38 million in PWA funds into Boulder Dam, which became, says his biographer Linda Lear, "his symbolic test case not only to establish the virtues of multiple-purpose basin-wide resource planning, but also to popularize the economic and social benefits of public power."[43] Harold Ickes's interest in multiple-purpose projects became significant for California beyond Boulder Dam.

Proponents of the embattled Water and Power Act continued to press the issue of public power, even though voters had defeated the concept three times. They had another opportunity when, in 1930, State Engineer Edward Hyatt finally submitted to the legislature the comprehensive water plan study, which had been under way since the Marshall Plan was proposed in 1921. The crisis of the depression, coupled with the fact that California was again suffering a serious drought, prompted interest in it by both Governor James Rolph and the legislature. Two companion measures proposing a multiple-purpose water undertaking for the Central Valley began to work their way through the legislature. Recognizing the tremendous cost of the project, state officials also contacted the Roosevelt administration to discuss federal assistance. Officials in the PWA and the Department of Interior advised them that the proposed plan would need stronger public power provisions to receive federal help. Public power advocates, led by long-time senator Herbert C. Jones, had been unable to amend one of the two measures to include public power just days before this news from Washington. Now the need for federal support enabled them to add provisions to the second bill—a preference to public agencies in purchasing power, a cancellation clause for private utility contracts, and provision for transmission lines and substations. With public power included to attract federal financial support, the Central Valley Project Act passed in July and was signed by the governor in August.[44]

Almost immediately a referendum campaign was mounted by public power opponents, including PG&E, and Governor Rolph called a special year-end election rather than waiting until the following June. Although public power proponents campaigned for the project on the power issue, the majority of its supporters focused on the project's other benefits: water development, federal financing, no state taxes, and creation of jobs. When the measure was submitted to the voters, six days before Christmas, the Central Valley Project Act was sustained by a margin of barely thirty-five thousand votes. Public power advocates, reported the *Sacramento Bee,* called it a "splendid victory . . . for the principle of public ownership" and a defeat of the private "power trust." In reality, the project was sustained by the timing of the election and because the people in counties that would benefit by the project got out the vote and those who would not benefit did not.[45]

The Central Valley Project Act established the California Water Project Authority (CWPA) to construct two dams on the Sacramento River in Shasta County, another on the San Joaquin River in Fresno and Madera counties, canals for irrigation, and electric power facilities. The CWPA was authorized to issue $170 million in bonds to pay for the project, but Californians counted on a substantial outright grant from the Works Progress Administration, plus federal agreement to purchase the bonds. But funds were not forthcoming. Similar projects in other parts of the country were in the works and not enough funding was available for all. In addition, Roosevelt decided, in 1934, to develop a national water resources plan before moving ahead with new undertakings. In June, his new National Resources Board took a number of projects under study, but its November report was noncommittal. Meanwhile, Californians pressed the matter. In 1935, they won a small victory when Congress passed a Rivers and Harbors bill that authorized, but did not appropriate, $12 million for flood control and navigation improvement on the Sacramento River. Encouraged, they worked toward getting the project included in Roosevelt's new public works program, for which Congress appropriated $4 billion in April. Through the summer, Californians worked to ensure an allocation for the project, and Harold Ickes assigned Commissioner of the Bureau of Reclamation Elwood Mead to study the project further.

In 1935, Mead recommended the Central Valley Project become a Bureau of Reclamation enterprise, and the CWPA agreed. Despite this decision, funding to begin was not in hand until the end of 1936, and construction did not begin until 1937. While Ickes saw the project as an opportunity to push forward federal government reclamation and resource management, it was more complex than other similar projects. Since the project area was well populated, an enormous number of water rights claims had to be settled, which required drawn-out negotiations. The Southern Pacific Railroad route through the Shasta reservoir site had to be acquired and the tracks relocated, and rights-of-way for project canals had to be secured. The Bureau also insisted on new engineering studies, which led to increasing storage at Shasta reservoir by 50 percent and substituting a new canal, the Delta-Menota, for the state's plan to pump water up the San Joaquin River bed. Consequently, delays plagued construction to the point that, after World War II began, the War Production Board shut down construction before it was completed and required the bureau of Reclamation to sell the available power from Shasta to PG&E.

In addition to other complexities, an ambiguous relationship between the bureau and the CWPA confused the public power issue. When the bu-

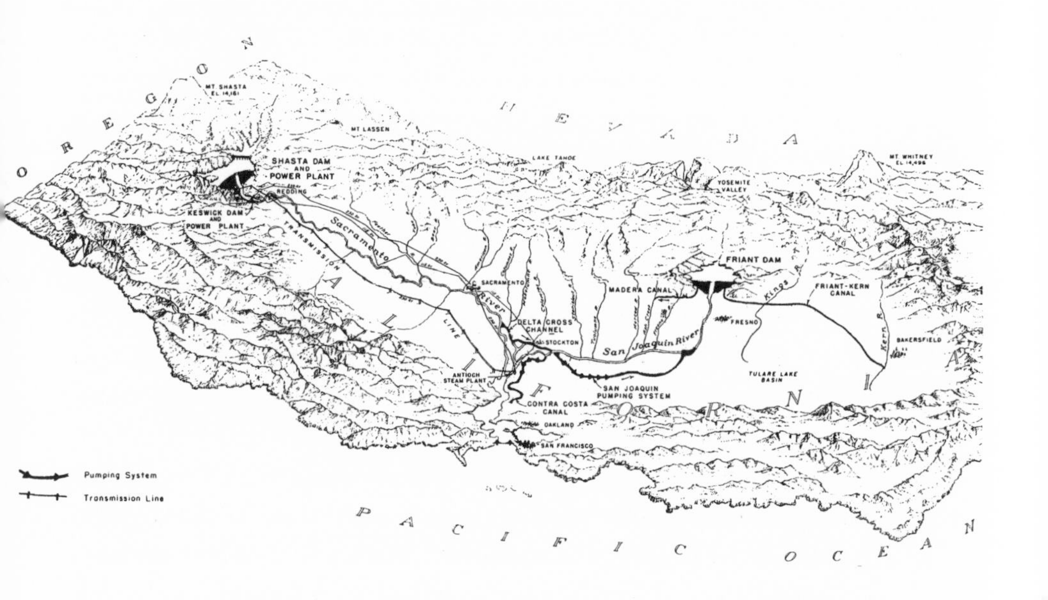

Map 12.1. Major features of the Central Valley Project. From U.S., Bureau of Reclamation, *Central Valley Project Studies; Allocation of Costs, Problems 8–9* (Washington, D.C., 1947), facing p. 183.

reau took over in 1935, the state agency assumed that the project eventually would revert back to its control. In 1935, it opened public power contract discussions with local agencies, but none were concluded, partly because authority for such negotiations remained fuzzy and partly because the agency still saw water as a more important issue than public power. In late 1935, following the death of A.F. Hockenbeamer, PG&E's president and a steady opponent of the Central Valley Project, the firm's new president James B. Black decided to seek cooperation with the CWPA and offered to purchase all the project's power. Through 1937 he persisted, proposing joint engineering studies to revise the company's plans in light of the project, even though the agency actively blocked the company's Feather River hydroelectric plans. The CWPA agreed, but at this point, the Bureau of Reclamation objected to cooperation with a private utility. State Engineer Edward Hyatt then moved ahead with a study that eventually proposed a major state public power effort.

While the CWPA continued to insist that power was secondary to water, the success of public power in the Tennessee Valley Authority (TVA) led Roosevelt, in 1937, to call for seven "little TVAs" around the country, one of which would be in the Central Valley. Culbert Olson, a strong public power proponent elected governor of California in 1938, thought Roosevelt's idea was lovely. Publicly supporting transformation of the project into a "little TVA," he drew on Hyatt's study and proposed selling the bonds originally authorized by the Central Valley Project Act to fund construction of transmission lines and steam-generating facilities. Neither the CWPA nor the legislature, however, supported Olson, and Ickes's Bureau of Reclamation, which saw the TVA as a threat, busily undertook reorganization and clarification of its own public power plans. The Bureau's effort, however, proved too little, too late for public power in northern California. When Olson left office, California's official promotion of public power ended, and project delays, plus the War Production Board's decision to force the sale of power to PG&E in 1942, undermined the bureau's ability to push the issue.[46]

Summary

After the war, PG&E's fight against the Bureau of Reclamation's plans to expand public power in the Central Valley Project gained support from the Farm Bureau Federation, Central Valley Association, and other groups who saw the effort "as a betrayal of the state water plan's intent to subsidize irrigation water from power sales."[47] In 1951, Congress finally forced the Bureau of Reclamation to accept a compromise contract that required PG&E to transmit or "wheel" the project's power to its handful of public power customers. Furthermore, the bureau was barred from building steam-generating facilities or distributing power directly to any future public power customers. The decision marked the end of major public power initiatives in California. Only the City of Sacramento had turned to municipal ownership after 1930, and it completed the process in 1946, without the help of the Central Valley Project. The structure of California's publicly regulated electric power industry at last was resolved.

Settling who should control waterpower resources and whether electric power should be developed by privately owned, publicly regulated utilities or by publicly owned ones had started with a collision between proponents of unfettered development of private waterpower and the rising natural resource conservation movement. Although California's hydroelectric people fought both state and federal regulation over their activities, they lost their struggle. But, during World War I, they worked with state regulators, conserving natural resources through power system interconnections and, in the process, strengthening both their own transmission networks and the position of proponents favoring an electric power industry comprised of regulated private utilities. Consequently, during the 1920s, when advocates of publicly owned utilities looked beyond the municipal to the state level, they discovered their opposition consisted of both the private utilities and their state regulators. The depression ended hope of the state water and power initiative. Even federal agency efforts toward public power gave way to the private utilities.

World War II, and the Cold War years that followed it, changed California dramatically. The state's population mushroomed, its economy prospered, and its publicly regulated utility industry successfully satisfied skyrocketing energy demand. It was a time of energy abundance and great confidence on the part of energy companies as well as government regulatory agencies. But headlong energy development focused public scrutiny on relationships between technology and the environment. The natural resource conservation movement evolved, during the postwar years, into environmentalism. By the 1970s, Californians faced increasingly difficult issues in energy, technology, and the environment.

CHAPTER XIII Energy in Abundance

PRIOR TO WORLD WAR II, development of California's petroleum and waterpower resources had made the state not only self-sustaining in energy, but also able to provide for a steadily growing population and expanding economy. Imported coal and oil products, relied on during the nineteenth century, ceased to be a meaningful part of the state's energy budget. Domestic petroleum became California's principal energy resource, supplanting coal, fuelwood, and animal power in industry and transportation. Natural gas became important, too, replacing kerosene, wood, coal, and solar collectors for domestic heating, and finding regular use in industrial heating processes. Finally, hydroelectricity provided between 80 and 90 percent of California's electric light and power needs and accounted for virtually all of its inanimate renewable energy resource consumption.

The region's prewar energy history also sketched an evolving interplay between technology and the environment. Over time, residents felt the impact of this interaction more and more, in no small part because their exploitative approach to the environment and their growing numbers intensified it. By 1850, the European-American settlers, who less than one hundred years before had constituted only a fraction of California's native

Photo 13.1. PG&E's oil-fired steam power plant at Moss Landing, along Monterey Bay, opened in 1950, its 340 MW capacity making it the company's largest steam plant. (Courtesy of PG&E)

population, already had domesticated the ecosystem unalterably, transfiguring grasslands and riparian woodlands in the creation of an extensive agroecosystem. Thereafter, the state's population doubled, on average, every twenty years. In not too many years, people's use of timber and fuelwood deforested many areas, gold mining silted up streams and rivers, hydraulicking caused a long-term environmental crisis in the Central Valley, and exploitation of petroleum and waterpower otherwise impacted the environment. Then, as people explored the meaning of resource conservation during the first part of the twentieth century, they began to discover complex linkages between oil, hydroelectricity, and the preservation of natural resources, and government's role in influencing the relationship between technology and the environment expanded.

World War II and the subsequent cold war era transformed California's economy, triggering unprecedented population and industrial growth. To keep up, energy industries expanded. They provided fuel and power in abundance, but the state's few years of energy independence during the twenties and thirties came to an end. Oil companies became part of a multinational industry, gas utilities reached into the Southwest and Canada to tap new resources, and electric power companies bought Columbia River hydroelectricity as well as adopted new methods of generation. The atom, harnessed to peaceful use through the generation of electricity, found its way into California's energy budget, and geothermal fields, discovered where hot spring resorts had once attracted thousands of visitors, offered another resource for electric power generation. None of this, of course, occurred without continued give-and-take between people, their technology, and the environment.

On Becoming a Postindustrial Society

At the outset of World War II, California's energy industries appeared to be in a good position to meet rapidly expanding power needs. Oil production had exceeded consumption during the previous two decades, and natural gas reserves were more than enough to meet expanded demand. A large surplus of electric power from Hoover Dam easily supported expansion in southern California, and a combination of planned steam and hydroelectric power projects by PG&E, plus completion of generating facilities at Shasta Dam, assured sufficient electric power for northern California. But the enormous wartime industrial expansion and influx of people almost overloaded the state's energy providers.

Between 1940 and 1946, some $35 billion poured into the state, fully 12 percent of all the war contracts let by the federal government. Established industries such as food processing, textiles, and small machine shops flourished. Southern California's aircraft industry, which had struggled during the 1930s to keep even one thousand workers busy, employed more than 280,000 men and women by the end of 1943, establishing a foundation on which the postwar aerospace industry would be constructed. Shipbuilding, primarily located in the San Francisco Bay Area, employed 300,000 workers in 1943, vitalizing the economy. This industry's enormous expansion, along with that of aircraft manufacturing, caused the first large-scale basic metals firms—those of steel, aluminum, and magnesium—to be opened in the state. At the same time, the war engendered an unprecedented number of huge military bases in California, many of which remained during subsequent years of international tension. Finally, the state's population expanded in proportion to wartime spending, swelling from 6.9 to 10.6 million during the 1940s alone.[1]

Some city and state officials feared that the war's end might bring a return of the depression, but the readjustment to peace instead brought continued economic growth. A positive psychological outlook imbued most business and government leaders. Annual per capita income, which had stood at $835 in 1940, rose to $1,752 by 1948. Pent-up consumer demand for durable products, which had been curtailed or severely rationed during the war years, propelled the economy. Benefits under the G.I. Bill led tens of thousands of ex-servicemen to enroll in a state college and university system that scarcely could expand fast enough to keep pace. The housing industry surged, fulfilling the needs of hundreds of thousands of young families; public schools were swamped by new enrollments. Fortunately, the state treasury, which had swollen during the war years, now spewed forth funds for highways and public education, while local communities poured tax dollars into public services and launched sophisticated growth campaigns. In response to these drives, what had been a flood of immigrants during the war years turned into a torrent during the 1950s and 1960s. With almost twenty million citizens, California surpassed New York, in 1962, as the nation's most populous state.

The cold war and national determination to perpetuate formidable military expenditures nourished California's prosperity. During the immediate postwar decade, over half the state's economic growth came from defense spending, and after 1957, it was augmented further by spending for the space program.[2] Southern California's aircraft industry began what became a postwar trend when it won 55 percent of the federal government's airplane contracts in 1948. At the same time, the wartime rocketry and aerospace research program at Pasadena's California Institute of Technology expanded into a new Jet Propulsion Laboratory. In northern California, the Lawrence Laboratory at Berkeley, and one in nearby Livermore, became one of the country's leading nuclear fission research centers, and Stanford University built successfully on years of work with the electric power industry, ascending as a national center for research and development in microwave electronics.[3]

Despite staggering federal defense spending cuts between 1968 and 1971, California's new industries survived and rebounded. By the late 1970s, California was one of the nation's most powerful and prosperous regions, no longer dependent on eastern capital, no longer reliant on extracting raw materials to be sent elsewhere for fabrication. Were it a nation, its economy would have been the eighth largest in the world. Its service sector had outstripped extractive and manufacturing industries, and the theoretical knowledge of people in its economy's technical and professional areas became a principal source for economic innovation. California's society had become a postindustrial one.[4]

New Sources for Oil and Natural Gas

California's oil industry strained under wartime conditions. Demand pushed state production to a new high, from 230.3 million barrels in 1941 to 326.5 million in 1945. Crude oil stocks, which stood at 145 million barrels in December 1941, were cut in half by mid-1945, despite reopening wells shut down during the 1930s and bringing new ones into production. California oil continued supplying the Pacific Coast region's domestic demand, but now it had to provide for the military's Pacific theater as well. The state's oil transportation network and refineries ran at capacity, production could not keep up, and supplementary supplies of both crude and refined oil had to be imported by rail from Texas and Rocky Mountain fields. Although proposals to accelerate importation by connecting Texas and California with a crude oil pipeline were turned down by federal authorities, knowledgeable observers realized that California would need such a connection if demand and growth continued in the postwar era.[5]

In anticipation of postwar growth, an intense but disappointing oil exploration campaign moved forward. The estimated reserves of several new pools, discovered in 1946, represented less than one-sixth of the state's annual production, and discoveries in 1948 were no better. Furthermore, postwar demand shifted in character. Western railroads switched from steam locomotives, fueled by heavy oil, to diesel-electric engines, which closed the largest part of the regional heavy fuel oil market and increased the need for diesel oil. At the same time, between 1945 and 1950, the demand for gasoline rapidly intensified as the number of automobiles in the state increased by 1.8 million, an increase of over 200,000 more than during the previous two decades. Because much of California's oil was crude fuel oil with marginal refining value, refiners began shipping heavy oil to the East Coast and importing lighter crude to refine into diesel oil or gasoline. Although the 1940s might appear to have been a period of transition from a time of substantial oil surplus to one approaching balanced regional supply and demand, the fast-rising markets for gasoline and other refined petroleum products, plus the declining heavy fuel oil market, actually pushed California toward becoming a net importer of oil.[6]

The increasing multinational integration of the oil industry ensured that California would continue to have enough oil to meet growing demand. The state's oil field production declined after 1953, but its oil industry was no longer isolated. Economies of scale and development of improved oil transportation technologies offset geographical peculiarities and other factors contributing to the region's earlier separateness. This changing situation led to the opening, in 1958, of a six-hundred-mile pipeline that connected the Los Angeles area to new oil fields at Four-Corners (the boundary juncture of Arizona, Utah, Colorado, and New Mexico). By the early 1960s, this and other western pipelines, plus overseas imports from Sumatra, Venezuela, the Middle East, and elsewhere, fully ended the Pacific Coast oil industry's regional isolation. Although new oil discoveries, along with development of improved secondary recovery methods, caused the state's own production to peak at almost 380 million barrels in 1970, Californians satisfied almost one-third of their oil needs with oil from outside their own borders; by 1975, fully 35 percent came from foreign fields (figure 13.1). Still, gasoline remained cheap, shortages were simply unknown, and California's automobile use climbed almost incomprehensibly into hundreds of trillions of miles traveled per year.[7]

The natural gas industry was tied closely to that of petroleum, particularly in southern California where over 80 percent of the gas produced was "wet" or "oil well" gas drawn off in conjunction with oil production. Before the mid-1930s, as much as 30 percent of wet gas had been flared, but expanding markets had reduced such wastage to 7 percent. By 1940, natur-

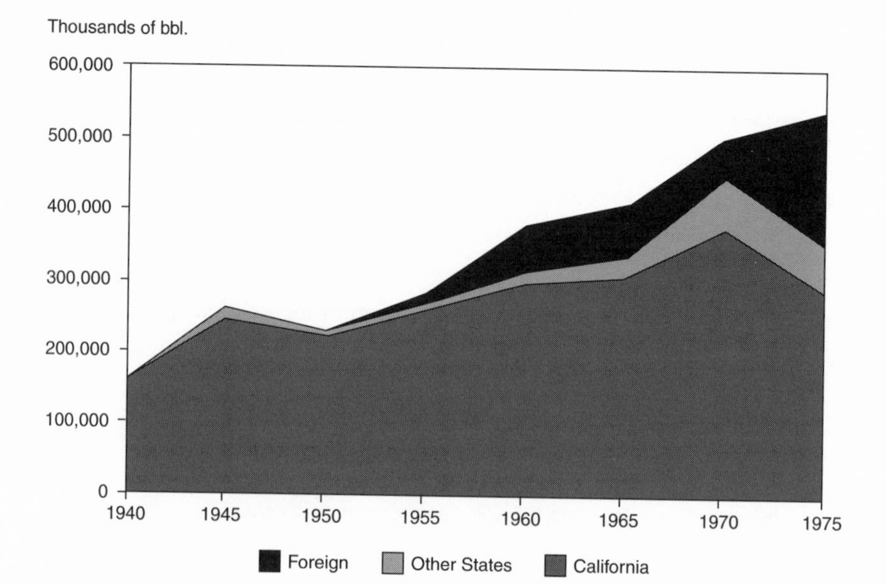

Figure 13.1. Sources of California Oil Consumption, 1940–1975. Source: *Supra,* Table C-20.

al gas use in California was substantially higher than that of the rest of the nation. Two-thirds of the state's homes burned natural gas for water heating and cooking, and about half used it for space heating. Almost a third of California's commercial businesses used gas, and many small industrial companies were equipped for gas only. These users were firm customers, with the highest priority for gas delivery. Large industrial users were provided service on an interruptible basis, so when demand exceeded availability, primarily during winter months, their gas could be curtailed, forcing them to switch to another fuel, almost always oil. Because wartime growth rapidly swelled the number of customers in all classes, the federal War Production Board ordered class restructuring so as to give war-production related customers the highest priority. Among them were new and expanded military camps, large housing projects, the rapidly expanded aviation and shipbuilding industries, thermal electric power plants, and vegetable dehydration plants.[8]

In 1942, the State Railroad Commission reported that existing gas facilities would be able to meet wartime demand, but this proved untrue. In northern California, where large dry-gas fields, like that at Rio Vista, provided about 70 percent of the area's needs and production could be regulated more closely to changes in demand, curtailments remained fairly small. But, after war authorities converted over to oil delivery the principal gas line connecting San Joaquin County with the San Francisco Bay Area, a new twenty-two-inch gas line had to be constructed from the Rio Vista field, where new wells were sunk. The wartime emergency also accelerated gas exploration that resulted in new discoveries after 1944 near Chico, Suisun Bay, and Winters.

In southern California, the gas situation was not as easily handled. Wet gas production increased uncontrollably with heavy wartime oil withdrawals. Much of the excess gas, from five billion cubic feet in 1940 to ninety-three billion cubic feet in 1945, was reinjected into oil fields to increase petroleum yield, but even so, insufficient facilities to store gas for heavy winter demand led to more and more curtailments to low-priority customers. In order to reduce curtailments, the nearly depleted Playa del Rey oil field near Santa Monica was converted, in 1942, to an underground gas storage field. An additional underground storage facility was developed in association with a depleted gas field at Goleta, west of Santa Barbara, and a pipeline connecting it to Los Angeles opened in 1944.[9]

Gas companies in southern California recognized that, if demand continued to grow after the war, it would soon outstrip natural gas reserves, so they turned outside the state for additional supplies. In 1945, the Southern California Gas and Southern Counties Gas companies cooperatively entered into a thirty-year contract with producers in New Mexico and Texas. The El Paso Natural Gas Company constructed a 990-mile pipeline to Blythe, on the Colorado River, where the California companies joined it to their own new 220-mile line to deliver the imported gas into their service areas. Finished in late 1947, the $70 million "Biggest Inch" pipeline delivered to southern California, by 1949, over 300 million cubic feet of gas per day. Completion was none too soon, for a severe winter gas shortage had struck southern California just the year before. Meanwhile, northern California's PG&E, realizing its reserves could not keep up with postwar growth, entered into a contract with the two southern California companies to share their newly imported gas. Because this was a short-term

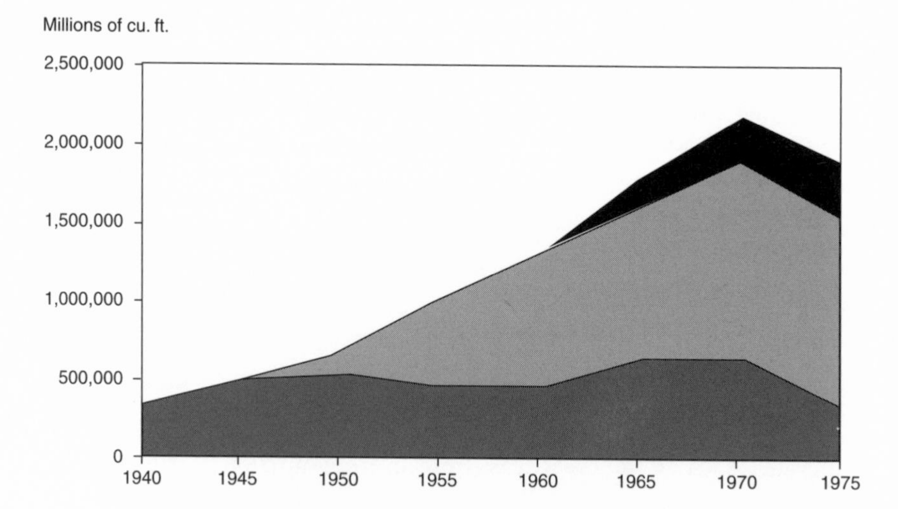

Figure 13.2. Sources of California Natural Gas Consumption, 1940–1975. Source: *Supra,* Table C-21.

arrangement, PG&E also contracted to purchase additional gas directly from the El Paso Natural Gas Company. In 1947, PG&E began construction of a five-hundred-mile-long "Super-Inch" pipeline from Needles, on the Colorado River, to the San Francisco Bay Area. At thirty-four inches in diameter, it was the largest gas pipeline in the nation and, by 1956, transported 500 million cubic feet per day.[10]

Even more quickly than Californians came to rely on imported oil, they found themselves dependent on imported natural gas (figure 13.2). By 1960, over 60 percent of the state's gas demand was fulfilled by imported southwestern gas. PG&E already had decided, in 1957, to turn to other sources and started construction of a pipeline stretching from the Canadian border through parts of Idaho, Washington, and Oregon. One hundred and fifty-one trillion cubic feet of Canadian gas were fed into the company's gas system in 1965, and double that amount in 1970. Moreover, to prepare for the future, PG&E slowed withdrawals from its northern California dry-gas fields to preserve those resources for some future rainy day. Like Pacific Enterprises, the parent company of Southern California Gas, PG&E also embarked on plans to bring additional gas from Wyoming and

Alaska. Although most residents of the six million California households using natural gas probably did not realize it, by 1975 over 80 percent of their steady and seemingly copious gas supply was imported from Canada or the Southwest.[11]

Moving Toward Steam Turbine Generating Plants

In order to meet the demands of wartime and postwar growth, California's electric utilities segued from hydroelectricity as the foundation of their power systems to thermal, or steam power. Although steam-powered generation in the state dated back to the midnineteenth century, hydroelectricity had come to provide the base electric power load. Even as increasingly efficient steam turbines began providing base load service in other parts of the nation, the high cost of imported coal and the unreliability of oil production led California utilities to limit steam power to "auxiliary" use, firming up power requirements during times of peak load. Oil shortage concerns on the Pacific Coast during World War I further discouraged power companies from erecting thermal generating plants. Instead of building thermal plants to meet increasing electric power demand, utilities pursued transmission line interconnections and hydroelectric development "as a conserver of oil." As one observer noted in 1918: "We know from past experience—aside from standby use—the development of electric power by steam can be practically eliminated. The conservation of oil by such elimination will be very important."[12]

Beginning in the 1920s, several drought years led electric utilities to look more favorably at steam generation. Most hydroelectric plants had been erected with minimal water storage facilities and depended on the run-of-the-river. In turn, sufficient stream flow depended on a deep, slow-melting Sierra Nevada snowpack. Between 1918 and 1926, annual precipitation averaged four inches below the normal twenty-four inches, and Californians, by then increasingly dependent on electricity, felt growing inconvenience as hydroelectric generation dropped. The first seriously dry winter came in 1918--1919, with precipitation averaging nineteen inches, but the industry consensus, reached as a result of wartime oil shortages, was for building more hydroelectric plants. In 1920–1921, precipitation again fell to nineteen inches. Consumer inconvenience mounted as state power administrator H.G. Butler called on city officials to halt electric sign and window display lighting, to cut back street lighting, and to stress to their citizens the moral obligation to economize electric power use. But the general opinion still held that saving natural resources—to wit, oil—by building more hydroelectric facilities was the clean, efficient way to resolve future power shortages. Not until the winter of 1924–1925 brought only twelve inches of precipitation did Californians' oil conservation stance began to change. Steam-generating plants took up the base load that year, with power companies, at least those with the water storage capacity to do so, "scrupulously conserving the water supply at hydro plants for peak loads."[13]

Drought served as a catalyst for a shift from hydroelectric to steam generation, and other factors secured the move. Although the electric power industry long had touted hydroelectricity as the key to cheap energy, infusing it with an almost spiritual quality, waterpower actually suffered disadvantages over the best thermal power. Plant construction costs were higher, design risks larger, land and water rights more complicated, and plant labor costs greater. More significantly, the cost of hydroelectricity rested in long-term interest payments on capital, while the cost of steam power was in fuel. In the eastern half of the nation, which possessed cheap coal, the early advantage held by hydropower gave way to improved thermal plants after 1914. Average thermal plant efficiencies increased to more than 5 percent between 1900 and 1914, while the country's best plants jumped to 10 percent. By 1930, average thermal efficiencies reached 15 percent, and the best plant efficiencies were over 25 percent. These gains translated into increased fuel efficiency, and power generated in average oil-fired thermal plants climbed from about sixty-five kWh/barrel of oil, in 1900, to more than three hundred kWh/barrel by 1930. Natural gas and coal showed comparable improvements. This caught the attention of the California industry, and the fuel conservation argument, which, since the 1910s, had urged hydroelectric over thermal power development, collapsed during the late 1920s in the face of flush oil discoveries that possessed enormous quantities of associated wet gas. Electric utilities now argued that flaring billions of cubic feet of natural gas every year was the greater waste of resources. Better to burn it under steam boilers for electric power generation.[14]

After the 1924 drought, southern California utilities moved more quickly to steam than those in the north, their shift speeded by their greater distance from undeveloped Sierra Nevada hydropower sites, their service area's faster population growth, and their ready access to ample wet gas reserves. SCE's completion of the state's first truly high-pressure, high-temperature turbine plant at its Long Beach steam power complex in 1927—two separate thirty-five-thousand kVA units with thermal efficiency twice that of any other plants in the state—exemplified the new direction. By 1929, southern California had almost as much thermal as hydroelectric generating capacity. Meanwhile, in the north, as hydropower projects started earlier in the 1920s reached completion, steam plants also became more attractive there. The Great Western Power Company installed its first steam turbine plant in 1929, the year natural gas pipelines from the lower San Joaquin Valley reached the San Francisco Bay Area, and PG&E began construction of the state's largest steam plant with two fifty-five thousand kVA turbines. Announcing that "Steam Will Lead in 1930: More Than Half of Total New Electric Generating Plant Capacity on Pacific Coast Will Be in Fuel-Burning Plants," *Electrical West* revealed that California's steam-generating capacity had risen from 407,000 kW in 1924 to 1,011,340 kW in 1930, an increase of 145 percent.[15]

"Under the circumstances which now prevail," observed PG&E's vice-president for engineering, A.H. Markwart, in mid-1930, "it is natural to question the future of hydro power in California." Waterpower development costs were rising, the state had suffered ten "subnormal or very dry" years, there appeared to be little hope of lower hydroelectric investment and operating costs, and further improvements in waterpower operating efficiencies seemed unlikely. Yet, as quickly as steam power had ascended, its climb was reversed. The Great Depression brought a decline in power consumption, normal and better than normal precipitation increased hydroelectric production levels, and the federal government began waterpower development with immense, multipurpose projects, the first at Hoover Dam. At the same time, hydroelectric plant labor costs were reduced through new techniques for automated operation, removing one of the serious disadvantages of hydropower over steam. Together, these factors halted most new steam plant construction. By 1934, Markwart and other utility leaders, shifting their attention to disposing of surplus power, reemphasized the marketing arm of their industry, and particularly pushed air conditioning.[16]

The change brought by the depression, however, did not last. In 1939, the anticipation of war again altered utility construction strategies. PG&E launched three new fifty thousand kVA steam plants, each built in cooperation with and sited alongside oil refineries on the eastern side of San Francisco Bay: Avon next to the Tidewater-Associated Oil refinery came on line in 1940; Martinez, adjoining the Shell Oil refinery, and Oleum, next to the Union Oil refinery, were finished in 1941. In southern California, municipally owned systems in Burbank and Glendale built two smaller steam plants, and San Diego Gas and Electric added a plant. At the same time, hydroelectric power projects were revived. PG&E rushed to complete a powerhouse on its Pit River system, received Federal Power Commission approval for the Cresta and Pulga plants on its Feather River system, and pushed forward another plant on the Bear River. The federal government installed another generating unit at Hoover Dam for the Los Angeles Bureau of Power and Light, plus two of four units in the Central Valley Project's Shasta Dam. Utilities also refurbished older steam plants, improved storage capacity and other features of older hydropower plants, and enhanced statewide transmission interconnections. Meanwhile, daylight savings time started in February 1942, conserving millions of kWh in its first year.[17]

The wartime emergency meant the abandonment of sales promotions, of course, but postwar planning quickly reestablished the "grow-and-build" strategy adopted over the years by power companies. Even before the war ended, an industry trade report indicated tremendous pent-up consumer demand for appliances, and a survey by the national Chamber of Commerce indicated that more than half of the U.S. population had savings with which they intended to buy appliances as well as automobiles. As postwar population growth accelerated, California utilities resumed sales campaigns and expansion of their generating capacities, but increasing complaints about service made it clear that they were struggling to keep up. Nationwide, material scarcities and labor strife impeded new powerhouse construction, and shortages of trained workers slowed line exten-

Photo 13.2. PG&E's Belden hydroelectric powerhouse on the Feather River was one of several modern plants completed after World War Two. Photo by author.

sions. For California's utilities, the situation was aggravated in 1947–1948, when drought again struck the state. An average statewide rainfall of fifteen and one-half inches shrunk the Sierra Nevada snowpack and storage reservoirs fell to less than 50 percent normal, and north of the Tehachapi Mountains, power curtailments lasted from February through April. The state legislature reinstituted daylight saving time, while an investigation by the State Railroad Commission, now the Public Utilities Commission (PUC), determined that the power industry was doing the best it could to meet demand.[18]

Power companies had to expand; the question was whether or not to emphasize thermal generation in their expansion plans. Steam power was very attractive, because of seemingly abundant cheap oil and natural gas and increases in thermal plant efficiencies, 40 percent in the newest and best units and 30 percent in average units. Furthermore, thermal plants could be brought on line quickly and sited near growing population centers. But rising postwar costs for natural gas and oil worked against this appeal, whereas falling postwar interest rates made hydropower more tempting. Even with perennial drought problems, hydroelectricity had been the key to California's cheap electric power, and some people within the industry wondered if becoming reliant on thermal plants would undermine their industry's ability to provide cheap electricity. Even though hydro plants took longer to build than steam, and even though there were a diminishing number of prime waterpower sites, hydroelectricity was a significant part of the culture of California's electric power industry. Top utility leaders, particularly in PG&E, had cut their teeth on hydropower, fought for it as sound policy, and believed in it. They found it hard to shift their emphasis to thermal power plant development, but the momentum for steam had been established by war, by drought, and by a positive history of increased thermal efficiencies that utility managers, nationwide, fully expected to continue. In 1947, backlogged orders for new steam turbine sets began to be filled, and twelve months later, thermal power accounted for three-fourths of the new generating capacity either under construction or being planned.[19]

California's electric power industry resolutely moved its base load away from hydroelectricity, shifting it to large, highly efficient steam turbine

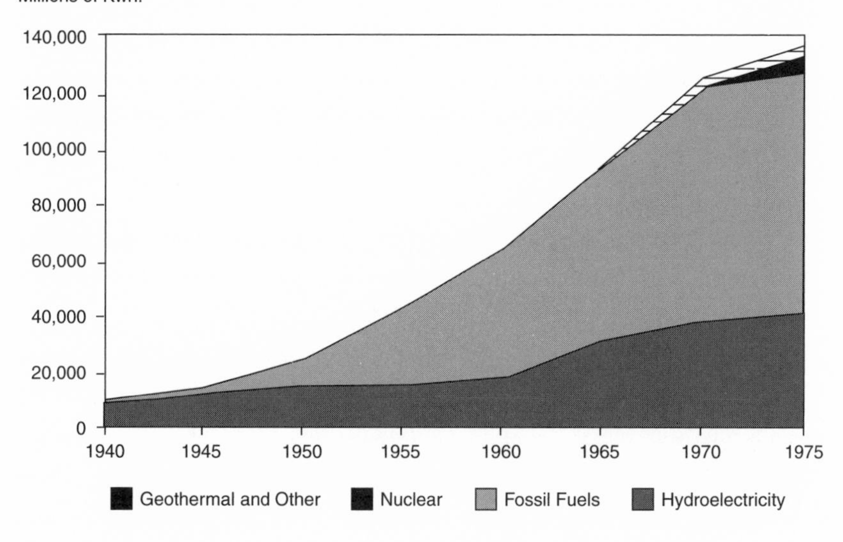

Figure 13.3. California Electric Power Generation, 1940–1975. Source: *Supra,* Table C-14.

generating plants. The state's utilities at last fell in line with the national industry. While the historic role of hydroelectricity and its cultural importance to the state's power companies motivated them to complete long-term waterpower projects, to upgrade existing plants, and even to start some new ones, much of the hydroelectric development after 1948 was part of large multipurpose projects sponsored by the Bureau of Reclamation and, later, the California State Water Project. Hydroelectric generation expanded from 11.5 million kWh to 17.4 million kWh between 1945 and 1960, but gas and oil-fired steam plants, which had supplied only 13 percent of California's electric power in 1945, provided 73 percent by 1960 (figure 13.3). In 1957, PG&E vice-president Walter Dreyer summed up the change: "For economic reasons nearly all future hydro in northern California will be designed to supply peaking power; consequently, the role of thermal-electric plants will be to supply power at high capacity factor, except during periods of heavy run-off."[20]

Accounts of utility expansion, in the early 1950s, emphasized power companies heroically meeting growing energy demands stirred by national security needs and by a rising standard of living. A subtle, almost permanent wartime atmosphere enveloped people everywhere in cold war America and cloaked the power industry's rapid expansion in a veil of incontestability. The PUC, traditionally concerned with rate setting, perceived new generating facilities as a necessity, and local government officials saw them, along with virtually any other sort of new construction, as welcome economic windfalls for fast-growing communities and the state. New power plants meant new tax revenues and jobs, and the general public saw the growing electric power network as part of the steady technological advance that they believed was responsible for America's victory in war, economic dominance in the world, and rising standard of living. Most U.S. residents revered, indeed were intoxicated by, science and technology during the postwar years. For the time being, their technological optimism obfuscated other issues. Virtually no one questioned the visual impact of power plants along scenic beaches, thermal pollution of the coastal and inland waters, the spewing forth of invisible air pollutants, or other environmental hazards intrinsic to the expanding electric power industry.[21]

Post World War II Development of Nuclear Power

The electric power industry's grow-and-build strategy needed more than the combination of hydroelectric and thermal power production to continue providing cheap and abundant electricity for California's rapidly growing population. Increasing reliance on imported and ever more expensive natural gas and oil made this clear. Consequently, the industry extended its interconnected transmission system to draw power from as far away as Bonneville Dam on the Columbia River and, perhaps more importantly, turned to two new energy resources: nuclear fission and geothermal steam. Nuclear power, suggests historian Roger Lotchin, was one of the most dramatic applications of the principle of using science and technology to overcome the constraints of regional geography. It created an early euphoria among some of its national supporters. Atomic Energy Commission (AEC) Chairman Lewis L. Strauss predicted, in a 1954 speech, "that our children will enjoy electrical energy too cheap to meter." In 1956, mathematician and AEC member John von Neumann repeated Strauss's prophecy, saying electricity, quite simply, would be free.[22] Geothermal steam, on the other hand, lacked such enthusiasm. It had been thought of as a potential energy resource in California since before the 1920s, but no one envisioned that it would be free.

Just after the war, California commentators on nuclear energy did not yet share the optimism expressed by Strauss and von Neumann. In 1945, a contributor to the University of California's journal, the *California Engineer,* briefly reviewed methods being considered to harness "atomic power" for peaceful energy production. His conclusion was cautionary: "there is no good ground for unbridled optimism about a cheap source of power to replace all others within the next few years. . . . no dynamo was built directly after Franklin flew his kite and Galvani saw his frog legs twitching; and explosive rivets were not seen until hundreds of years after the invention of gunpowder." A year later, R.G. Spiegelman contributed to the same journal a more optimistic view, noting that "harnessing the atom to the wheels of industry" had great potential and was not too far distant. But L.A. DuBridge, president of the California Institute of Technology, steered back, in 1948, to the less sanguine judgment: "large-scale use of atomic energy . . . may still be thirty to fifty years in the future. . . . [and] it is unlikely that atomic energy will compete with coal, oil, gas or hydroelectric plants in the foreseeable future."[23]

Nevertheless, California's need for new ways to generate electricity and its participation in wartime atomic bomb research translated into both strong interest and important political influence over nuclear energy development. The war had favorably positioned the state in the nuclear research field. The University of California's Lawrence, Livermore, and Los Alamos laboratories, crucial to the Manhattan Project, became centerpieces of the postwar nuclear program, and atomic research flourished at the California Institute of Technology and Stanford University. The emerging ideological climate of the late 1940s thrust the atom onto center stage in the cold war. From its beginning in 1947 through 1957, three Californians made up one-sixth of the membership on the congressional Joint Committee on Atomic Energy (JCAE). Pasadena's Congressman Carl Hinshaw, who understood the energy squeeze felt by Californians, was passionately supportive of nuclear power. Representative Chet Holifield, Montebello, eventually became the JCAE's chairman, and he placed the issue of peaceful atomic power development on the congressional agenda in 1953. Finally, William Knowland, Senate minority and majority leader, represented California as a whole on the committee. The JCAE, observes historian Michael Smith, led the way in convincing the AEC to redefine peaceful nuclear energy as "a necessity, rather than a threat, to national security."[24]

California's electric power industry also found internal reasons to be favorably disposed to undertake nuclear power development. Nuclear energy was a good fit for large electric utilities. It possessed a central power-generating format, great growth potential, and capital intensity. Moreover, they risked little by investing in nuclear and other new generating technologies. As privately owned, publicly regulated utilities they were prevented from making non-electric investments, but the PUC guaranteed them the ability to recoup their outlays for new generating facilities through their customers. Liability, the one obstacle that might have held back private utilities from participating in nuclear energy, was resolved when the JCAE drafted the 1957 Price-Anderson Act, which exempted the power industry from lawsuits stemming from injuries suffered in a nuclear accident.[25]

When the AEC issued an invitation to the nation's electric power industry to study reactors, PG&E quickly agreed to participate. PG&E and Bechtel Corporation, a giant California-based construction firm, had been exploring nuclear power with the AEC since 1951, participating in a number of early studies. In 1954, the utility teamed with General Electric Company to build the state's first licensed commercial nuclear test plant at Vallecitos, near Livermore. Even before Vallecitos became operational in 1957, PG&E looked forward to nuclear plants being integrated competitively into its existing generating system. Based on its Vallecitos experience, the firm's leaders decided, in 1958, to install the state's first major commercial nuclear plant at Humboldt Bay, four miles south of Eureka. Opening in 1963, the 60,000-kW plant was modeled on Vallecitos and sited as a third production unit alongside two existing steam units. Southern Californians did not have a nuclear power plant until 1968, when SCE opened the state's first truly large-scale reactor (435,000 kW) at San Onofre, half-way between Los Angeles and San Diego. A decade later, nuclear fission generated less than 4 percent of the electricity used by Californians (map 13.1).[26]

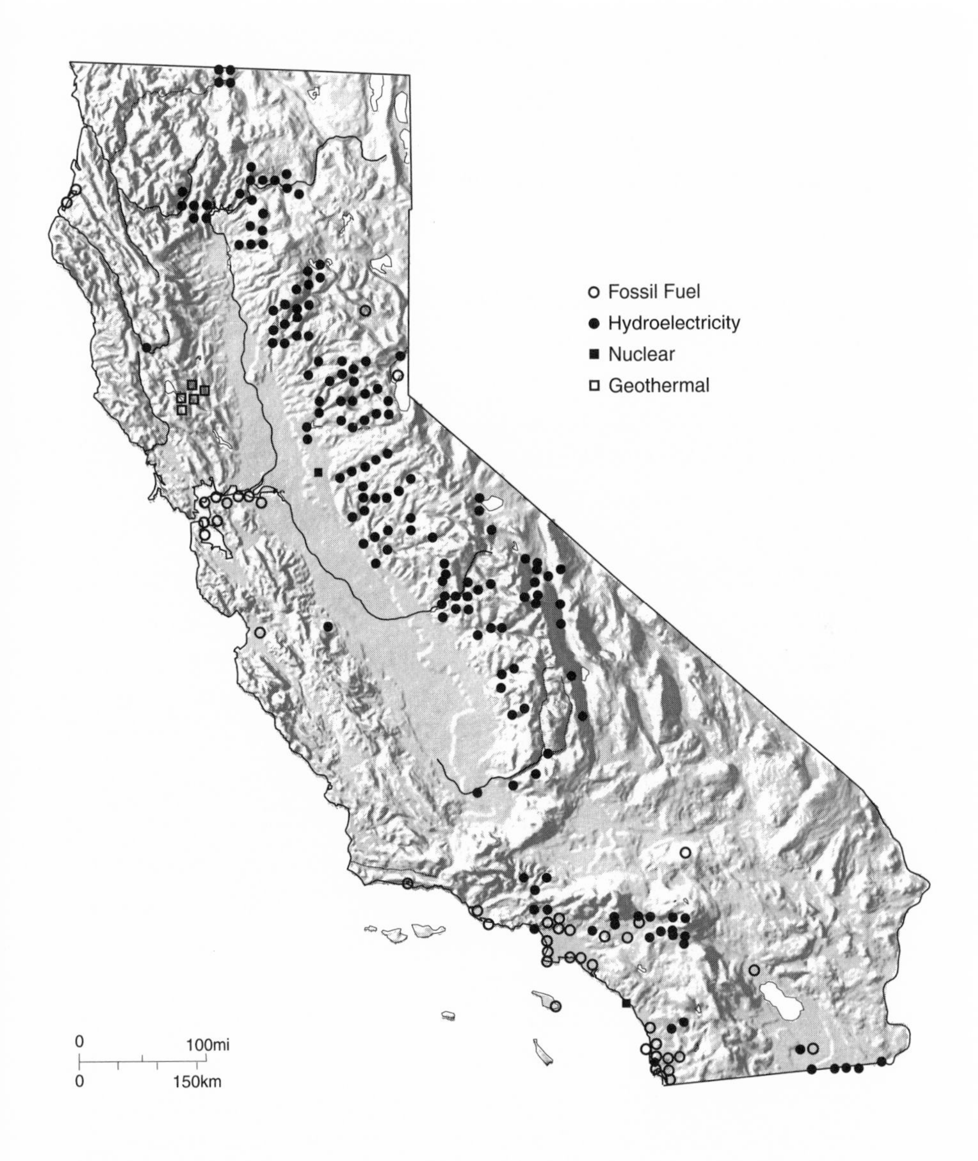

Map 13.1. California Electric Power Plants, 1979. Adapted from California Energy Commission, *1979 Biennial Report* (Sacramento 1979), facing 1; and David Hornbeck et al., *California Patterns: A Geographical and Historical Atlas* (Palo Alto, CA: Mayfield Publishing, 1983), 115.

Through the late 1960s, the general public largely shared the nuclear industry's enthusiasm for atomic power. Like many other Americans, Californians embraced Walt Disney's film depiction of "Our Friend the Atom." The AEC mounted a major advertising campaign on behalf of nuclear power, and the Price-Anderson Act assured the public that atomic power was safe. Because of the cold war, people did not question the industry's shrouding atomic power in a cloak of official secrecy. For most Californians, even the PUC, nuclear power was not much different than steam power. People trusted their government and the power industry. They simply had no idea that nuclear energy was based on little empirical evidence, involved massive technical uncertainties, and exposed society to substantial safety and environmental hazards.[27]

Development of Geothermal Energy

As nuclear power acquired a foothold in the state, the potential of geothermal energy in Sonoma County caught PG&E's attention. During the nineteenth century, resorts appeared at hot springs scattered throughout California, the Geysers Resort Hotel next to Big Sulphur Creek at the foot of Geyser Canyon among the more famous ones. Although no actual geysers existed in the canyon, the resort's hot springs, plus vapor vents and fumaroles in the canyon, became a magnet to tourists. Meanwhile, in Larderello, Italy, soon after 1910, entrepreneurs tapped "volcanic steam" to operate low-pressure steam turbines, which, by 1915, were producing sufficient electric power to supply a number of towns in the Tuscany region. Word of this unique method for generating electric power gradually spread, and steam vents in Geyser Canyon caught the interest of a Californian. In the Fall of 1919, John D. Grant of Healdsburg secured an option to develop geothermal power on the resort's property.[28]

Grant formed the Geyser Development Company in 1921. Because established oil-drilling firms feared destruction of their equipment by natural steam explosions, he was unable to get one to undertake the drilling he planned. Undaunted, he moved ahead on his own, drilling three wells in 1922 and 1923 and hooking a thirty-five kW turbo generator to Well No. 1, "which brilliantly lighted up the hotel, cottages, bath house, and surrounding grounds." His project attracted some attention from the electric power industry, as well as earth science researchers. Arthur L. Day, Director of the Geophysical Laboratory in Washington, D.C., spent the summer of 1924 studying the canyon's characteristics, and General Electric Company engineers ran several tests on the steam potential of Wells No. 1 and No. 2. Enough steam power existed in the area, said some investigators, to generate light and power for two or three good-sized communities; moreover, the power could be developed for $75 per kW, compared to $250 per kW for hydroelectricity. The relative nearness of Geyser Canyon to San Francisco Bay Area towns and cities added another attractive feature for power generation.[29]

Reports by visiting engineers greatly encouraged Grant, and, in 1925, he reorganized his company to sink five more wells. In order to draw on the expertise of engineers with experience in the electric power industry, he retained John Debo Galloway as chief engineer for the project. Galloway, born in San José in 1869, and a graduate of Rose Polytechnic Institute in Terre Haute, Indiana, had entered the civil engineering profession in San Francisco in the early 1880s. His work spanned every aspect of civil engineering, he was a prominent consultant to PG&E and its predecessor companies over the years, and his publications earned him high respect within his profession. Intensely concerned that energy resource development was the key to state growth, he had involved himself in state water politics during the 1920s and was a strong proponent of hydropower. He was the ideal person to head up Grant's geothermal project, and they moved ahead, in 1925, to drill additional wells from which they hoped to capture enough steam to power a 10,000-kW plant. Unfortunately, Grant was unable to raise the necessary $2 million required for the project, and as flush oil discoveries in the state drove down prices for oil and natural gas, investors lost interest in Geyser Canyon.[30]

In 1953, with natural gas and oil prices rising, B.C. McCabe founded Magma Power Company and renewed efforts at the Geysers. He acquired extensive development leases along Big Sulphur Creek and in Geyser Canyon, and, in 1955, drilled Magma No. 1, sited among the wells drilled thirty years earlier by John D. Grant. The next year, Dan McMillan, an associate of McCabe, formed the Thermal Power Company, and the two firms cooperatively drilled five more wells. Having proven again the area's steam thermal capabilities, McCabe and McMillan negotiated a contract, in 1958, with PG&E in which the utility would build a generating plant and purchase steam from Magma and Thermal at a price based on the average cost of oil and natural gas. PG&E Unit 1 went on line in June 1960, using a thirty-eight-year-old, 12,000-kW salvaged marine turbine generator. Three years later, PG&E opened a second 14,000-kW generating plant. Meanwhile, Magma and Thermal continued development along a fairly narrow zone extending from Geyser Canyon to the north side of Big Sulphur Creek, but Union Oil Company's success, in 1966, with a well some distance from the canyon greatly expanded the region's geothermal possibilities. A year later, Magma and Thermal merged their holdings with those of Union Oil, turning over field operations to the larger company. PG&E continued its effort by adding two more 28,000-kW units north of Big Sulphur Creek, and wider exploration began throughout the area.[31]

Since the Geysers was the first geothermal development in the nation, questions concerning control of the energy resource were inevitable. Should they be considered water resources under state control or mineral resources under federal control? The federal Geothermal Steam Act of 1970 and subsequent court rulings determined that geothermal resources should be treated as a mineral, title remaining with the federal government for lands granted under the Stock-Raising Homestead Act of 1916, which included the Geysers. During the early 1970s, accelerating oil prices further stimulated geothermal development, and the first federal geothermal lease sale was held in 1974, with Shell Oil Company purchasing parcels in the Geysers field. PG&E also continued its development of generating facilities, opening six more 55,000-kW plants between 1971 and 1973 and bringing a 110,000-kW unit on line in 1975, the first of another series of larger plants. The next ten years saw considerable geothermal development at the Geysers, with several other drilling companies and three more electric utilities becoming involved. By the end of the 1970s, the Geysers geothermal field was generating about 2 percent of California's electric power, all of which was used in the northern California market (map 13.2).[32]

Summary

World War II and the two decades which followed it proved to be an extremely prosperous period for California. Although, even before 1950, the region's own energy resources were unable to meet demand, California's

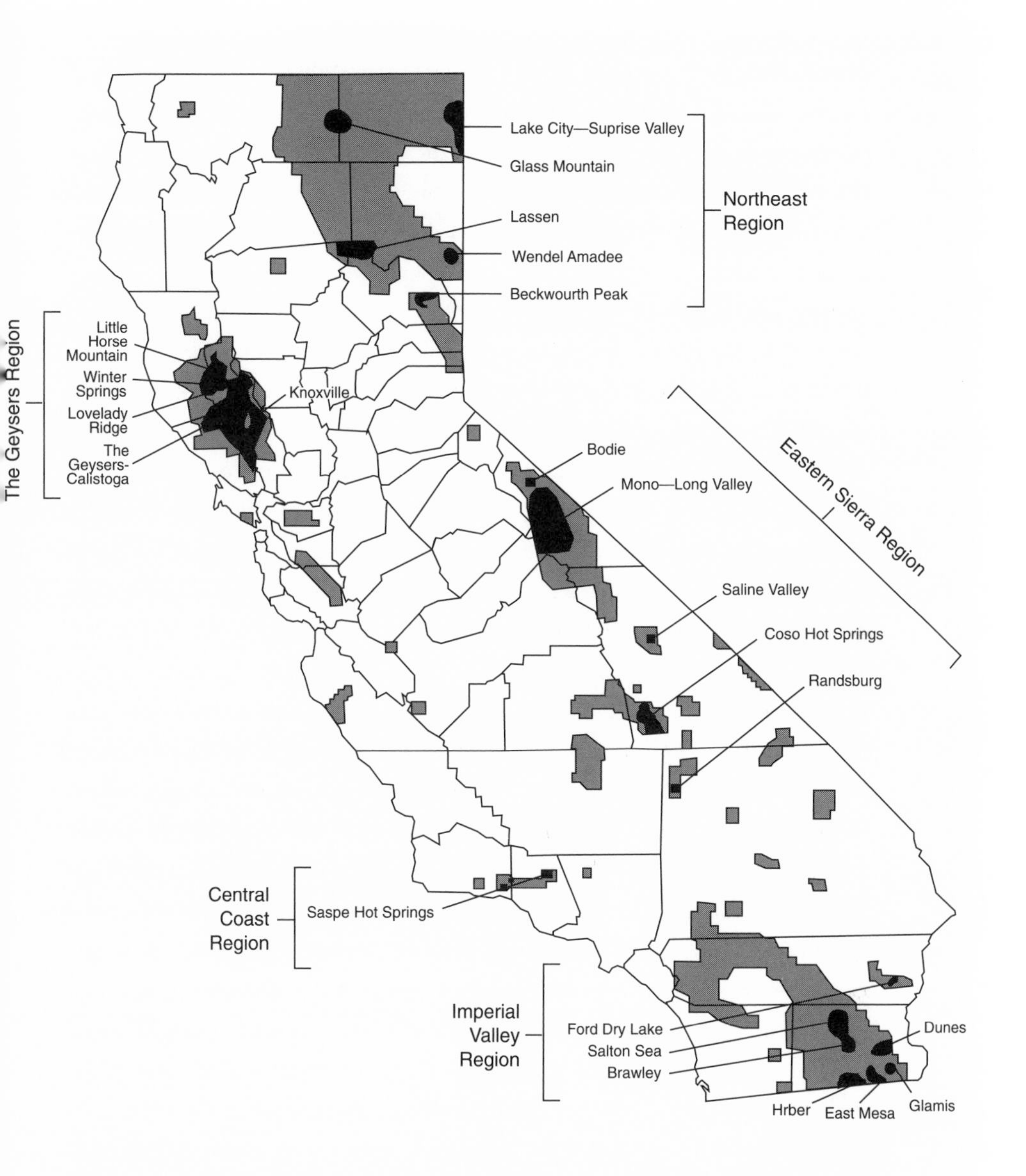

Map 13.2. California Geothermal Resource Areas, 1978. Source: California, Office of Planning and Research, *Report on the State of Geothermal Resources Task Force* (Sacramento, June 1978): 3.

regional isolation melted away as improved transportation and communication technologies toppled geographical barriers. As the state's population climbed rapidly from 6.8 million toward 20 million, its energy industries expanded the resource base on which they drew. In addition to increasing the state's own oil, gas, and hydroelectric production, they imported energy from various locations in western North America, as well as from distant corners of the world, and they began nuclear and geothermal electric power generation. Oil from the Middle East to Texas provided gasoline to propel Californians across their freeway landscape. Hydroelectricity from the Colorado and Columbia rivers, as well as from California's own streams, pumped precious water through a growing system of aqueducts to irrigate millions of acres in the dry San Joaquin Valley and to quench the thirst of arid but booming southern California. Natural gas from fields in New Mexico to Canada fired cooking ranges and home heaters, and helped to generate the electric power which ignited the state's postindustrial society.

Encouraged, indeed expected by state and local government to keep up with growth, oil companies and gas and electric utilities busily built new facilities and encouraged energy use. To be sure, their rapid expansion had an enormous impact on the natural environment, but neither they, state and local officials, nor the general public seemed to notice much. The PUC monitored electric power lines for fire and safety hazards but showed no interest in other environmental issues. When a few citizens of the central coast town of Morro Bay wrote letters to their local newspaper, in 1954–1955, opposing the aesthetic loss over the local economic gain from a new power plant, they were ignored.[33] If there were technical problems, such as air pollution, nuclear radioactivity and waste disposal, people, including legislators, fully expected scientists and engineers to resolve them. For most people, technology remained a tool for conquering environmental obstacles and harnessing natural energy resources for practical use. For the time being, negative environmental impacts, resulting from energy production and consumption, were part of the price of progress. Yet that very progress ultimately led Californians to develop new attitudes about their environment. During the 1960s and 1970s, the calculus by which Californians made energy choices changed dramatically.

CHAPTER XIV Energy and Environmentalism

CAREFUL OBSERVERS SAW the intimate association between energy and the environment long before the 1970s. Reliance on fuelwood during the nineteenth century had much to do with early deforestation and people's concerns about it. Waterpower for hydraulic mining created an environmental crisis in the Central Valley, and the turn-of-the-century conservation movement revealed linkages between hydroelectric development and the preservation of natural resources. California's population was relatively small in those early days. Though the natural resource conservation movement, inspired by John Muir, had rooted deeply in California, the state's resources seemed boundless. Its citizens found ways to deal with the problems that beset them without probing too deeply into connections between their lifestyles and the natural environment.

California's great transformation after 1940, however, altered this situation. Staggering population growth filled in the state's open landscape. Cities sprawled across rich farmland, and beaches disappeared beneath crowds of suntanned bodies, highways, and buildings. Trails, roadways, and resort developments imprinted foothills, mountains, and deserts. By the 1970s, California contained few corners of untrammeled space. Quiet meadows and streams became rare treasures, and an evolution in environmental values occurred among its citizens. As the state's landscape changed, the Sierra Club moved the conservation movement beyond preserving, developing, and enjoying scenic resources. Affluent and wasteful lifestyles of the 1950s and 1960s inspired emergence of the environmental movement. As part of the growing culture of popular dissent against the seemingly ubiquitous technocratic idea of progress, environmentalists turned against the rationalist vision of a technologically dominated and despoiled environment. They soon found themselves in conflict with producers, who operated under free market ideology and paid little attention to the environmental impact of their activities.[1]

Air pollution and offshore oil development were among the first energy-related environmental issues to gain public attention. During the 1950s and 1960s, these issues provoked a merging of environmentalism and government planning, first at the local level in communities such as Santa Barbara, and then statewide through nonprofit groups such as California Tomorrow. With the 1973 Arab oil embargo and subsequent decade of rising energy costs, a whirlpool of changing conditions and opinions brought established, modernist doctrine about energy development into open opposition with environmentalism. Nuclear power stood at the vortex, with government increasingly arbitrating between colliding interests. As a result, innovative approaches to dealing with energy and related issues emerged. By the end of the 1970s, a new relationship between energy and the environment—one characterized by diversified energy production and energy conservation—began ascending in California as a paradigm for other regions, as well as for the nation.

Air Pollution Control Measures

Air pollution was a familiar phenomenon, especially where coal was burned. People tolerated it and tried to contain the worst of it through smoke abatement programs, but a series of postwar air pollution crises ended American passivity about the problem. In 1948, in Donora, Pennsylvania, nineteen people died and one-tenth of the population was affected severely when a cloud of sulfur dioxide smoke enveloped the steel-mill town for a week. In 1952, a "killer fog" in London, England, caused four thousand deaths, and a similar fog in New York City killed two hundred people a year later. The federal government responded in 1955 by enacting a National Air Pollution Control Act to fund research about the problem.[2]

Although Californians, who had little coal to burn, may not have felt threatened by killer fogs, air pollution was endemic to the Los Angeles basin. There, surrounding mountains and sunshine created an inversion layer of warm air that trapped cooler air, blowing inland from the Pacific Ocean, as well as smoke and other pollutants generated by basin residents. In 1903, the first of several local ordinances was adopted to abate chimney smoke. During the 1930s, severe seasonal pollution came from smoke spewed by the oil-burning smudge pots used to prevent frost damage in orchards. By the 1940s, the basin's air pollution had evolved into an almost constant eye-burning haze. Local health departments and newly formed air pollution control districts used police powers to go after petroleum refiners and oil-burning industries, and succeeded in lowering sulfur dioxide pollution by requiring offending plants to reduce emissions. Still, the problem of "smog," as the haze came to be called, worsened.[3]

In 1950, A.J. Haagen-Smit, a professor of biochemistry at the California Institute of Technology, discovered that Los Angeles's smog was caused largely by the sun's rays reacting photochemically with automobile emissions. This was a distressing revelation for Californians, because the automobile already had become an essential part of their lives. The state's postwar population boom was accompanied by a comparable growth in the number of automobiles. California's population grew from 9.3 to 15.7 million between 1945 and 1960, and climbed to twenty million by 1970. Motor vehicles increased in number from 3.1 to 8.5 million between 1945 and 1960, reaching 13.6 million by 1970. At the same time, the number of highway and freeway miles grew to over fourteen thousand and that of city and county roads to more than one hundred thousand. The number of highway miles driven annually by Californians soared, rising from about six billion to fifty billion per year between 1945 and 1960, with many more if local miles are counted.[4]

The sheer growth in numbers of automobiles and miles driven ensured a worsening air pollution problem, but the situation was exacerbated as a result of the way in which American automotive and oil industries had addressed the problem of engine knock during the 1920s and 1930s. Historian

August Giebelhaus recounts how petroleum companies raised fuel octane by adding tetraethyl lead to gasoline. This technological fix eliminated the "loud 'ping' and subsequent loss of power" called knocking. High octane, leaded gasoline encouraged automobile manufacturers to produce faster, heavier cars with higher compression engines that provided increasingly lower fuel economy. Octane and compression ratios climbed steadily after the 1930s, fuel economy dropped, and as long as gasoline remained cheap, consumers joined industry leaders in adopting a "bigger is better" attitude about automobiles. Unfortunately, the larger, more powerful cars that burned more gasoline per mile driven also emitted more pollutants.[5]

Through the 1950s, air pollution control districts, university researchers, meteorologists, and other experts studied the smog problem. Even the automobile industry examined it. In 1959, the California legislature passed the nation's first antismog law, directing the State Department of Public Health to establish air quality standards followed by limits on automobile emissions. Thus began a series of technological fixes. A State Motor Vehicle Pollution Control Board began testing crankcase and exhaust emission devices, and it required that approved devices or systems be installed on all new automobiles sold in the state after 1965. The pollution control technologies significantly reduced hydrocarbon and carbon monoxide emissions, but they did not eliminate nitrogen oxides. In 1963, the federal government enacted its first Clean Air Act. It was followed by two more Clean Air Acts, in 1967 and 1970, that developed air-quality standards and a program for implementing them. In order to meet new state and federal emission standards, plus deal with nitrogen oxides, the auto industry introduced the catalytic converter. Leaded gasoline, however, fouled the converters, and the oil industry responded by developing nonleaded gasolines.[6]

In 1970, Governor Ronald Reagan announced to Californians: "We have turned the corner on smog."[7] Although air quality improved over the next several years, continued growth in the number of automobiles, and the propensity of Californians to drive to every destination, prevented the most populous urban areas from meeting air-quality standards. In 1977, amendments to the federal Clean Air Act provided for economic sanctions against states not meeting federal standards, and the Environmental Protection

Agency ordered states with severe automobile pollution to develop annual vehicle inspection programs by 1 July 1979. Even though California's emission control standards were stricter than the federal ones, the idea of invading the California motorist's inviolable private sanctuary caused the state's legislators to balk at imposing vehicle inspections. They failed to meet the federal deadline, and the EPA stopped giving permits for new pollution-emitting industrial plants. In 1980, legislators again failed to adopt a vehicle inspection plan. This time, the EPA withheld some $850 million in federal highway funds. Pinching back federal largesse had the desired effect, and California adopted an acceptable inspection plan in 1983.[8]

Catalyst for National Environmental Control Policies

Linkages between the environment and energy also appeared in connection with offshore oil drilling, and, following World War II, the offshore oil issue in Santa Barbara served as a rehearsal for later, more protracted controversies over nuclear power. The first offshore drilling occurred in the 1890s, in Summerland, just below Santa Barbara, a picturesque community nestled on the coastline between the sea and towering Santa Ynez Mountains. California's oil industry was born nearby during the nineteenth century, and, coincident with expanding oil activities, Santa Barbara also developed as a resort community. For the most part, no friction existed between the area's petroleum and recreational focuses, even after enactment of federal and state leasing laws in 1920 and 1921 caused a local oil boom. In 1925, however, natural disaster began to shift the balance. A devastating earthquake destroyed most of downtown Santa Barbara, and wealthy community members seized the opportunity to enhance the city's resort focus by rebuilding in what they saw as a tasteful architectural style reflecting Santa Barbara's Hispanic past. Aesthetics ascended in importance for Santa Barbarans, and conflict emerged between people's regard for the area's natural beauty and the oil industry's activities. The first antioil protest occurred in 1929, over drilling a field within the city limits.[9]

"Historic battle lines" were drawn, pitting the "oil industry (with the tacit approval of state and federal government)" against "the awakening involvement of the Santa Barbara citizenry to protect the environmental uniqueness of their community."[10] Until 1945, this emerging conflict was tempered by depression, war, the fact that most oil drilling occurred onshore, and ambiguity about federal and state government jurisdiction over offshore oil. That year, however, citing national defense needs, President Harry Truman proclaimed federal control over all offshore oil deposits. Two years later, the Supreme Court sided with the Truman administration in the *United States v. California*. This seemed to open the way to offshore drilling, but, in 1953, the state regained jurisdiction when newly elected President Dwight D. Eisenhower disavowed federal offshore claims. The same year, Congress, supported by the oil industry, enacted a Submerged Lands Act that gave states title to submerged lands within the three-mile limit, and an Outer Continental Shelf Lands Act that granted federal control beyond the three-mile limit. In response, Santa Barbarans became increasingly concerned about future oil development, and they called for establishment of an oil-free sanctuary to buffer their community.[11]

Santa Barbara's quest for a sanctuary faced difficult odds. The cold war created an atmosphere of national need that led both federal and state governments to support the oil industry's desire for offshore development. State and federal concern focused on working out the three-mile limit as it pertained to the Santa Barbara Channel Islands and getting the lease process started. Nevertheless, Santa Barbarans presented a strong lobby in the state legislature and succeeded in getting a sanctuary to protect shoreline recreational and residential areas from the impact of oil development. The Shell-Cunningham Act, enacted in 1955, granted tidelands leasing and regulation to the State Lands Commission and incorporated a sixteen-mile strip of the tidelands as the Santa Barbara Oil Sanctuary. Two years later, Santa Barbara annexed the sanctuary, thereby effectively gaining power to prohibit drilling within its waters via the city's zoning authority. The California Supreme Court blocked a state effort to reverse the city's action.[12]

Between 1957 and 1966, all the tidelands outside the Santa Barbara Oil Sanctuary had been leased. From the shore, one clearly could see the oil platforms rising two hundred feet high in waters north and south of the sanctuary. As the nation's oil appetite grew, pressure for further offshore drilling increased. In 1965, the U.S. Supreme Court at last resolved lingering confusion over state and federal jurisdiction boundaries, ruling that

Photo 14.1. Cleanup following the 1969 Santa Barbara oil spill lasted for months. Photo by Steve Murdock. (From the Robert Easton Collection, courtesy of Special Collections, Library, University of California at Santa Barbara.)

the state's three-mile limit included a three-mile area around the Channel Islands. At the same time, the federal government took steps to assure monitoring and abatement of coastal water pollution. The door was opened to federal leasing on the outer continental shelf, and the Santa Barbara County Board of Supervisors responded by passing a resolution that requested Congress to establish an outer continental shelf oil-free preserve consistent with the north-south boundaries of the existing tidelands sanctuary. They gained a small victory with Congressional approval of a Federal Ecological Preserve, a two-mile buffer zone extending seaward beyond the sanctuary. But sale of leases and drilling did not slow. Three hundred and sixty-three thousand acres were leased in 1968, and drilling began almost immediately.[13]

By 1969, a total of 925 wells operated along California's coastline under both state and federal regulation. The massive Santa Barbara oil spill that began on 29 January 1969, came from the blow-out of Union Oil Well A-21, the fifth well sunk on the new federal leaseholds. Although the well was capped quickly, a fissure on the ocean floor eventually released 235,000 gallons of petroleum, creating an 800-mile-long oil slick that fouled California beaches from Santa Barbara to San Diego, interrupted commercial fishing, and killed over three thousand birds. The crisis mobilized virtually all Santa Barbarans in the effort to protect their environment from oil pollution, and it compelled the California State Lands Commission to impose a moratorium on drilling new wells on state-owned submerged lands. Moreover, it made clear that America's dependence on oil had extremely serious environmental consequences. It served as a catalyst for the larger state and national environmental movement, helping to stimulate passage of the 1969 National Environmental Policy Act, the 1970 California Environmental Quality Act, and the state's 1972 Coastal Zone Conservation Initiative, Proposition 20, all of which mandated prior consideration on various levels of environmental impacts resulting from human activities.[14]

Viewing California's Society Ecologically

Struggles over air pollution and coastal oil drilling were crucial elements in California's rising environmental movement. Private citizens, interest groups, and professional people—among them public health officials, doctors, local politicians, university professors, and city and county planners—coalesced around these and other issues. In 1960, Alfred E. Heller, an independently wealthy conservationist and publisher of the *Nevada County Nugget,* teamed up with Samuel E. Wood, bureaucrat, planner, and staff to the state assembly's Committee on Conservation, Planning, and Public Works. They formed California Tomorrow, a nonprofit educational organization. Really a partnership between Heller and Wood, plus a tiny board of directors comprised of influential friends and acquaintances, the organization had no members. Heller, as president, bankrolled it; Wood quit his legislative staff position to be executive director.[15]

In 1962, Heller and Wood published *California Going, Going . . . ,* a small booklet that highlighted a variety of concerns about the state's future. In his historical sketch of California Tomorrow, John Hart notes that the *San Francisco Chronicle* and other major newspapers gave the booklet front-page coverage, that Governor Edmund G. (Pat) Brown, who had shown no concern for conservation issues before, "made a point of it" in his 1962 gubernatorial campaign against Richard Nixon, and that, not coincidentally, the Planning and Conservation League, California's first and principal formal conservation lobby, was founded the year after it was published. *California Going, Going . . .* was the first truly comprehensive indictment of California's environmental decline, and it eventually saw twenty thousand copies in print.[16]

California Tomorrow became the catalyst for environmentalism in the state. Through booklets and their *Cry California* magazine, Heller and Wood made California Tomorrow a forum wherein business people, government leaders, writers, and private citizens openly discussed environmental and planning issues ranging from pollution, off-shore oil drilling, and suburban sprawl to deforestation, energy, and overpopulation. They built a bridge between environmentalists and planners, over time merging the California Center on Environment (1969) and the Planning and Conservation Foundation (1975), and affiliating with the California Environmental Intern Program (1976–1980). In the process, California Tomorrow launched the nation's first venture to focus on better planning as the essential solution to environmental degradation.[17]

Although planning was their watchword, after 1962, Heller and Wood lost track of the governor's Office of Planning work on a State Development Plan that had been mandated by the legislature in 1959. When Governor Ronald Reagan's administration finally produced a report on the plan in 1968, it did not contain a single one of the various specifics called for by the legislature. Heller was outraged, and in a *Cry California* editorial, called the report a "pile of mush . . . intended as a barricade against the future." Angry with the state and angry with himself for not keeping watch over the process, he was disconcerted particularly that many professional planners saw the report as proof that statewide planning was impossible. He, therefore, persuaded California Tomorrow to develop a plan for the state, one that would sketch out the necessary contents for a comprehensive state plan and show that it could be done.[18]

A California Tomorrow planning task force, chaired by Harvey Perloff, the dean of Architecture and Urban Planning at the University of California at Los Angeles, began work on the project by determining four underlying causes of California's plight. According to John Hart, the task force found two basic causes to be the lack of economic and political strength among individuals. Too many Californians were poor, and too few were willing to participate in what seemed to be an unresponsive political system. The task force determined that the other two causes were California's population size and distribution and its wrongheaded patterns of resource consumption: too many people, living in the wrong places, consuming resources wastefully.[19]

With this causal framework in place, the task force developed a two-part plan for California and presented it at conferences and seminars across the state. The first part laid out what Californians already had, and suggested what California would be like if things were allowed to go on unchanged. It was not a happy picture, for, by continuing the present course, the basic causes of California's predicament were never addressed. The state's urban, rural, economic, social, and environmental problems simply would get worse. The plan's second part presented an alternative future, one that might exist if Californians changed what they were doing and addressed the underlying causes of their troubles. It recommended state-wide land use planning; a state infrastructure plan; a population policy; development of regional general purpose governments and community councils in addition to city and county government; California environmental standards for clean air and water, toxics, and auditory and visual pollution; affordable housing; and accessible open space. In the same way that scientists were emphasizing the interrelationship of all things in nature, the California Tomorrow task force focused on environmental quality in the broadest possible sense, viewing California's society ecologically.[20]

The *California Tomorrow Plan* was published in 1972, and over twenty thousand copies were sold during the next ten years. It was perhaps a logical outcome of California Tomorrow's work. Since its initial firebell on the environment, California *Going, Going . . . ,* California Tomorrow had poured a decade of effort into urging Californians to take notice and to act in changing the course of their state. As historian Samuel P. Hays observes, California became a leader in environmental policy development, "originating policies in coastal-zone management, environmental-impact analysis, state parks, forest-management practices, open-space planning, energy alternatives, air-pollution control, and hazardous-waste disposal." Environmentalists across the nation borrowed California's innovations, and California Tomorrow inspired similar planning efforts in Delaware, Vermont, and other states.[21]

Energy Crisis

The 1973 energy crisis punctuated environmentalism and planning in California. Symbolically, the crisis began with the Yom Kippur War and the oil embargo placed by Arab nations against the United States and other supporters of Israel. Yet, at least a year earlier, Californians began realizing they faced an energy problem. In 1972, Lester Lees, professor of environmental engineering at California Institute of Technology, wrote in *Cry California* that the nation's 5 percent annual growth rate in energy demand over the previous decade was unsustainable. California's was 8.9 percent. In May 1973, three of the leading contenders for the next year's gubernatorial election addressed the subject. As the *California Journal* reported, Assembly Speaker Bob Moretti suggested that a threatened summertime gasoline shortage in the state was being engineered by the oil industry to achieve higher profits and lower pollution control standards. State Controller Houston Flournoy called Moretti a demagogue, and said energy companies "cannot continue to satisfy our ravenous and growing appetites for energy at prices we can live with." Finally, Lieutenant Governor Ed Reinecke postulated a scenario of drastic energy shortages, a position he garnered from Professor Lees and other experts at his own two-day conference on energy. The *Journal* concluded: "California has not yet been hit by what could be called an energy 'crisis,' but such a crisis could develop in the next decade unless the supply of energy is substantially increased or the growth rate of demand slowed, or both."[22]

In October, the Arab oil embargo enraged the public, and subsequent increases in the cost of fuel, controlled by the Organization of Petroleum Exporting Countries (OPEC), further incensed them. For the first time in history, soaring gasoline prices and real shortages threatened to slow down the suburban, freeway-commuter lifestyle of California's automobile-based society. Because oil fuel generated much of the state's electric power, utility bills also rose steeply. Furious consumers blamed the oil companies, evidenced prejudice against people of Middle Eastern ancestry, and indicted public utilities and government. On the national level, presidents Nixon, Ford, and Carter in turn struggled with the energy issue, Carter describing the situation by 1977 as "the moral equivalent of war." Two years later, it only seemed worse, as the Iranian revolution precipitated further cutbacks in oil imports and a "second" energy crisis. Lines at California gas pumps became even longer.[23]

Amidst the energy crisis, Californians discovered they faced a formidable conundrum. They all relied on and freely used energy, but many of them also were staunch environmentalists. On one hand, they shared what political theorist Langdon Winner described as the "Grand Consensus": energy and efficiency meant growth and progress. They had accepted a social contract agreeing that energy producers would "supply safe, reliable, abundant energy at a reasonable price," and the rest of society would "pay the going rate and let producers go about their business."[24] So, when the system faltered, their easiest course of action seemed to be to blame the crisis on OPEC, the oil companies, or government failures of one sort or another. On the other hand, many Californians already had begun facing the fact that the very exploitation and use of energy resources that so benefited them conflicted sharply with their desires for a healthy environment. Thus, they discovered the more difficult task: the necessity to evaluate their own beliefs and behavior. Those who served on the frontlines against air pollution and opposition to offshore oil drilling, plus participants in California Tomorrow, were among the first to confront this conflict of values. Millions more followed during the 1970s.[25]

In fact, the energy crisis came at a critical juncture in the evolution of environmental movement in California. Successes in coastal protection, pollution control, environmental policy planning, and a generally increasing desire among people for greater participatory democracy made environmentalism a political force in the state. The Sierra Club was just entering the field of environmental activism full force, and a number of new environmental groups, such as Friends of the Earth, and legal nonprofit groups, such as the Environmental Defense Fund and Natural Resources

Defense Council, added strength to the movement. Energy *per se* became a pivotal issue, one seen differently by energy companies and environmentalists. California's energy regime saw the energy plight as a problem of supply. Thus, companies sought to live up to their part of the social contract by accelerating offshore oil development, experimenting with new oil-recovery methods, seeking natural gas deregulation, and building more thermal as well as nuclear power plants. Environmentalists, however, scrutinized this and other supply-side assumptions. In doing so, they turned against, among other things, nuclear power.[26]

Nuclear Power and the Seismic Hazards

Supporters of nuclear power as a long-term solution to diminishing fossil fuels had pushed forward the technology further and faster than that of any other new potential energy resource. Although "the storage and disposal of radioactive waste" remained a principal unresolved problem, one which was cumulative in nature, California policy makers saw it as "a question of costs rather than constituting an insuperable technological barrier." The eventual development of fusion technology, they thought, would overcome the fission waste problem.[27] Faith prevailed that technology would resolve the drawbacks to nuclear power, and California and the rest of nation moved ahead to make nuclear generation part of the electric power network. By 1970, nearly one hundred nuclear plants were operating, under construction, or on order in the United States. But life on the Pacific Rim embodied a distinctive geological characteristic that helped to disrupt nuclear power in California: earthquakes.

In 1961, Pacific Gas and Electric Company announced plans to site a nuclear plant on Bodega Head, a narrow, curving peninsula that formed the outer edge of the scenic northern California fishing village of Bodega Bay. Local people immediately objected that the plant would destroy the area's scenic beauty. Protest widened with the formation of the Northern California Association to Preserve Bodega Bay, whose executive secretary, David E. Pesonen, left his position as part-time conservation editor for the Sierra Club to lead the fight. After failing to stop the project before the Public Utilities Commission, the group took its objections to the Atomic Energy Commission. The AEC did not consider aesthetic issues, only those of radiological safety. Therefore, in order to gain standing before the AEC, the Northern California Association, recounts nuclear power historian J. Samuel Walker, focused on "whether or not the proposed plant presented seismic hazards that made it unsuitable for nuclear power."[28]

Each side turned to experts to resolve the seismic hazard issue, only to discover that seismological science provided no clear answers. Geologists came into sharp conflict with each other over the potential seismic hazard at the site. While they all agreed that a small fault, one related to the well-known San Andreas Fault, passed directly beneath the site, they could not agree on the larger safety question. Some structural engineers felt construction techniques might effectively mitigate any hazard, but even they knew the real test could come only with an earthquake. Although PG&E engineers designed structural safeguards to protect the nuclear reactor against major ground displacement, opponents would not yield from the side of extreme caution. Then, in March 1964, a horrendous tremor, measuring 8.6 on the Richter scale, struck the southern Alaska coast. The devastation wrought in Anchorage and other Alaskan port communities helped sway the AEC. In October, AEC staff reported against nuclear reactors being designed with untested and untestable aseismic construction techniques. The AEC concluded: "Bodega Head is not a suitable location for the proposed plant at the present state of our knowledge."[29]

Although the Bodega Bay incident did not slow development of nuclear power on a national scale, within California the seismic safety issue posed a serious setback to the industry. It empowered nuclear opponents throughout the state and awakened many other Californians to nuclear safety issues. During the late 1960s, earthquake concerns dashed plans for three more nuclear plants: Corral Canyon, near Malibu in Los Angeles County; Point Arena, above Bodega Bay on the Mendocino coast; and Tulare, in the San Joaquin Valley. In 1973, the U.S. Geological Survey discovered a major offshore fault close to PG&E's just completed Diablo Canyon plant near San Luis Obispo. The Nuclear Regulatory Commission (NRC), which had assumed the AEC's nuclear power regulatory functions, ordered the company to make extensive seismic safety modifications, a process that long delayed the plant's opening. Three years later, in 1976, active earthquake faults were discovered near PG&E's Humboldt Bay facility, on the state's far north coast, and the NRC ordered it closed down. David Pesonen, by this time state chairperson of Californians for Nuclear Safeguards, argued that PG&E had known about the faults all along.[30]

In 1976, the seismic safety issue, along with the fundamental and unresolved problem of radioactive waste disposal, led California environmentalists to place Proposition 15, a Nuclear Power Plants Initiative, on the state ballot. The measure proposed stringent precautions against nuclear accidents and requirements for safe disposal of wastes. If adopted, it effectively would have halted all nuclear power development in California. Thus, not surprisingly, the state's energy regime opposed it, and a fierce election campaign ensued. Although voters ultimately defeated the measure two-to-one at the polls, the legislature so feared its adoption that, on the eve of the election, pronuclear and environmentalist legislators joined forces to enact three less draconian nuclear safety bills. In truth, the legislative measures had virtually the same effect as the initiative would have had, for they placed a moratorium on state approval of new nuclear plants until the federal government solved nuclear-waste-disposal and spent-fuel recycling problems. Legislators hoped a quick technological solution to the nuclear-fuel-cycle issue would be found. Of course, after Proposition 15 lost, pronuclear representatives regretted they had supported the three bills.[31]

State Energy Conservation

The Bodega Bay imbroglio went beyond energizing California's antinuclear movement. Public opposition to the project originally was rooted in objections that any power plant, thermal or nuclear, would have harmed the aesthetic beauty and the ecology of the bay. This proenvironmental position spread among residents of communities throughout California, and the electric power industry saw it as a major obstacle to their building new power plants.[32] In 1965, the California State Resources Agency established a Power Plant Siting Committee to assist utilities and satisfy public concerns. The committee released a policy statement to guide its activities:

> It is the policy of the State of California to ensure that the location and operation of thermal power plants will enhance the public benefits and protect

> against or minimize adverse effects on the public, on the ecology of the land and its wildlife, and on the ecology of state waters and their aquatic life. . . . Because sites of power plants often are critical to the interests of the state, plant owners should meet with the state's Secretary of Resources early in the planning stage to review and define proposed locations.[33]

Eleven proposed sites were reviewed and approved by the committee during the next six years, but the thrust toward state planning, inspired by California Tomorrow, encouraged the legislature to ask that the Resources Agency plan long-term for new power plants. The 1970 Power Plant Siting Coordination Act stipulated that the agency work out with utilities at least a twenty-year plan for siting sufficient power plants to meet the state's projected population growth.[34]

The same year the Power Plant Siting Coordination Act was passed, the legislature, prompted by the Santa Barbara oil spill, also enacted the California Environmental Quality Act (CEQA). It required that, for every major construction project or government proposal that might significantly affect the environment, an environmental impact report be completed and submitted for approval to the appropriate public agencies. For California's utility companies, CEQA was a problem. Already complaining that they had to deal with as many as thirty federal, state, and local agencies to gain approval for location, design, and construction of new power plants, they did not want to have to comply with the new act. Instead, noted John Zierold, the Sierra Club's state lobbyist at the time, "they wanted to create in the Resources Agency a one-stop shopping center for power plant siting."[35]

Energy now became an important concern in the state legislature. In 1970, Assembly Speaker Bob Monagan asked that the Assembly Science and Technology Advisory Council advise the legislature on issues concerning California's future electric power needs. In June 1971, in a fifteen-page report, the council urged that the state, among other things, move away from fossil-fuel electric power generation, designate a single agency to determine power plant siting, and establish a State Energy Conservation Commission to reduce energy waste. It emphasized that minimizing environmental impacts should be of paramount concern in meeting state energy needs. In order to gain more information, the Assembly Planning and Land Use Committee commissioned the Rand Corporation to study further the power-plant-siting issue, as well as ways to slow the growth rate of electricity demand.[36]

In 1972, two unsuccessful power plant siting bills were introduced in the Senate, and five Assembly bills died in the Planning and Land Use Committee while it awaited completion of the Rand reports. In October, the studies were finished, but before the committee could consider them, voters approved the Coastal Zone Conservation Initiative, Proposition 20, which created a commission to plan for long-range conservation of the state's coastal zone. Combined with passage of CEQA, this led the Resources Agency to rein in its power plant siting committee. Resources Secretary Norman Livermore, Jr. said such dramatic environmental protection advances had occurred since the 1965 power plant siting policy was established that his agency could not continue its activities without authorizing legislation. Meanwhile, more electric power need studies were completed by the Public Utilities Commission, the Resources Agency, Stanford Research Institute (SRI—on behalf of the state's five largest electric utilities), and the Assembly Science and Technology Advisory Council.[37]

The studies differed on a number of issues. The PUC warned of electric power shortages if immediate legislative action was not taken to facilitate construction of new generating plants. SRI endorsed the PUC's study and urged rapid expansion of the state's nuclear generating capacity. The Rand Corporation, however, encouraged measures to reduce the electricity demand growth rate. It called for adopting energy conservation measures and slowing down nuclear power development in favor of cultivating geothermal and solar energy, strategies that SRI impugned. The advisory council's report offered legislators guidelines for a proposed State Electric Power Authority, which would deal with facility siting, study feasible conservation techniques, and research electric power issues unique to California, such as those of seismic safety and geothermal resource development. Finally, the Resources Agency report straddled the line between energy development and environmentalism:

> Over the last decade, there has been inadequate concern for a continuing sufficient supply of our prime energy resources—oil and gas. Other forms of energy were likewise treated with indifference. Actions taken by federal, state, and local governments have, in effect, reduced our energy supplies. The off shore drilling moratorium on state lands, coastline petroleum sanc-

tuaries, emission standards that reduce engine efficiency, and other restrictive measures, such as the recently passed Coastline Initiative (Proposition 20), have a serious influence on the State's energy situation.[38]

In 1973, at the instigation of pronuclear Senator Alfred Alquist and environmentalist Assemblyman Charles Warren, the legislature tried to resolve the conflict between energy development and environmental protection. Alquist secured Senate passage of SB 283, a bill to create a one-stop power plant siting procedure. Encouraged by the Rand study, Warren sought to regulate utilities more stringently and to encourage energy conservation, and worked closely with the Planning and Conservation League. Over strong utility opposition, he steered through the Assembly AB 1575, a measure to create an Energy Resources Conservation and Development Commission. Each bill faced almost certain death in the opposite chamber, until Alquist drastically amended SB 283 to resemble Warren's measure. Douglas Foster, writing for *Cry California* three years later, described the Warren-Alquist State Energy Conservation and Development Act as "a tenuous marriage of convenience at best, . . . a complicated compromise, with all parties hoping a new high-powered agency might—given enough good leadership, expert staffing, scientific research and, perhaps, blind luck—figure some way to solve the state's energy problems."[39]

Despite indignant protests in the Senate that Alquist had prevented committee consideration of the Warren bill by making it his own, and despite objections from angry utility interests, the new measure squeaked through both houses in September. On 2 October however, Governor Ronald Reagan returned the bill unsigned, persuaded to do so by legislators who were upset over the lack of Senate hearings and by heavy lobbying from opponents of the measure—the Resources Agency and the PUC, General Electric (a major nuclear power company), and business and manufacturing interests, all of whom sided with investor-owned utilities. Alquist took the high moral ground and accused the governor of "fiddling while the state's energy crisis burns." Reagan countered the senator's tongue lashing by appointing his own Energy Planning Council and by encouraging the legislature to give him a new bill after a full-hearing process. The Arab oil embargo, triggered during the third week of October by President Nixon's authorization of a $2.2 billion weapons airlift to support Israel in the Yom Kippur War, virtually ensured a revised version of the Warren-Alquist bill would be forthcoming. When it again reached the governor's desk in May 1974, Reagan signed it.[40]

Stalemate, Three Mile Island, and Proponents of Nuclear Power

Born amid crisis, the California Energy Commission (CEC) derived from the sort of thinking California Tomorrow inspired about environmental and resource planning. Charles Warren believed California's energy problem stemmed from a poverty of rational resource management. He wanted the CEC to take an ecological approach to planning, and he made this the agency's primary charge. It should examine the implications of its energy forecasts, he said, "in terms of, among other things, critical environmental and other resources of the state, public health and safety, capital requirements for new facilities, and costs to consumers." The CEC's other powers logically flowed from this: energy resource conservation, power plant siting and certification, research and development of "alternative" energy technologies, emergency electric load curtailment, and self-financing through a tax on electricity sold to California consumers. All in all, Warren hoped the CEC could slow the state's annual electricity demand growth rate, which hovered around 8 percent, and expand research and development on alternative energy resources.[41]

Utilities did not see the CEC's charge in the same way. Rather, they saw it as the one-stop power plant siting agency they had wanted all along. They expected that it would continue the practice of determining the state's future power plant needs from their own, PUC-certified, long-range forecasts. They also assumed it would accept their current forecast. Based on the assumption that the 8 percent annual growth rate in electricity demand would continue into the 1990s, this forecast buttressed their grow-and-build plans to expand generating capacity from the existing 35,000 megawatts to about 105,000 megawatts by 1993, over half the new capacity to be nuclear. Finally, they expected that the CEC would deal with environmental issues expeditiously. It would resolve all power plant siting issues in a single proceeding, and its judgment would be "conclusive on all environmentally related questions involved in siting, including land use, air and water quality, and aesthetics."[42]

From the moment the State Energy Resources and Development Act took effect in January 1975, the CEC found itself embattled and embroiled in the head-on collision between energy development and the environment. As California Tomorrow writer Douglas Foster observed, the first commissioners appointed by newly elected Governor Jerry Brown turned out to be "primarily development oriented" and "by and large pro-nuclear," thereby upsetting the environmentalists.[43] During its first year, the CEC missed a couple of legislative deadlines for issuance of regulations concerning energy waste reduction, which garnered early criticism from the legislature. Then, it entangled itself in controversy by choosing to meet a legislative deadline for its first energy forecast by adopting, without revision, the PUC's predictions. The fracas worsened in early 1976, when it dismissed the head of its forecasting and assessments division. He claimed he was fired after revising the PUC energy demand projections so that the forecast did not support the need for nuclear power plant construction. Assemblyman Warren charged that the PUC-based report that the CEC submitted to the legislature was "giving the utilities a license to steal." "Damn it," he said, "if the commission is always going to side with the utilities, the way the Public Utilities Commission does, I might as well repeal my goddamned act that created the thing."[44]

The CEC's position on nuclear power served as a lightning rod for the public conflict between energy development and environmentalism. During the spring 1976 ballot campaign, three commissioners publicly opposed the Nuclear Power Plants Initiative. The nuclear safety laws passed by the legislature just before the election to head off the impact of Proposition 15 specifically exempted seven nuclear plants: Humboldt, near Eureka, which the NRC ordered shut down before year's end; San Onofre's one operating and two unfinished units, halfway between Los Angeles and San Diego; Rancho Seco, which just had been completed near Sacramento; and two units under construction at Diablo Canyon, near San Luis Obispo. After Proposition 15 failed, nuclear supporters anticipated that the CEC would exempt from the legislative measures—which did not go into effect until January 1977—the two-unit Sundesert plant planned for construction near Blythe and the Colorado River in Riverside County.[45]

There seemed to be reason to think the CEC would exempt the Sundesert plant. The San Diego Gas and Electric Company already had invested over five years and $100 million in planning the project and had filed official notice of intent to construct in June. It believed it had no feasible alternatives to meet the forecasted growing demand for electricity in its service area. Except for Ronald Doctor, the only true environmentalist on the CEC and, ironically, a nuclear engineer, the commissioners appeared likely to support an exemption. After the election, however, Governor Brown began answering complaints about the commission's generally proenergy development point of view. He appointed pronuclear commissioner Richard Tuttle to a judgeship, replacing him with Emilio "Gene" Varanini, who had helped draft the CEC legislation as a former aide to Assemblyman Warren. A few months later, commissioner Bob Moretti quit, charging the commission had become narrowly environmentalist. "If the lights ever go out," he warned, "the price California will pay will be enormous, economically and socially." To take Moretti's place, Brown appointed Suzanne Reed, his own energy advisor from the Governor's Office of Planning and Research.[46]

Through 1977, Sundesert continued to dominate issues before the CEC. The state's energy regime appeared confident that the plant remained a "pipeline project" for approval, and that, despite growing criticism, nuclear power soon would be made acceptable in California. Utilities personified a remarkably successful technological system that had gained enormous momentum. They showed no sign of altering their basic assumptions or modifying their past behavior. For example, at PUC hearings between 1975 and 1979, PG&E based its rate requests on forecasts that, as usual, assumed regular, massive additions to generating capacity. The entire hearing process seemed designed to ignore persistent testimony by the Environmental Defense Fund that stood PG&E's forecasts and assumptions on their heads. Indeed, in the midst of it and even though the Sundesert issue remained unresolved, in August 1977, PG&E filed its own notice of intent to construct a 2400-megawatt nuclear power facility in the Central Valley, near Waterford in Stanislaus County.[47]

But environmentalists kept pressure on the CEC and individual commissioners. For example, former antiwar activist Tom Hayden, who headed the Campaign for Economic Democracy and who had Governor Brown's ear, took note of pronuclear Commissioner Alan Pasternak's activities. After informants advised Hayden that Pasternak had represented a very pronuclear view at a Tennessee "Workshop to Outline Acceptable Future Energy Systems," Hayden reminded him he was being watched: "It is the duty of an Energy Commissioner, whatever his or her personal preferences, to give the utmost scrutiny to whether nuclear energy can be made safe at all. The nuclear safety legislation passed in 1976 makes this quite clear." Hayden then added, "Instead of fulfilling your duties, you seem to be advocating a strategy of enhancing public acceptability of nuclear power by going through various motions."[48]

Finally, in January 1978, the CEC made its ruling. Behind-the-scenes staff work on household energy conservation projects, investigations into alternative energy resources, and energy education strategies had begun to inform forecasting methodologies that seriously questioned electric power demand projections. It seemed clear that San Diego Gas and Electric could achieve the same ends for which it proposed building the Sundesert nuclear plant more efficiently and cheaply by reducing energy waste through conservation, power pooling, the upgrading of existing conventional fossil-fuel plants, and/or by the use of renewable resources, such as solar, wind power, cogeneration, and geothermal. Practical alternatives existed. The project was denied.[49]

The pronuclear camp was infuriated. Republican Attorney General Evelle Younger removed his deputies from representing the CEC on nuclear issues. PG&E filed suit, claiming the state's nuclear safety laws obstructed achievement of federal policies, and a San Diego coalition of labor, business, and pronuclear groups sued to recover economic and environmental damages allegedly suffered as a result of the moratorium. Bills were introduced to override the CEC decision and to abolish the commission, threats were made against its budget, and Senator Alquist proposed replacing it with a Department of Energy with a single director. "As co-author of the legislation that created the state Energy Commission, sometimes I feel like a Dr. Frankenstein who has created a monster," he grumbled. "In three years of existence, this agency, this bureaucracy, has done nothing. They have yet to authorize the construction of a single power plant, either nuclear or conventional."[50]

Although it seemed plain that the CEC would uphold the nuclear power moratorium, environmentalists continued to battle against nuclear power within the state. On 29 March 1979, the Three Mile Island accident in Pennsylvania galvanized California's nuclear opponents. New safety bills were introduced into the legislature, and the state's few nuclear plants were scrutinized more carefully than ever. Snags at the Sacramento Municipal Utility District's (SMUD) Rancho Seco plant, which was of the same design as the Three Mile Island facility, suddenly became vital public news. The plant had gone on line in October 1974 and was problem-plagued from the start. It was shut down for thirteen of its first eighteen months in operation, and in 1979, PG&E sued SMUD for not fulfilling its power delivery contracts. Opposition to the plant from environmentalists became a part of daily life. Finally, a decade later, SMUD voters approved a nonbinding referendum to shut the plant down, and the utility complied. It had one of the worst lifetime reliability records in the industry, operating at an average of only 39 percent of maximum output, compared to a generously estimated industry average of 63 percent.[51]

The other major nuclear target was PG&E's Diablo Canyon facility. Though it was initially endorsed by the Sierra Club, which, in 1964, agreed that Diablo Canyon was "a satisfactory alternative site" to destroying the nearby Nipomo Dunes, local groups, such as the Anti-Diablo Mothers for Peace, soon formed to oppose it. By the early 1970s, Sierra Club dissenters had splintered off to form the more militant Friends of the Earth, which roundly opposed nuclear power, and opposition to Diablo began building. In June 1977, following the defeat of Proposition 15, the Abalone Alliance, a statewide network of antinuclear and environmental groups, organized to protest the project. Borrowing adversarial techniques from similar protests staged in Germany and by the Clamshell Alliance at Seabrook, New Hampshire, in April 1977, the alliance tried to occupy the site in August. A second attempt, in 1978, brought six thousand people. Then, after Three Mile Island, opposition became increasingly intense. In June 1979, forty thousand demonstrators rallied at nearby Avila Beach to stop the plant, and a fourteen-day blockade in 1981 involved perhaps twelve thousand protesters and led to the arrest of over nineteen hundred. The incident prompted Langdon Winner to conclude that proponents of nuclear power seemed bent on demonstrating "that with enough police, national guard troops, and attack dogs, it will be possible to install nuclear power plants."[52]

New Energy Paradigm

As the 1970s closed, rapidly rising oil prices caused a resurgence of offshore oil drilling conflicts, and the curtailment of the nuclear option led the electric power industry to pursue coal-fired power plants. PG&E planned to build Fossil I and II, a 1600-megawatt coal-burning plant in northern California that would receive its daily ration of more than seven thousand tons of coal by rail from Utah. With Southern California Edison, the Nevada Power Company, and the City of St. George, Utah, PG&E joined in an even grander scheme to build a 2,000-megawatt coal-fired plant just north of Las Vegas and another 500-megawatt plant in the Warner Valley near St. George. Two slurry pipelines would deliver crushed coal mixed with water from strip mines in Utah's Alton Coal Field, between Bryce Canyon and Zion National Park, to the Harry Allen-Warner Valley Energy System.[53]

But California's energy future no longer rested only in the hands of its energy companies. Although the CEC worked with the old energy regime's supply-side approach to energy issues and plotted the nuclear and coal generating plant scenario for the year 2000, it did not believe the plants would be necessary. The protracted struggles over offshore oil drilling, the linking of environmentalism and planning, nuclear power, and the energy crisis itself had thrust forward the environmental movement as a major force in California political life. Energy industries, because of their modernist, grow-and-build approach to expanding demand, and because of the inherent environmental impact of their activities, bore the brunt of environmental activism. The convergence of energy and environmentalism wrested nuclear power from California's energy budget, immersed political leaders in energy planning, and initiated steps toward a policy of energy conservation. The tiny West Coast office of the Environmental Defense Fund made it possible to imagine an array of alternative energy futures by computerizing PG&E's own planning models. It then shared the computer program with the CEC and PUC. By weighing energy conservation and a

Map 14.1. California Nuclear Scenario for the Year 2000. A California Energy Commission plan for 23,500 megawatts of new nuclear capacity based on utility projections of 3.5 percent annual growth in electric power demand. Adapted from California Energy Commission, *1979 Biennial Report* (Sacramento 1979), 25; and David Hornbeck *et al.*, *California Patterns: A Geographical and Historical Atlas* (Palo Alto, CA: Mayfield Publishing, 1983), 115.

Map 14.2. California Coal Scenario for the Year 2000. A California Energy Commission plan for 29,300 megawatts of new coal-fired plants based on utility projections of 3.5 percent annual growth in electric power demand. Adapted from California Energy Commission, *1979 Biennial Report* (Sacramento 1979), 23; and David Hornbeck et al., *s* (Palo Alto, CA: Mayfield Publishing, 1983), 115.

multitude of other variables, environmentalists successfully challenged the basic assumptions on which the utility's coal-fueled generating plant construction plans were based. In 1980, Fossil I and II and the Allen-Warner Valley coal plants were abandoned.[54]

A new energy paradigm began to emerge in California. Energy industries found themselves confronted and embarrassed by a dedicated and organized activist population. With the prodding of environmental organizations, the regulatory persuasion of the PUC and CEC, and legislative insistence from the state and the federal governments, California energy providers began to respond to the interrelationship between energy and the environment in new ways. The integration of energy conservation techniques and alternative energy resources into the existing energy systems began to alter California's energy landscape in earnest during the 1980s, revealing radical new possibilities for meeting energy needs.

CHAPTER XV The Soft Energy Path

IN 1971, David Brower, longtime Sierra Club director who had resigned from the organization to form Friends of the Earth, went to northern Wales to participate in saving Snowdonia National Park from the encroachments of copper mining. There he met Amory Lovins, a young physicist and environmentalist who had left Oxford University to fight the mining venture. Brower persuaded Lovins to work with Friends of the Earth, where he turned his attention to energy issues. He quickly determined that prevailing energy systems were enormously inefficient. When one looked at the consumption rather than the supply end, the waste of energy was stunning: only 30 percent of the energy in coal used to generate electricity reached consumers; only 15 percent of the energy in gasoline actually propelled the average car. Lovins was sure that energy users cared more about the services energy provided than how much actually was used, and he thought the idea that energy and the economy had to grow in tandem was simply wrong. He decided that conservation techniques, such as electrical cogeneration and building insulation, combined with harnessing renewable energy resources, such as solar and wind power, offered an alternate "soft energy path" to the orthodox "hard path" followed by energy industries.[1]

Lovins's soft path found ready champions among devotees of appropriate technology, technologies possessing qualities of simplicity, cheapness, environmental sensitivity, and accommodation to small-scale application. First popularized by E.F. Schumacher's *Small is Beautiful* and by *The Last Whole Earth Catalog,* the idea of appropriate technologies for energy production offered an increasingly attractive alternative to traditional energy practice. Organizations emerged, such as the New Alchemy Institute, located in Massachusetts, and the Farrallones and Portola institutes, both located in the San Francisco Bay Area. These groups melded environmentalism with energy concerns, gathering and disseminating widely scattered materials on solar energy, wind power, waterpower, biofuels, and conservation techniques, as well as designing buildings and carrying out research projects. They advocated that people decouple from centralized energy systems and called for development of appropriate technologies as part of a larger effort to sustain the philosophy of a decentralized society.[2]

Soon after his 1974 election, California Governor Jerry Brown, an enthusiast of Schumacher, asked architect and president of the Farrallones Institute Sim Van der Ryn to suggest ways state government might implement Schumacher's small-is-beautiful philosophy. Van der Ryn suggested establishing an Office of Appropriate Technology (OAT). Shortly thereafter, Brown appointed him state architect, and then officially sanctioned the appropriate technology movement by organizing OAT. Van der Ryn opened OAT's doors in May 1976, with a budget of $25,000. Over the next two years, he helped make OAT and California, in writer Hal Rubin's words, "an acknowledged leader in renewable energy investigations." OAT undertook a variety of projects: reusing waste heat from power generation in the state capitol area, training solar technicians among low-income families and minorities and women in the construction trades, researching alternatives to septic tanks and central-sewer systems, developing "Local Energy Initiatives" with local governments and community groups, and even putting a renewable energy trailer on the road to demonstrate such things as solar water heating, cooking, and wind power. Inspired, some local governments, such as Santa Clara County, even formed their own offices of appropriate technology. After three years, Van der Ryn returned to the Farrallones Institute, but OAT thrived. By 1981, it had a staff of almost forty and a budget of nearly $3.2 million, and stood as an symbol of California's emerging new energy paradigm.[3]

Facing Changing Energy Realities

Not surprisingly, advocates of the existing energy regime belittled Lovins, the soft energy path, appropriate technology, and agencies like OAT. People and societies had come to think of the historical pattern of rising energy use and growing waste as normal. Like addicts, they felt their prosperity and health depended on continuing their energy binge, so they waved aside environmental, social, and political risks. They persisted in responding with traditional solutions to OPEC's control over international petroleum pricing and the need to meet apparent growth in consumer energy demand: mine more coal, pump more oil, find more gas, erect more power plants. Alternatives seemed impractical or fanciful or both. By turning down White House thermostats and declaring a moral equivalent of war on energy while sitting in front of a fireplace wearing a sweater, President Jimmy Carter defined energy conservation as "freezing in the dark." And virtually every energy expert pooh-poohed the idea of developing and using solar, wind, biomass (plant and animal material and waste), and other renewable energy resources, General Electric Vice President Bertram Wolfe calling Lovins's vision an "energy Shangri-la."[4]

Frustrated with the rising tide of environmental opposition over their efforts to bring new energy facilities on line or tap new resources, energy experts reeled at having amateurs attack their basic competency. They saw themselves as having toiled successfully for a century, both in the private sector and within cooperative government agencies, to supply energy for an always growing consumer demand. Managers and engineers in companies such as Pacific Gas and Electric and Southern California Edison perceived themselves as daring, resourceful entrepreneurs struggling to harness nature's white coal in order to serve a grateful public. Oil company workers saw themselves as hardy roughnecks venturing into wastelands to draw forth nature's bounty for the benefit of humankind. Now their efforts were belittled and their expertise and abilities directly challenged. They felt humiliated. Bewildered, many of them responded defensively.[5]

In 1981, for example, Chevron Corporation tried to still critics by holding special weekend retreats around the state to which they invited college professors to mingle with and hear their experts in hopes that the professors would return to their campuses and talk sense to students. At other convocations and in government or institutionally sponsored committees, expert participants ignored any alternatives to the traditional, hard energy path. For them, there was simply no point in talking about anything except nuclear power, coal, petroleum, or synfuels. Anthropologist Laura Nader was struck by "the number of taboos" held by participants at gatherings in which she participated and by colleagues on the National Academy of Sciences Committee on Nuclear and Alternative Energy Systems. She later wrote: "Solar was never mentioned by anybody other than myself, literally not mentioned. . . . The social and political consequences of nuclear power were not discussed. Nobody used the word safety. These were all taboo areas. The fact that we were making decisions that closed off options to the next generations was considered irrelevant."[6]

Nevertheless, government and energy managers, particularly in electric utilities, found they could neither ignore the environmentalist-soft path coalition nor avoid facing changing energy realities. Not only were citizens groups stymieing efforts to promote nuclear power, but the rising price of household energy, which started during the late 1960s, actually had slowed the electricity consumption growth rate. At the same time, thermal efficiency limits halted the electric power industry's development of larger steam turbine plants, a process which historically had offset rising fuel costs. These factors combined to convince utilities, especially PG&E, to reevaluate and scale back long-term plans to meet new load demand, to pay attention to alternative strategies suggested by adversarial groups, such as the Environmental Defense Fund, and to shift from their historical grow-and-build strategy and supply-side management to slower growth and demand-side management. Following this trend, the industry's national research arm, the Electric Power Research Institute (EPRI) in Palo Alto, founded in 1972 to head off a congressional effort to begin a federally controlled research and development program, started sponsoring work during the late 1970s on various unconventional power generation technologies: fuel cells, integrated gasification combined-cycle plants, and cogeneration, the latter exploiting the steam used in industrial processes, such as food canning, to turn electrical generators. In doing so, EPRI abandoned the emphasis it first had given to nuclear power breeder-reactor research.[7]

As utilities stepped away from their decades-long grow-and-build ideology, they also began cooperating with the California Energy Commission. Both the industry and the CEC looked initially at alternative fuel resources, such as coal and oil shale. However, the CEC soon concluded that coal, with all its environmental hazards, offered only a short-term solution to the state's energy situation, and oil shale and other synthetic fuels could be, at best, long-term solutions to diminishing world petroleum reserves. Because energy consumption growth rates would continue more slowly in the future and because conventional energy resources would be unable to eliminate dependency on imported oil without greatly hurting the environment, the CEC concluded that energy conservation and alternative energy resource and technological development offered the only reasonable strategies. Thus, the CEC helped steer California's energy regime toward the soft energy path.[8]

The CEC joined with OAT in helping local communities initiate energy conservation programs. It formulated the highest state standards in the nation for energy efficiency in newly constructed buildings, electric appliances, and commercial and industrial equipment. It worked with the legislature to get funds for innovative energy conservation projects and tax credits to encourage homeowners and businesses to invest in insulation, weather stripping, and other energy conservation measures. The CEC and Public Utilities Commission worked with utilities and local governments to develop demand-side load management strategies, energy-saving rate structures, energy audit programs, and retrofit of outdoor lighting. Meanwhile, utilities began joining with local energy advocacy groups at community symposiums and fairs, promoting energy conservation with displays and information. PG&E, soon a national conservation leader in the industry, initiated low-interest conservation loans to customers and a zero-interest financing program, ZIP, for those who installed packaged residential energy conservation measures.[9]

The federal government did not seem, at first, to promote such changes in the energy industry. The CEC found itself at odds with persistent federal advocacy for coal and nuclear power. Although the National Energy Conservation Act of 1978 and federal tax credits supported the CEC's conservation strategy, national support of alternative energy development remained weak. President Carter's National Energy Plan focused largely on stabilizing oil imports. It called for a transition to alternative fuels only within the hard energy path. Thus, the Power Plant and Industrial Fuel Act restricted industrial and utility boiler use of oil and gas in new facilities, requiring a switch to alternative fuels, primarily coal; the Natural Gas Policy Act deregulated gas to stimulate domestic discovery and production; and a windfall profits tax on the oil industry was aimed at providing funds for massive subsidies in a synthetic fuels program. Carter's Crude Oil Equalization Tax, supported by the CEC because it would have raised oil prices, failed in Congress. Convinced a largely conventional energy future was infeasible, the CEC opposed any centralized federal energy program that might override regional initiatives. It concluded that California's "mixed resource strategy is inherently more resilient and robust than the national strategy."[10]

Ironically, the one measure of Carter's plan that proved crucial to stimulating the alternative energy strategy supported by the CEC and that forced utilities to change policies dramatically—the 1978 Public Utility Regulatory Policies Act (PURPA)—was ignored in the CEC's 1979 discussion of the National Energy Plan. Among PURPA's most important provisions was the requirement that regulated public utilities purchase the electricity produced by small power producers and cogenerators at a price equivalent to the utilities' production costs. It also exempted small producers and cogenerators from most federal and state regulations. Although some utilities outside California took legal action against the law, in 1983 the U.S. Supreme Court upheld PURPA and the implementing regulations promulgated by the Federal Energy Regulatory Commission. California legislation, between 1979 and 1981, added to PURPA's impact by establishing a revolving fund for demonstration projects to convert biomass into usable energy, authorizing bond issues to assist generation projects that reduce fossil and nuclear fuel use, and providing for accelerated depreciation, tax credits, and other encouragement for the undertaking of various alternative energy projects.[11]

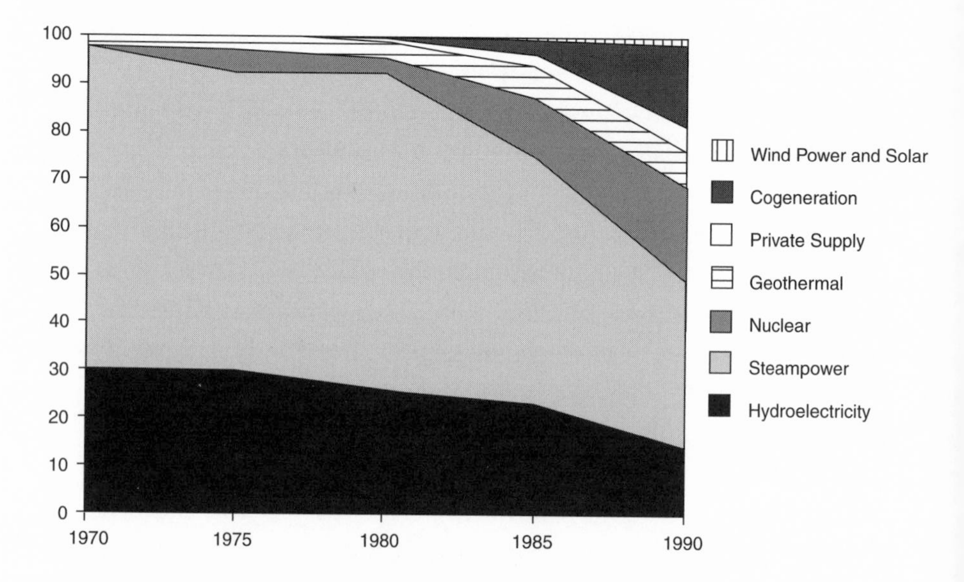

Figure 15.1. California Electric Power Generation by Percentage of Resource, 1970–1990. Source: *Supra*, Table C-22.

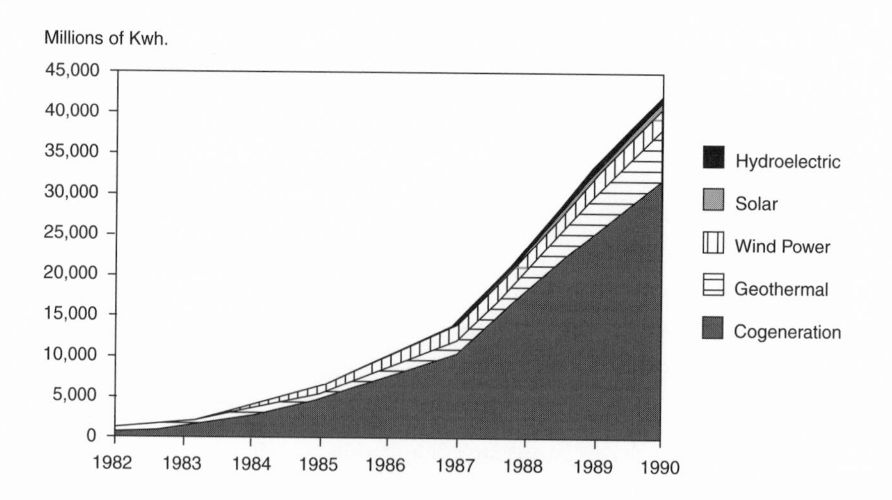

Figure 15.2. California Electric Power Production Under PURPA, 1982–1990. Source: *Supra*, Table C-23.

In 1979, the PUC began implementing PURPA by authorizing PG&E to purchase electricity from independent producers at the utility's "full avoided costs," essentially the cost of conventional fuels as forecasted by the state. Almost overnight, market forces engendered by PURPA and other government economic enticements lured independent hydroelectric, wind power, and solar entrepreneurs into the energy field. They were joined by industrial cogenerators, biomass fuel-generating companies, and geothermal firms. In 1985, California's nonutility electrical-generating capacity exceeded that of any other state in the nation, and the PUC decided that the state's overall generating capacity was sufficient to carry it into the twenty-first century and stopped accepting new applications under the program. By 1990, as the bulk of approved PURPA projects came on-line, their combined production fed more than forty-one billion kWh of electricity into California's electric power grid (figures 15.1 and 15.2). Although the renewable energy resources of waterpower, wind power, and solar energy provided less than 10 percent of this amount, they attracted the most attention.[12]

The Resurgence of Small Power Projects

Hydro

Dozens of small pre-1940 hydroelectric plants still existed in the Sierra Nevada during the 1970s, constructed by individuals, lumber companies, mines, and resort owners. Brothers Dan and Russell Worman, for example, homesteaded one hundred acres in the foothills east of Fresno, around 1918, and opened a small sawmill. To operate it, they installed a tiny hydroelectric plant driven by a three-foot Pelton wheel salvaged from an abandoned mining site. In 1981, the Worman mill still operated with its original equipment. On Pyramid Creek, near the summit of Highway 50 to Lake Tahoe, another little Pelton wheel plant, probably constructed around 1930, provided power to the Twin Bridges Resort until the mid-1970s and remained extant in 1981. Further north, near the Plumas County town of Graeagle, the California Fruit Exchange moved a 1919 hydro plant from a closed-down gold mine and installed it, in 1935, to power its box manufacturing mill. From a concrete diversion dam on Gray Eagle Creek, water flowed through a 2,150-foot-long wood flume into 1,220-foot-long wood stave pipe, then through an 1,800-foot-long steel penstock that delivered it to a Pelton wheel driving an Allis-Chalmers generator. In 1941, the Graeagle Lumber Company took it over, then the Placerville Lumber Company, in 1958. Both provided power for themselves as well as for nearby Graeagle. The plant was shut down after utility lines reached the town in 1964, but it remained in place until 1982.[13]

Within weeks after PURPA was enacted, entrepreneurs in California, New England, and other parts of the country with waterpower sites began filing for small hydroelectric projects. By 1982, over eighteen hundred applications had been submitted, one fourth of them in California; the region's projects represented 5,000 megawatts, twice the Diablo Canyon nuclear plant's capacity. The first Californians to do so were Barbara and George Mallett, who had in 1973 bought a resort in Weaverville, located in the Trinity Alps between Redding and Eureka. Years before, other owners had installed a small 60-kW hydro plant to provide power for the resort, and, in 1979, the Malletts sought their license under PURPA to bring the plant up to a capacity of 600 kW. They worked their way through a variety of bureaucracies: state and federal fish and game departments; the state's Water Quality Control Board, Office of Historic Preservation, and the Public Utilities Commission; and some thirteen federal agencies finally leading to the Federal Energy Regulatory Commission (FERC). With approvals in hand, they then negotiated PG&E's first PURPA contract in 1980. A month after the contract was signed, the Mom & Pop Power Company went on line. The Malletts expected to recoup their $190,000 investment in three years.[14]

While Barbara Mallett shared her family's experience by organizing the Small Power Producers Association of California, a support group to assist other minihydro developers, other entrepreneurs launched a free-for-all competition for waterpower sites. Some, like Henwood Associates, Inc., were family affairs like the Malletts. The Henwoods, led by engineer sons Mark and Ken, acquired the old Graeagle system, in 1980, with their father Carroll, also an engineer. They refurbished the Pelton wheel, generator, and 1919 Woodward Governor and went on line under PURPA on 28 December 1981, three days before their temporary construction financing ran out. Others, like Joseph M. Keating, who filed under PURPA on the historic Pyramid Creek hydro site, and Terence O'Rourke, CEO of Consolidated Hydroelectric, Inc., a subsidiary of Consolidated Petroleum Industries, Houston, Texas, filed on many sites. O'Rourke, for example, hired Ott Water Engineers in Redding to search for unclaimed sites statewide. Identifying five hundred, they narrowed them down to one hundred, and had filed preliminary permits on seventy-four by the end of 1981. A year later, Consolidated's first plant opened on Bailey Creek. O'Rourke contracted with Dongfang Electric Machinery Works of Deyang in China's Sichuan Province for the 670-kW low-head Francis-type turbine and generating equipment, and invested $2.5 million. By 1984, some thirty small hydro projects had been approved by various state bureaucracies.[15]

Although the rush to develop minihydro plants addressed energy needs by harnessing a renewable resource, it also raised environmental issues. As journalist Matt Herron noted: "Organizations that initially had welcomed small hydro as a benign counterforce to oil tankers and nuclear reactors now regrouped to fight what they perceived as a menace to the natural environment. Dick Roos-Collins, the small-hydro specialist for Friends of the River, termed it 'hydromania—the worst nightmare in 30 years for river people. . . .'"[16] Roos-Collins and other environmentalists claimed filings were especially excessive on some rivers. The North Fork of the Feather River accumulated forty separate claims, and on the upper San Joaquin River, where reservoirs already immersed fifty-one stream miles and water diversions partially dewatered another ninety-eight miles, "new proposals would inundate 16 miles more and dewater another 42."[17] But Barbara Mallett argued that small-hydro developers should not be blamed for the sins of earlier, disastrous, big hydroelectric projects, and Terence O'Rourke pointed with pride to the $400,000 in voluntary environmental mitigation work that went into Consolidated's Bailey Creek system. When the two sides finally met in Sacramento, in 1982, they discovered that they agreed on most issues, yet continued small-hydro development through the decade resulted in continued conflict with river conservationists.[18]

PURPA-inspired development of small hydroelectric systems also came at a time of resurging interest in public ownership of hydroelectric resources. Beginning in 1970, hydroelectric projects, which had received fifty-year licenses under the federal Water Power Act of 1920, began coming up for renewal. Under the law's provisions, municipalities had preferential rights on such sites, and communities desiring or having publicly owned electric systems had the opportunity to acquire already established hydroelectric plants. As licenses came up for renewal, municipalities filed their claims and utilities went to court. The City of Santa Clara, for example, filed in 1974 to take over the four plants comprising PG&E's Mokelumne River system, when the company's license expired in 1975. In the ensuing battle, FERC upheld the 1920 municipal preference provisions in 1980, then reversed itself in 1983 by ruling it only applied to new sites. In 1985, the Washington, D.C. Federal Appeals Court overturned the second FERC decision, thereby reaffirming municipal preference. The next year, however, saw Congress succumb to pressure from PG&E, other utilities, and the Edison Electric Institute by passing new legislation which gave preference to existing licensees. PG&E finally received its renewed license in 1987.[19]

While PURPA did not cause these battles, *per se*, passage of Proposition 13, in 1978, which rolled back and restructured property taxes in California, and Proposition 4, in 1979, which limited state government spending, combined with PURPA to give municipalities reason to seek their own hydroelectric resources. As Morgan Hill Councilman Neil Heiman said, in 1982, when his city approved a feasibility study to install a hydro system at nearby Anderson Dam: "With the passage of Propositions 13 and 4, we are living on fixed income. We have to find a way to pay our utility bill." As a result, a large number of municipalities, irrigation districts, and water districts took under study or filed on small waterpower sites. Minihydro entrepreneurs suddenly discovered competition for sites on which they filed. O'Rourke's company, for example, found that various municipalities duplicated some of his firm's filings. Consolidated Hydroelectric immediately launched a legal war against the municipalities' claims, winning most cases and losing only some fourteen sites. In the end, most new public power proposals failed to move beyond feasibility studies, and the renewed interest in public ownership faded.[20]

Wind

Wind power, like minihydro projects, gained popularity under PURPA and also drew on a history of technological developments. The idea of attaching small DC generators to windmills dated at least to the 1890s, and the Army Signal Corps experimenting with wind power. The Windcharger and Jacob companies' wind turbines were popular throughout Great Plains and Rocky Mountain states during the 1920s and 1930s, but virtually none were sold in California. In the 1940s, Palmer Coslett Putnam built a gigantic 1,250-kW two-blade wind turbine at Grandpa's Knob, Vermont, which lost a blade after eleven hundred hours of operation and was never repaired. Putnam's work inspired Federal Power Commission engineer Percy H. Thomas to study large-scale turbines for several years, but his work attracted little attention during the postwar era of abundant, cheap energy. In 1954, University of California geography professor James J. Parsons concluded that windmills for water pumping were fine but that wind power would otherwise "be expected to make but a very small contribution to Western energy requirements."[21]

Americans had little interest in wind power, but in Europe, where DC wind turbines operated successfully before World War II, interest was greater. In Germany, Ulrich Hutter worked with Ventimotor, a government-owned company founded in 1939, and continued on his own after the war to build a couple of small two-blade AC turbines plus a 100-kW turbine, the latter operating through 1968. Meanwhile, a largely self-educated Danish utility engineer born in 1887, Johannes Juul, began to experiment with wind rotors in Denmark in 1948. After experience with two- and four-blade designs, Juul determined that "it would be best to build a windmill with three blades." In 1956, he constructed a 200-kW, three-blade turbine that operated successfully until 1967.[22]

In both Europe and the United States, wind power gained renewed attention after the first energy crisis jolt in 1973. A federal wind power program soon developed. Centered at the National Aeronautics and Space Administration's Lewis Research Center in Cleveland, the government contracted primarily with aerospace firms for research and development. After purchasing designs for Ulrich Hutter's large turbine in 1974, NASA contractors scaled up Hutter's design in its MOD turbine series: MOD-0 at 100-kW, MOD-1 at 2,000-kW, and MOD-2 at 2,500-kW. In 1982, PG&E installed an experimental Boeing MOD-2 atop the hills along Interstate Highway 680, north of the Carquinez Strait, and the company reported that it generated 2.1 million kWh during the next two years. The turbine rotor diameter was about three hundred feet, and each blade weighed ninety-nine tons. The machine soon proved to be unreliable: it failed every 20.6 hours on average. In just over a decade, aerospace companies were out of the business, the NASA program ended, and PG&E's MOD-2 was felled with dynamite in 1988. During the same period, the West German government funded a similar large-scale research program, headed by Ulrich Hutter, which also resulted in ruin.[23]

As large-scale wind power work met with failure, successful wind power developments came from other sources. Guaranteed sale of power under PURPA prompted a number of entrepreneurs in and out of California to undertake windmill designs. The Canadian firm DAF Indal, Ltd., entered the state, in 1980, with a 50-kW egg-beater turbine, which the California Department of Water Resources placed in a test program at the San Luis Reservoir. FloWind Corporation of Pleasanton acquired an egg-beater turbine design from the U.S. Department of Energy, and other companies began producing two- and three-blade turbines: Unique Windpower (San Francisco), Fayette Manufacturing (Stockton), GE3 (San José), U.S. Windpower (Livermore), and Aero Turbine (Denver). Ready backers quickly appeared among wealthy individuals who leaped to cash in on state and federal alternative energy tax credits and accelerated depreciation benefits that lasted through 1986. They invested heavily in wind power manufacturing firms as well as wind farms, which sprouted in California's windy coastal mountain passes—particularly, Altamont Pass between Stockton and Livermore and the Tehachapi Mountains north of Los Angeles.[24]

Like minihydro plants, wind turbines at first appeared to be environmentally benign. Indeed, the Sierra Club looked into funding a study of wind power along California's coastal zone. But, as towers appeared in more and more locations, environmentalists, as well as the general public, became concerned. In the Altamont Pass area, environmentalists spoke up when, between 1984 and 1989, several hundred raptors that soared in the mountain's natural updrafts died in collisions with spinning wind turbine blades. The avian-death problem proved extremely difficult to mitigate, and it played a significant role in defeating plans for development of a 458-turbine wind farm at Tejon Pass in southern California's Tehachapi Moun-

Map 15.1. California Wind Energy Resources. Adapted from California Energy Commission, *1979 Biennial Report* (Sacramento 1979), 53; and David Hornbeck *et al.*, *California Patterns: A Geographical and Historical Atlas* (Palo Alto, CA: Mayfield Publishing, 1983), 115.

tains. For the general public, wind turbines particularly produced a strong NIMBY (Not In My Back Yard) response. In Benecia, a single eighty-foot tower, privately erected in 1981 under the city's encouragement of renewable energy efforts, raised public outcry. City residents demanded a public hearing to argue that it was unsightly, unsafe, and should be removed. But residents near the wind farms complained the most: of the throbbing noise of turning rotors, of television and FM radio interference, and of the general visual blight towers brought to the rural landscape. Perhaps the most celebrated NIMBY complaint came, in 1989, from Palm Springs' celebrity Mayor Sonny Bono, who railed against the ugly turbines in nearby San Gorgonio Pass as mere tax write-offs.[25]

Sylvia White, a professor of urban and regional planning at California Polytechnic Institute, Pomona, summed up the early feelings of frustrated environmentalists, feelings that lasted through the decade:

> Unscrupulous entrepreneurs are raping the environment in the name of energy conservation. Bulldozers are following their noses toward the least sign of a breeze—scouring, eroding, flattening, trenching and extinguishing species as they go. Access roads are cut across fragile soils, and cement pads are installed. . . .
>
> It's time that we outgrew our romantic notions and opened our eyes to the need for development controls on alternative-energy projects.[26]

Many observers were sure that as soon as tax credits and depreciation benefits ended, wind power in California would collapse. It did not. By the mid-1980s, windmill manufacturers had worked many of the bugs out of their turbine designs and problems in connecting wind plants to utility systems largely were solved. While weaker firms fell by the wayside, companies such as U.S. Windpower steadily were reducing the capital costs of installed turbines. Meanwhile, a rebirth in Danish wind power had refined Johannes Juul's three-blade design, and California wind farm operators soon discovered that Danish turbines were among the most efficient in the world. As early as mid-1982, Danish-built turbines accounted for 44 percent of the forty-four hundred operational turbines in California and 62 percent of that year's new installations. By 1984, 90 percent of the state's new installations were Danish-built, and American manufacturers improved their designs and retrofitted older turbines based on the Danish mills. By the end of the decade, a variable-speed turbine had been developed by U.S. Windpower, and the company had completed an advanced turbine feasibility study with EPRI and three major utilities, which concluded that new state-of-the-art turbines could soon produce electricity for $.05 per kWh or less, $.02 less than the best mills installed in 1987. Even as oil prices dropped during the decade, thereby lowering utility payments for generated power to PURPA energy producers, turbine technical advances plus lower capital costs meant wind power would continue competing effectively.[27]

As the 1990s opened, almost sixteen thousand wind turbines with a capacity of over 1,500 megawatts provided more than 1 percent of California's electric power from wind farms extending over more than twenty-seven thousand acres. Fully half of them were located in Altamont Pass, "the wind energy capital of the world," the bulk of the rest at San Gorgonio Pass and in the Tehachapi Mountains. Although the rate of new installations slowed with the expiration of federal and state tax credits, the number of operating turbines and their cumulative capacity continued growing. Both PG&E and SCE, which had discovered that the cyclic nature of wind power actually fit with their high summer demand cycle, announced that wind farms would be among their first choices if they needed to build additional power plants, the Sacramento Municipal Utility District planned to construct its own wind power facility, and U.S. Windpower's president, Dale Osborn, reported that even "a $.03/kWh turbine is not out of reach." Spurred by PURPA and federal and state tax incentives, California had become the world's largest laboratory for wind power, and its newly born wind power companies were beginning to undertake joint ventures in Spain, England, Japan, and elsewhere.[28]

Solar

The general public often spoke of solar energy in the same breath as wind power, and, like wind power, solar energy has a largely unnoticed history. Well before World War II, Professor F.A. Brooks of the University of California's Davis Agricultural Experiment Station, as well as staff members of the university's extension program, carried out research and development with solar water heaters and cookers. Entrepreneurs Frank Walker and William J. Bailey successfully commercialized solar water heating in southern California. Outside the state, Smithsonian Institution Secretary C.G. Abbot experimented with solar heaters during the 1920s and 1930s, and the Massachusetts Institute of Technology carried on a solar house research program from 1938 to 1959. Bell Telephone Laboratories worked on the "sunshine battery" during the early 1950s, NASA also sponsored research and development on photovoltaics, and the U.S. Army Quartermaster Corps experimented with solar furnaces. Moreover, knowledge of this work and other worldwide solar research activity was shared at the First World Energy Symposium in Phoenix, Arizona, in 1955, and through the Association of Applied Solar Energy. At the outset of the 1970s, however, few people knew of these activities.[29]

Solar power's historical invisibility was a mixed blessing. On one hand, as historian Ethan B. Kapstein pointed out, in 1981, the fact that solar technology "vanished without a trace" spoke to its great strength. It was a technology which could "operate quietly and efficiently, without leaving a legacy of pollution, radioactivity, or raped landscapes." On the other hand, as federal and state support for solar energy during the 1970s reached entrepreneurs, solar equipment designers and manufacturers learned little from the past. For example, the problem of hard water clogging solar collector pipes, as well as difficulties with corrosion and dirt, had been solved years earlier. Unfortunately, for many buyers, makers of new solar energy systems were still rediscovering both the problems and the solutions on a hit-and-miss basis. In 1979, the CEC and state Department of Consumer Affairs cooperatively established a toll-free hot-line number for complaints, project manager Kathryn Ramsay explaining "there are a lot of people just getting into [the business] who don't understand the basics of solar." More generally, noted Kapstein, "if only some time had been dedicated to studying the solar homes designed years ago," a great deal of effort and money would have been saved during the 1970s.[30]

Makers and users of solar devices soon overcame the technical problems, solar's popularity grew, and it became an essential component of the soft energy path. Magazines and books appeared, such as *Northern California Sun* and *The Mother Earth News Handbook of Homemade Power*, offering easily made to technically sophisticated designs and plans for solar water heaters, solar cookers, and passive solar houses. For the appropriate technology movement, solar was especially attractive because of the possibility for decentralizing energy production. Hot water heater collectors, even photovoltaic collectors, could be on anyone's rooftop, completely uncontrolled by utilities. Freeing people from dependency on central station electric power networks and making human settlement patterns more flexible became, for many solar champions, a crucial social benefit of harnessing the sun's energy. In 1977, David Pelka, chair of the Energy-Science Department at Northrup University in Los Angeles, argued: "Because it is decentralized, solar energy allows for greater individual initiative . . . it speaks to some traditional American values . . . to an earlier, more simple time when a man could be self-sufficient and provide for himself and his family."[31]

Pelka's vision appealed to backyard inventors, tinkerers, do-it-yourselfers, and other solar boosters, but it did not persuade the general public to adopt solar energy. Commercially available solar collectors remained costly and beyond the reach of most homeowners. They lacked back-up systems, required owner maintenance, and, in the minds of many consumers, were inconvenient. Unless one was well-off or a true convert, solar energy was seen as a luxury. Indeed, contrary to Pelka's view that solar energy enhanced individual initiative, the concept of a solar society implied for many people a community of cooperation that ran against the grain of individualism. The cooperative tenor was evident in Sim Van der Ryn's 1979 proposal that the 1,271-acre Hamilton Air Force Base in Marin County, sold as surplus by the military, be transformed into an ecologically balanced solar village. His plan envisioned a community with on-site solar industries providing some twelve hundred jobs, an internal solar vehicle transit system, retrofitted passive and active solar housing, community gardens and aquatic agriculture, and other cooperative features.[32] It did not get off the ground.

Federal and state tax credits and PURPA, not philosophical vision, brought solar energy into full bloom in California. By 1980, state and federal tax incentives had stimulated almost eighty thousand residential solar installations, the majority to heat swimming pools. The CEC was urging passive and active solar space conditioning, photovoltaic electric generators, solar thermal electric generation, sun drying fruit by agriculturalists, and industrial and commercial solar process heating. The state also supported development of a municipal solar utility program, in which local governments would finance, lease, or sell solar systems and provide other support to their residents. Local governments pushed solar forward, a handful of counties actually mandating solar domestic water and pool heating for new and/or retrofit housing construction. The City of Davis became a nationally recognized model for its comprehensive program addressing solar energy, conservation, bicycling, tree planting, and other energy issues. In response to these efforts, a new California solar industry accounted for over half the nation's production of solar collectors and photovoltaic materials. OAT reported, in 1982, that about seven thousand California companies made, distributed, or installed solar collectors. Solar energy was the primary focus of fifteen hundred of them.[33]

Some skeptics observed that when the big utilities figured out how to meter the sun, solar energy would become a reality. They were not far off the mark. In 1974, the federal government launched a large solar energy initiative, starting with a $17 million commitment and peaking at more than $500 million per year in the early 1980s. As with wind power, it focused on large-scale projects, and research contracts were funneled into high-tech companies across the country. SCE's Solar One, a gigantic solar thermal electric generating plant in the Mojave Desert which opened in 1982, received $120 million for project development. Covering 130 acres near Barstow, 1,818 highly reflective mirrors or heliostats—each with a surface area of over 400 square feet—automatically tracked the sun, concentrating its rays onto a central nickel, iron, and chrome alloy boiler mounted atop a 300-foot tower. Superheated steam turned a generator below, producing some 10,000 kW for eight hours a day during the summer and four hours during the winter, plus 7,000 kW from a thermal storage system for about four hours after sunset. In 1983, PG&E, Rockwell International, and ARCO Solar, a subsidiary of Atlantic Richfield Company, teamed up with federal support to build a more advanced 30,000-kW solar thermal system with 2,000 heliostats—each with over 1,000 square feet of surface area—a 400-foot-high "power tower," a boiler filled with liquid sodium, and a heat transfer system.[34]

PG&E and ARCO Solar also decided to try photovoltaic electrical gen-

Photo 15.1. Near the Mojave Desert town of Barstow, Southern California Edison Company opened a ten-megawatt Solar One plant in 1982, then the world's largest solar-powered electrical generating facility. (Courtesy of Southern California Edision Company.)

eration. In 1984, they opened a 177-acre photovoltaic farm in San Luis Obispo County, which contained several hundred thirty-four by thirty-six-foot rectangular panels mounted on twenty-foot-high pedestals. The "mechanical toadstools" automatically tracked the sun, photovoltaic cells on the panels converting sunlight directly into some 6,000 kW of electricity. With oil prices still high and photovoltaic cell improvements reducing cost of power generation steadily downward to $1.50 per kWh in 1980, it seemed a prudent project to undertake. However, plans to expand the plant to two thousand pedestals covering five hundred acres never materialized. Oil prices kept declining, and even though photovoltaic electricity dropped to about $.35 per kWh by 1988, its cost remained six times higher than that of conventional electric power generation. Even federal and state tax credits and PURPA benefits could not overcome this.[35]

In 1990, ARCO Solar sold the San Luis Obispo County farm to Carrizo Solar Corporation, which decided to dismantle it the following year. Said Carrizo co-owner Mike Elliston: "There is more money from selling the solar panels than from selling the electricity." The outlook dimmed for photovoltaics. Not only had oil prices fallen, but federal research and development support had reached a low point, dropping from $150 million in 1980 to $24 million in 1988. President George Bush's 1991 National Energy Strategy hardly brightened the picture, for it virtually ignored solar research and development, devoting only two out of twenty-eight pages to renewable energy resources. Although the industry tried to fill the dollar gap, it was constrained because the U.S. world market share for photovoltaics dropped during the 1980s from 86 percent to below 30 percent. While research and development continued by PG&E, other large energy companies, universities, and EPRI, the success of photovoltaics, like that of other soft-path energy resources, depended on the price for traditional fuels, as well as on research dollars.[36]

Cogeneration, Biomass, and Geothermal

Cogeneration, biomass projects, and geothermal plants generated over 90 percent of the 41 billion kWh that PURPA power producers fed into California's electric power grid in 1990, and cogeneration and biomass accounted for three-quarters of this. Like other "new" entries into the energy field, cogeneration and use of biomass fuel also had a past. Cogeneration in California dated back to the turn of the century, when Standard Oil ran cogeneration plants on its Kern County to San Francisco Bay Area oil pipeline in 1902, and privately owned commercial buildings in San Francisco generated electricity as a by-product of steam heating systems during the 1910s and 1920s. Similarly, using forest residue, peat, or garbage as fuel, and retrieving methane gas from agricultural or other waste were old ideas. A hundred years earlier, steam-powered electric plants in north coast logging territory began using sawdust fuel, and Sacramento and San Joaquin river delta peat had been molded into fuel briquettes, used directly as fuel, and studied for conversion to alcohol or gas. With the exception of limited sawdust fuel use by logging companies, however, none of these ideas had been able to compete with cheap fossil fuels. After the 1920s, both cogeneration and biomass fuel use largely disappeared under the weight of the central-station electric power industry.[37]

By 1990, however, California led the nation in cogenerated electric power. Almost half of PG&E's one hundred largest industrial customers—such as Dow Chemical, Crown Zellerbach, and Union Oil Company—began cogenerating electricity. Gilroy Foods garlic dehydration plant, Paul Masson Vineyards wine processing and bottling facility, and other smaller enterprises also became cogenerators. By 1986, in oil-rich Kern County, where raising steam by gas-fired generators and injecting it into wells to enhance oil recovery had become common practice after the initial energy crisis, PURPA begat over twenty cogeneration plants with a combined capacity of 500 megawatts. Kern River Cogeneration Company sold steam to Texaco and electricity to both Texaco and SCE from its 300-megawatt Omar Hill plant, "the biggest operation west of the Mississippi," according to its manager Charlie Myers. The CEC estimated that converting all of Kern County's enhanced oil recovery operations into cogeneration could provide 5,600 megawatts of new electric generating capacity, and easily swamp the state's utilities with excess power.[38]

PURPA's alteration of energy economics, coupled with support from the CEC, also brought back biomass energy. In the Sierra Nevada town of Burney, east of Redding, Ultrasystems launched a $21 million, 11,000-kW steam turbine power plant in 1982, to which the firm would transport forest residues for fuel, selling three-quarters of its electricity to PG&E. In 1983, Royal Farms, in the San Joaquin Valley town of Tulare, undertook the nation's first project to generate electricity using swine manure biogas fermented in a covered lagoon. Owner Roy Sharp anticipated PG&E paying as much as $36,000 per year for his excess electricity. Other CEC projects included use of peach pits by Tri-Valley Growers to fuel its plant boilers and cotton-gin trash for agricultural process heat at the J.G. Boswell Company. In 1992, a 47-megawatt biomass burning straw, Bermuda grass, orchard prunings, and wood chips came on line east of Los Angeles in Riverside County's Coachella Valley.[39]

Closely connected to use of biomass fuel, a PURPA-instigated "waste-to-energy" movement also emerged during the mid-1980s. The state initially began addressing the waste problem, in 1977, by creating the Solid Waste Management Board. A number of legislative measures followed this, including SB 1395, the first "garbage-to-energy bill." By 1986, with CEC support, thirty-nine waste-to-energy projects had been proposed or undertaken by cities, counties, and garbage haulers. But five already had been canceled, and twelve seemed indefinitely stalled for various reasons. Of the remaining twenty-two, one was operating in northeastern Lassen County, another was under construction in Los Angeles County, and twenty were undergoing financing or permit and siting approvals. Once on line, the twenty-two facilities were expected to be able annually to convert twelve million tons of solid waste into as much as 920 megawatts of electric power. Within three years, however, only four plants had actually materialized, the rest dying on the drawing boards.[40]

"What happened, basically," said Rod Miller, with the Sacramento-based public interest organization Californians Against Waste, "is the public rejected the concept of having the incinerators in their neighborhoods. . . . they don't want the problem of solid waste turned into an air pollution problem." Even though the operational waste-to-energy plants released emissions below federal and state air quality standards, the industry faced the NIMBY syndrome and a tough "emissions-driven market." People preferred to recycle garbage rather than burn it, an idea that was pushed hard by the Sierra Club and groups, such as the Northern California Recycling Association. The Sierra Club policy on waste was plain: first, reduce wastes produced; second, recycle; and, third, only then incinerate to produce energy. Yet, waste-to-energy was too sound an idea to abandon, in both environmental and energy terms, so the industry turned to solving their pollution problem by improving incineration technology. The Combustion Power Company in Menlo Park, on the San Francisco peninsula, developed an extremely low-emissions-producing "fluidized bed combustion system," and, by 1990, they operated six plants, one in southern California and five in the San Francisco Bay Area.[41]

While cogeneration and biomass generating projects provided most PURPA power to California's utilities, PURPA also greatly accelerated geothermal development. Prior to the 1980s, geothermal power generation was centered in northern California, around the Geysers, and PURPA stimulated several new facilities there, including plants constructed by the Sacramento Municipal Utility District and the Northern California Power Authority, the latter serving public power municipalities such as the City of Santa Clara. By 1991, generation capacity at the Geysers exceeded 2,000 megawatts, making it the largest geothermal development in the world. But PURPA also instigated geothermal development elsewhere in the state. In 1980, two 10-megawatt plants were dedicated in southern California's Imperial Valley, one jointly built by SCE and Union Oil Company, the other by the Desert Valley Company, a subsidiary of Magma Power, which had pioneered the Geysers during the 1950s. In 1982, Earth Energy, Inc. completed its Salton Sea Unit No. 1, and San Diego Gas and Electric broke ground for its 45-megawatt binary geothermal plant in Heber. Within ten years, fifteen Imperial Valley geothermal facilities tapped five geothermal fields with a generating capacity of 379 megawatts.[42]

As the 1980s came to a close, geothermal was one of the brightest spots in California's renewable energy picture, not only for generating power but, in some locales, for district space heating. New geothermal fields were being tapped. The California Energy Company built nine plants, with a combined capacity of 240 megawatts, near China Lake in the Mojave Desert. Two plants went on line in Mono County in 1990, a 30-megawatt facility in Lassen County joined two smaller plants already there, and Trans-Pacific Geothermal Corporation contracted with Magma Power to develop the Surprise Valley field in Modoc County. Meanwhile, cooler geothermal fluids were tapped by the community of Susanville, in northern California, and by the City of San Bernardino, which received CEC funding to convert its waste-water treatment plant to geothermal heat in 1983. San Bernardino expanded its initial project into one of the largest geothermal district-heating systems in the country, providing space- and water-heating to twenty-seven major public and commercial buildings. Yet, even though thirty-one new geothermal well drilling permits were granted by the California Department of Conservation's Division of Oil, Gas, and Geothermal Resources in 1990, a dark cloud loomed over geothermal's future.[43]

In 1987, PG&E noticed a decline in steam production at its Geysers facilities. It first delayed and then canceled expansion of its capacity because of insufficient steam. In 1989, it decided to sell one of its units, and two years later the Department of Water Resources closed its plant near Clear Lake. Geysers steam production declined at as much as a 25 percent rate, and pressures that had been 514 PSI in the 1960s, fell to 300 PSI by 1990. Some new and deeper wells seemed only to bring steam with increasing amounts of corrosive hydrogen chloride and noncondensible gases, and instead of producing power at the field's gross capacity of 2,093 megawatts, 1992 output reached only 1,226 megawatts. For PURPA producers, this problem was exacerbated by declining oil prices and the start-up of PG&E's Diablo Canyon nuclear plant. The utility's payments for their geothermal steam production, based on the price of oil, dropped precipitously, from $39.50 per megawatt in 1985 to $15.15 in 1988. The CEC launched an intensive study of the field's steam pressure problems, but the long-term prospective for continued geothermal energy at the Geysers remained unclear.[44] In a larger sense, problems at the Geysers reminded energy planners that, in their arena, the give-and-take between technology and the environment includes a myriad of ever-changing factors.

Summary

By the end of the 1980s, a diversified, innovative, and less centralized energy paradigm ascended in California. It had taken shape slowly, growing in counterpoint with the environmental movement. Some of the more progressive institutions that moved it forward either changed their focus or did not survive. California Tomorrow's interest in and influence concerning the interplay between energy and the environment withered, its attention turning to other issues, such as ethnic diversity and education. Governor Jerry Brown's successor, Republican George Deukmejian, abolished OAT in 1982, an action that was symptomatic of the appropriate technology movement's failure to exercise enough political power to entrench itself securely in the establishment. At the same time, suggests historian Carroll Pursell, OAT's demise "was also an example of the triumph of hegemonic culture over deliberate subversion by a truly oppositional culture." As developments in gigantic solar generating plants, photovoltaic farms, and wind farms illustrate, California's established energy regime "showed a remarkable ability to expropriate emerging technologies for their own benefit."[45]

Although California's new energy paradigm was not as radical as some people might have liked, its emphasis on alternative methods of electric power production and energy conservation provided a softer, more environmentally benign, and remarkably efficient energy path. Between 1973 and 1988, observed political scientist Dan Mazmanian, California's population increased 37 percent and its economy grew 46 percent, while energy consumption increased only 8 percent. By altering its electric power energy supply mix, California conserved fossil fuels. By adopting the strictest standards in the world for building insulation and electric appliances, along with other conservation measures, the state greatly slowed growth in energy consumption. Ironically, California's utilities found themselves swamped by an unprecedented surplus of power, which resulted from an enormous expansion of their own capability to meet power demand by the addition of new PURPA-inspired generating capacity and by reduced consumer demand as a result of energy conservation. This, coupled with falling worldwide oil prices and abundant natural gas supplies, brought a hiatus in California's energy revolution toward the end of the 1980s, and utilities backed away from conservation in order to sell more, not less, electricity.[46]

The chronic issue of air pollution, however, reignited California's efforts to meld environmental and energy planning and policy making. Significant efforts to control air pollution had been achieved through better automobile technology, as well as controls on industrial emissions. In southern California, the powerful South Coast Air Quality Management District (SCAQMD) attained major air quality improvements over the years, but, there and elsewhere, smog continued. Meeting federal air quality standards, promulgated by the 1977 Clean Air Act, seemed almost impossible. Local and state agencies undertook a variety of measures, frequently at the behest of citizens. Bicycle lanes and special paths appeared for people who chose to reject the automobile. Bus systems were improved, and expensive light rail lines constructed in communities such as Sacramento, San Diego, and San José. Finally, measures that seemed somewhat restrictive to some citizens came with car-pooling strategies: the addition of car-pool and bus lanes to freeways, construction of ride-sharing parking lots near suburban freeway access points, and increases in bridge tolls.[47]

Californians only grudgingly accepted measures that manipulated their use of the automobile, for such actions seemed tantamount to attacking individual freedom. When, by the end of the 1980s, some agencies revealed plans to charge parking fees for drivers who did not car-pool to work, the shopping mall, or other public places, prickly opposition appeared. Even though officials believed this was the only way to get people to use mass transit and bring metropolitan areas up to air quality standards, many people saw these latest plans as a distinct threat to democratic and laissez-faire tradition. Journalist Joanne Jacobs observed, in 1990:

> It's one thing to persuade Californians to recycle cans and bottles or ask for paper bags at the grocery store. But no degree of enviro-sadism will get Californians to take the bus. . . . Most people value very highly the mobility, convenience and freedom of hopping in the car and driving where they want to go, when they want to go. Raise gas taxes (they went up Saturday), charge for parking, boost tolls, and we'll pay more, but we'll still drive.[48]

This sort of resistance, plus California's seemingly never-ending population growth, led even SCAQMD to conclude, during the 1980s, that federal air quality standards could not meet existing deadlines, and the federal Environmental Protection Agency concurred. However, notes Dan Mazmanian, "environmentalists took the EPA and SCAQMD to court to compel the development of a strategy to bring the [Los Angeles] basin into compliance with federal standards." As a result, SCAQMD released an air quality plan in 1989 that called for a regional reduction in electric power generation that was 30 percent lower than the CEC and PUC planning target for the year 2010. This forced "the key players in the state's energy and air pollution control arenas," says Mazmanian, "to confront and reconcile their differences."[49]

In 1990, a statewide group of environmentalists, government planners, and utility representatives recommitted California to energy conservation. During the early 1990s, PG&E, SCE, and San Diego Gas and Electric Company each turned toward demand-side management, adopting major new conservation programs that ranged from energy audits and customer rebates on energy-saving appliances to low-energy lighting retrofit services for business customers. Commentator Peter Asmus suggested: "Utility executives all across the country are shaking their heads in disbelief, wondering whether their California counterparts have finally gone off the deep end." But they had not. They still sought to control their world, and lobbied hard to put their stamp on legislation promoting conservation and renewable energy production.[50] As always, people stood at the nexus of the ever-changing interplay between technology and the environment.

Conclusion

POINT, COUNTERPOINT. The interplay between technology and the environment, orchestrated by people with sundry values and interests, is the key feature of any society's energy experience. Sometimes subtle, sometimes raw; sometimes reciprocal, sometimes collisional—the give-and-take becomes unmistakable in modern societies, where a wide technological sweep can make possible the successful exploitation of complex and versatile environments. This process is revealed plainly in California's energy history, as is the recognition that every energy choice made by people requires weighing environmental endowments and available technologies. Moreover, California's experience reveals that each energy choice, in turn, has its own impact on the environment, and, often, on the process of technological development.

Californians exemplified the American propensity to exploit and commodify the environment under the aegis of corporate capitalism. Thus, general characteristics of California's experience are part and parcel of the complexion of American society. Yet the specifics of its energy development diverged in many ways from that of the rest of the nation, in large part because its resource endowment was different. Moreover, Californians lived in relative autonomy from their eastern compatriots, especially during the nineteenth century, and this led them to distinctively harness regional energy resources. They developed varied and sometimes singular ways to use waterpower, to briquette lignite coal and peat, to burn petroleum and straw as fuel for raising steam, and to manufacture gas from oil. They sought to employ tidal power, developed active use of solar energy, and made unique contributions to urban transportation and agricultural technology. At the same time, their lack of good regional coal and reliance on expensive imports of this essential industrial fuel helped to stifle familiar manufacturing development. Californians became, perhaps, the earliest Americans to yearn for what, during the last part of the twentieth century, people began calling "energy independence."

At the turn of the century, California's usable energy resource base changed dramatically, and the state became independent of meaningful energy imports for almost fifty years. Petroleum ascended as the principal fuel for transportation, agricultural field work, and regional industries; natural gas and natural gas liquids became popular domestic and industrial fuels; and electricity, generated principally by waterpower, supplanted virtually all other forms of illumination, provided agricultural and industrial power, and enormously impacted domestic life and energy use. By the 1930s, the state's twentieth-century energy triad—oil, gas, and waterpower—was firmly established, use of renewable resources beyond waterpower faded away, and, after a long struggle, residents were close to working out who would control their waterpower resources. Yet California's energy budget still did not bear much resemblance to that of the rest of the nation, where fuelwood continued to be important and coal played a crucial role. Nor did California's newfound energy independence stimulate the traditional manufacturing growth for which so many of its citizens continued to hope.

Then World War II dramatically affected California. An influx of new investments, industry, and people altered the state, and by the war's conclusion California had developed an uniquely new economic base. During the postwar era, aerospace companies matured, driving forward southern California's economy; a research-based electronics industry, spawned by long experience developing hydroelectricity, propelled northern California forward. Moreover, the war excited more than the transformation of the regional economy and society. The growth it engendered also led to a renewed dependency on imported energy resources. The emergence of new industrial sectors and concomitant population increase forced petroleum companies and gas and electric utilities to search out and import enough energy to meet accelerating demand.

By 1970, California's energy condition seemed to be in line with that of the rest of the nation. Its prosperity appeared inextricably linked to energy development. Its energy regime followed the same grow-and-build strategy found elsewhere in the country. Its energy resource budget, save having no coal, followed the national pattern. But its citizens' concern about air pollution, offshore oil drilling, and nuclear power exposed an undercurrent of environmentalism that came into direct conflict with the energy regime. As energy shortfalls struck the nation, the state once again diverged from the national course. Californians and their policy makers balked at national energy plans. Acting on a profound concern for the deteriorating state of their natural environment, they took an independent course. Grassroots activists and legislators halted installation of nuclear power plants, urged development of renewable energy resources, and took up energy conservation. Resources used and then abandoned in past years were rediscovered, and new technologies harnessed them in different ways. Federal and state legislation helped to stimulate investment in conservation and development of these and other untested energy resources and technologies.

In taking up appropriate technology and the soft energy path, California forged a new energy paradigm. Hard energy industries and the oil, gas, and waterpower triad were not replaced, but the introduction of alternative energy resources, new production methods, and conservation strategies substantially refashioned them. California became a laboratory for environmentally astute energy strategies and technologies. There soon trickled from it to the world a stream of new energy ideas and techniques, and the California Energy Commission undertook an export technology program, in 1986, to support the new renewable energy industries. In effect, the CEC came to act as an international economic development agent for renewable energy companies, among other things, granting funds to assist them in competing against German, Japanese, and other foreign firms. The resulting good health of California's new energy industries elevated the state as a leader in the battle against the worsening greenhouse effect of the earth's atmosphere caused, in large part, by carbon-dioxide gas emissions. Because of its extensive research into solar panels, fuel cells, and advanced gas turbines for clean, efficient, and cost-effective electric power generation, the state made possible viable alternatives to the exclusive use of fossil fuels.[1]

Loss of Individual Independence in Making Energy Choices

If nineteenth-century booster Titus Fey Cronise, wheat farmer W.S.M. Wright, or oil man George S. Gilbert stepped into the Golden State as it prepared to enter the twenty-first century, they would be staggered by the quantity, but probably not by the diversity, of energy resources used by their fellow Californians. Nuclear power, photovoltaics, and steam turbine and geothermal electric power plants would be new to them; but only the applications of waterpower, windmills, solar heat, biomass, petroleum, natural gas, and coal might be unfamiliar. They would understand their fellow Californians' discomfort in relying on imported oil, for they had depended on imported coal. They would be empathetic with citizens' nervousness about far-off places often dictating fuel prices, and they would know how such things could determine the speed at which regional energy resources might be developed. After all, they had burned wood instead of expensive British coal, and they had seen the ready availability of eastern kerosene slow the development of their own early efforts to tap California's oil. They would understand, too, that resources could run out, for they had seen their skimpy coal mines close and had worried over shortages of fuelwood. They might even understand that oil prices could suddenly rise because nations had gone to war in the Persian Gulf.

Yet Cronise, Wright, and Gilbert probably would be confounded by other changes. Although they had experienced some dependency on a few coal importers and Standard Oil's early monopoly over kerosene, their compliance with such energy supply systems paled before the subservience of modern Californians to the control over energy production and distribution exercised by several gigantic corporations with worldwide ties. Making the matter more entangled, these huge corporations have their hands in every new energy development, work closely with universities and government in every aspect of energy research, and are loath to relinquish any portion of the energy field. Cronise, Wright, and Gilbert would have a very difficult time finding a modicum of the individual independence, which they had enjoyed in making energy choices. Central state commissions formulate energy policy, give direction for the future, and interact with the powerful corporations and a myriad of other government agencies, legislative committees, and public and special interest groups. Although a state utilities commission, plus a dozen other state agencies and another layer of federal agencies, play a role in regulating energy companies and consumer prices, the individual faces formidable obstacles to try anything new or different.

Particularly confounding to Cronise, Wright, and Gilbert would be the number of state and federal bureaucracies, as well as public interest groups, concerning themselves with the environment and its exploitation. Environmental concerns had touched the lives of nineteenth-century Californians and even impinged on energy matters, but rarely in a restrictive way. Now impact on the environment appeared to dominate every energy choice. How would these time travelers have reacted to the anathema against burning coal for manufacturing or against burning wood in stoves and fireplaces in geographic basins like that of Lake Tahoe? What would they make of permit requirements for cutting wood, setting up a windmill, or building a small waterpower system on a nearby stream to operate machinery or generate electricity? One could not even mount a solar water heater, much less build an energy-efficient house, without facing a permit bureaucracy. How, indeed, would Cronise, Wright, and Gilbert react to changes in sea life tied to the warming of ocean waters by power plant cooling systems, to air pollution caused by automobile emissions, and to global warming linked directly to the burning of fossil fuels?

Yet, despite their bewilderment at the complexities of the energy-technology-environment relationship in modern California, Cronise, Wright, and Gilbert certainly would sympathize with Californians' concerns about energy and related issues. Because they understood shortages, they probably would support calls for energy conservation. They would understand individuals, citizens groups, and trade associations advocating all sorts of energy resource policies and development positions and railing against unfair utility rates, for their generations had done the same. They would probably even understand the rhetoric of government bureaucrats, corporate and interest group representatives, politicians, university professors, and individual Californians who speak of implementing alternatives to the current energy system or accomplishing a transition to a decentralized energy future. They would probably applaud them, for they were dreamers, too, and they would see plainly the drawbacks of the gradual sacrifice of individual Californians' energy independence to an increasingly sophisticated, leviathan-like energy network.

Lessons from Energy History

We make energy choices and establish policies from a variety of factors, ranging from ideological, philosophical, and cultural to economic, environmental, and technological. Their omnipresence is precisely why nineteenth-century Californians like Cronise, Wright, and Gilbert would be able to understand and empathize with our present energy concerns and conditions. But the fluctuating relative importance of such factors in the making of decisions over both short and long spans of time also explains why these time travelers sometimes would be perplexed by what they encountered and why learning from the past is tricky.

Some lessons from energy history are plain. Fossil fuels are exhaustible. Just as California's coal reserves were spent in some fifty years and its oil and gas reserves depleted substantially in another fifty years, fossil fuels everywhere on earth constantly are being used up. We inevitably will reach the time when worldwide reserves will near exhaustion. Although people may lose sight of this reality, it nevertheless remains true, and one only can argue when, not if, fossil fuels will run out. Reasonable people probably would be safest in assuming it will be sooner rather than later, if for no other reason than to be prepared for the inevitable. Such a position is especially important for societies that persist in opting for economic growth as the way to the good life and as the solution to social ills.

Most insights from history, however, are not as straightforward. For example, general economic betterment over time correlates, to some extent, with increased energy consumption. Per capita energy consumption in

California followed an upward trend for 150 years, moving from between 150 million and 200 million BTUs in 1850–1860 to some 280 million BTUs in 1970–1980 (figure C.1). Significant dips in per capita consumption appear to coincide with periods of economic depression (the 1870s, 1890s, and 1930s). Other declines not explained entirely by economic slowdowns can be attributed to improved efficiency in production and energy use. A variety of technological improvements between 1870 and 1900 contributed to more efficient energy conversion in steam boilers, water wheels, kerosene lamps, and agricultural field equipment. Similarly, since 1900, steady technological improvements to steam turbines, electric power generating and transmission systems, electric lights and appliances, and internal combustion engines have slowed dramatically the growth of per capita consumption despite an enormous and almost unbroken climb in overall energy use. In all this, California's experience is not unlike that of the United States as a whole (figure C.1).

This apparent linkage between economic growth and energy use still governs some energy planning, but, as California's experience during the 1970s and 1980s reveals, economic growth and energy consumption are not linked in some immutable fashion. Therefore, if our ideological position remains opposed to a relatively static economic state, one which might force a redistribution of wealth among our citizens, this does not mean we need to continue using more and more energy in order to achieve the economic growth and opportunity needed to stave off redistribution. Rather, it may mean that we must work very hard indeed at more efficient use of energy through improving production and consumption technologies, harnessing renewable resources, developing safe nuclear power production, and implementing widespread and effective conservation measures. Of course, we have been doing some of these things all along, and more recently, by following the soft energy path, we seem to be taking up the rest. Indeed, California provides a model of energy conservation and diverse resource development. But, it remains very difficult for us not to place too much emphasis and hope on some single known resource, such as solar energy or nuclear fusion, or dream unrealistically of some magic, unknown resource just waiting to be discovered, that will let us get on with lives as usual.

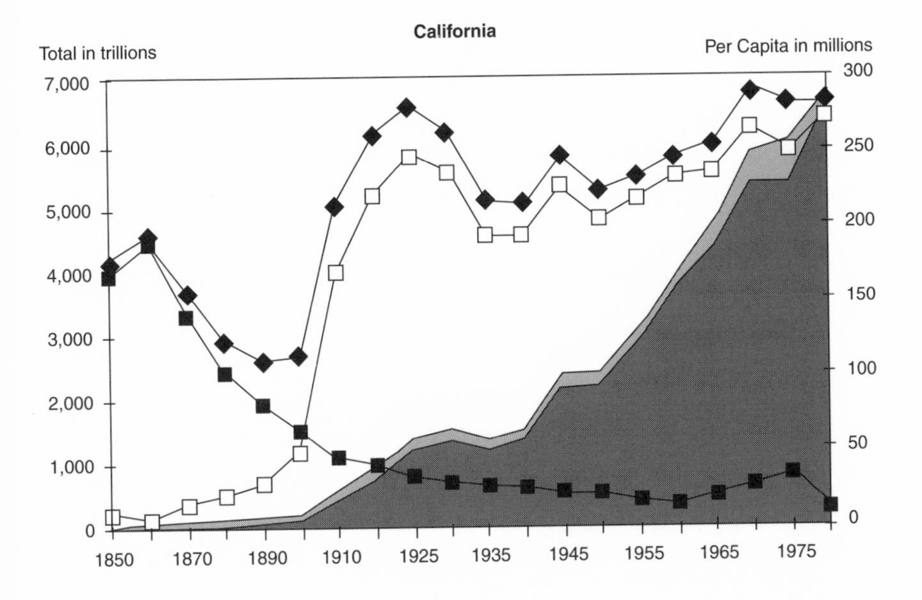

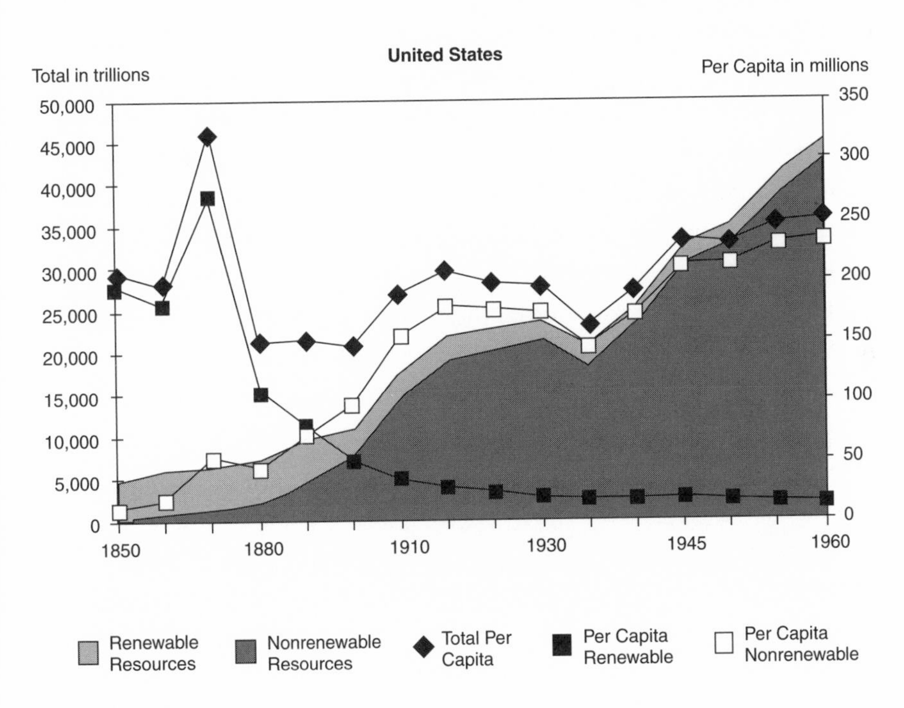

Figure C.1. California and United States Total and Per Capita Energy Resource Consumption in BTUs, Selected Years from 1850–1960. Sources: *Supra,* Table B-3; U.S. data drawn from J. Frederic Dewhurst and Associates, *America's Needs and Resources, A New Survey* (New York: Twentieth Century Fund, 1955), and Sam H. Schurr and Bruce C. Netschert, *Energy in the American Economy, 1850–1975* (Baltimore: Johns Hopkins Press for Resources of the Future, 1960).

Whatever today's policy makers decide, there do appear to be some pitfalls about which they might be wary. Our general optimism about energy is perhaps the greatest trap, for it often has blinded us to realities. Titus Fey Cronise, we might recall, was convinced that California could and would build great waterfall cities in the Sierra Nevada foothills, and he and many other Californians were sure, despite geologists' warnings, that good coal would be found in sufficient abundance within the state to spawn a great manufacturing society. Other boosters talked of oil and gas reserves that would run from rocks like rivers and last forever. Still others spoke of sufficient hydropower to meet energy needs cheaply for centuries. More recently people have invested similar hopes in nuclear, solar, and geothermal energy, as well as in energy conservation and decentralization. Utopia always has seemed just around the corner, when it comes to energy futures.

Policy makers also must watch out for our unquestioned acceptance of massive technological and corporate systems. While they relieve individuals from having to spend time chopping wood, fiddling with solar collectors, building windmills, pedaling bicycles, and otherwise being inconvenienced, they also undermine individual and even small group independence. While economies of scale can certainly provide efficiencies, it is also true that they become vulnerable to catastrophic failures, require substantial investments to ensure reliability, and by their very existence mitigate against employment of relatively simple, localized, and possibly more appropriate technologies, which might accomplish the same tasks with equal efficiency and greater sensitivity to long-term environmental well-being. Finally, they demand a vast regulatory bureaucracy to protect society against unacceptable or illegal practices. As Cronise, Wright, and Gilbert no doubt would say of modern society, regulation can appear to get rather out of hand.

Concern for the natural environment, the source of much regulatory activity in recent years, also will be of continuing importance for policy makers, particularly if we fully meld environmental planning and policymaking with that of energy. Would the environmental regulations and labor laws in place today financially or otherwise have prevented the construction of many, if not most, of the pre-World War II hydroelectric plants on which Californians greatly depend today? Likewise, what might have been the impact on earlier developments in California's oil industry under current regulatory restrictions? Policy makers might find it prudent to consider that environmental concerns have the potential to prevent tapping sorely needed resources and to slow other societal developments which might depend on the experiences gained in resource development. Conversely, it might be wise to remember that enacting regulations against flaring natural gas connected to oil production helped to harness its beneficial use, and entrepreneurs do adjust to changes over time, as the builders of minihydro projects during the 1980s revealed in their successful dealings with the modern bureaucracy.

Finally, in a world shrinking in terms of time and distance because of advances in communication and transportation technologies, California's regional differences with other parts of the world now may appear less important as determinants of its energy future than in the past. Yet the state's present leadership in solar, wind power, and geothermal energy production seems inextricably linked to regional geographic characteristics, much as its hydroelectric system developed as it did because of the location of rivers and streams. Therefore, policy makers also must consider that regional peculiarities will continue to play a role in California's quest to achieve a mixed energy resource base for the twenty-first century. What the energy past plainly reveals is that the constant interplay between technology and the environment, along with a host of other factors, has always made up a crude calculus of ever-changing advantages by which we make energy-related choices. California's recent energy experience leaves little reason to believe this has changed and every reason to challenge our creative thinking for the future.

APPENDIX A California Energy Resource Production Statistics

ble A–1. California Energy Production by Resource in Physical Quantities and Equivalents, 1850–1980[a]

r	*Coal and Coke*	*Oil*	*Natural Gas*	*NGL[b]*	*Hydro-electricity*	*Fuel Wood*	*Hydraulic Power*	*Wind Power[c]*	*Animal Power*	*Nuclear Power*	*Geothermal Power*
50	—	—	—	—	—	443	2	294	173	—	—
55	—	—	—	—	—	786	22	735	640	—	—
60	—	—	—	—	—	1,129	42	703	1,142	—	—
65	61	—	—	—	—	1,403	58	698	784	—	—
70	142	—	—	—	—	1,677	74	970	768	—	—
75	167	—	—	—	—	1735	81	1,158	738	—	—
80	237	41	—	—	—	1,793	87	1,179	716	—	—
85	72	325	—	—	—	1,742	106	1,069	791	—	—
90	111	307	41	—	—	1,691	125	1,125	1,137	—	—
95	74	1,245	110	—	17	1,633	126	1,031	1,257	—	—
00	177	4,320	41	—	86	1,574	127	934	1,049	—	—
05	47	34,276	148	—	300	1,588	123	883	730	—	—
10	11	77,698	2,765	—	879	1,602	125	588	814	—	—
15	10	91,147	21,891	13	1,885	1,610	72	358	492	—	—
20	2	103,377	66,041	48	2,568	1,619	57	290	368	—	—
25	1	232,492	187,789	303	5,171	1,501	25	169	205	—	—
30	11	228,100	334,789	835	6,804	1,388	16	91	108	—	—
35	1	207,832	284,109	552	7,751	1,296	10	53	65	—	—
40	1	223,881	351,950	660	8,781	1,225	4	21	38	—	—
45	—	326,482	502,442	814	11,619	1,040	4	12	26	—	—
50	—	327,607	558,398	1,190	14,809	900	3	7	19	—	—
55	—	354,812	538,178	1291	14,555	715	—	—	—	—	—
60	—	305,352	517,535	1203	17,444	500	—	—	—	—	—
65	—	316,428	660,384	995	30,523	773	—	—	—	300	300
70	—	375,291	667,689	890	37,867	928	—	—	—	3,100	600
75	—	296,035	299,535	838	40,780	1,031	—	—	—	6,120	3,956
80	—	343,276	404,651	787	17,531	1,134	—	—	—	8,905	5,287

a. Physical quantities: Coal and Coke = 000 tons; Oil = 000 bbls; Natural Gas = 000,000 cubic feet; NGL = 000,000 gallons; droelectricity = 000,000 kWh; Fuelwood = 000 cords; Waterpower, Wind power, and Animal power = 000 tons of coal; Nuclear and othermal = 000,000 kWh.

b. Includes natural gasoline and other natural gas liquids.

c. Includes sail and windmill power.

Sources: *U.S. Minerals Yearbook;* annual reports of the California State Mineralogist; Federal Power Commission data; California, *tistical Abstract,* 1970, 1986, and 1990; and *infra,* Tables C–2, C–3, C–4, C–5, C–6, C–8, and C–14. Coal equivalents for waterpower, wind ver, and animal power based on Sam H. Schurr and Bruce C. Netschert, *Energy in the American Economy, 1850–1975* (Baltimore: Johns pkins University Press, 1960), 485–487.

Table A–2. California Energy Production by Resource in Trillions of BTUS 1850–1980[a]

Year	*Coal and Coke*	*Oil*	*Natural Gas*	*NGL*[b]	*Hydro-electricity*	*Fuel Wood*	*Hydraulic Power*	*Wind Power*[c]	*Animal Power*	*Nuclear Power*	*Geotherm Power*
1850	—	—	—	—	—	9.3	—	7.7	4.5	—	—
1855	—	—	—	—	—	16.5	.6	19.5	16.8	—	—
1860	—	—	—	—	—	23.7	1.1	18.4	29.9	—	—
1865	1.6	—	—	—	—	29.4	1.5	18.3	20.5	—	—
1870	3.7	—	—	—	—	35.1	1.9	25.4	20.1	—	—
1875	4.4	—	—	—	—	36.4	2.1	30.3	19.3	—	—
1880	6.2	.2	—	—	—	37.6	2.3	30.9	18.8	—	—
1885	1.9	1.9	—	—	—	36.5	2.8	28.0	20.7	—	—
1890	2.9	1.8	—	—	—	35.4	3.3	29.5	29.8	—	—
1895	1.9	7.2	.1	—	1.6	34.2	3.3	27.0	32.9	—	—
1900	4.6	25.1	—	—	7.9	31.7	3.3	24.5	27.5	—	—
1905	1.2	198.8	.2	—	14.8	30.8	3.2	23.1	19.1	—	—
1910	.3	450.6	3.0	—	32.6	31.1	3.3	15.4	21.3	—	—
1915	.3	528.7	23.5	1.0	51.5	31.2	1.9	9.4	12.9	—	—
1920	.1	599.6	71.0	3.9	90.1	31.4	1.5	7.6	9.6	—	—
1925	—	1,348.5	201.9	24.6	117.1	29.1	.7	4.4	5.4	—	—
1930	.3	1,323.0	359.9	67.8	124.0	26.9	.4	2.4	2.8	—	—
1935	—	1,205.4	305.4	44.9	126.3	25.2	.3	1.4	1.7	—	—
1940	—	1,298.5	378.3	53.6	134.6	23.8	.1	.5	1.0	—	—
1945	—	1,893.6	540.1	66.1	174.9	20.2	.1	.3	.7	—	—
1950	—	1,900.1	600.3	96.7	192.2	17.5	.1	.2	.5	—	—
1955	—	2,957.9	578.5	104.9	170.8	13.9	—	—	—	—	—
1960	—	1,771.0	556.4	97.7	187.9	9.7	—	—	—	—	—
1965	—	1,835.3	709.9	80.8	324.8	15.0	—	—	—	3.2	3.2
1970	—	2,176.7	728.5	72.3	403.0	18.0	—	—	—	33.0	6.4
1975	—	1,717.0	322.0	68.1	439.6	20.0	—	—	—	66.0	42.6
1980	—	1,991.0	435.0	63.9	189.0	22.0	—	—	—	96.0	57.0

a. British Thermal Unit values computed from physical quantities of each energy source produced.
b. Includes natural gasoline and other natural gas liquids.
c. Includes sail and windmill power.
Source: *Supra*, Table A–1. BTU conversions based largely upon Schurr and Netschert, *Energy in the American Economy.*

Table A–3. California Energy Production by Resource as a Percentage of Total State Production, 1850–1980[a]

Year	Coal and Coke	Oil	Natural Gas	NGL[b]	Hydro-electricity	Fuel Wood	Hydraulic Power	Wind Power[c]	Animal Power	Nuclear Power	Geothermal Power
1850	—	—	—	—	—	43.0	.2	35.7	21.0	—	—
1855	—	—	—	—	—	31.0	1.1	36.3	31.6	—	—
1860	—	—	—	—	—	32.4	1.5	25.2	40.9	—	—
1865	2.2	—	—	—	—	41.2	2.1	25.6	28.8	—	—
1870	4.3	—	—	—	—	40.7	2.3	29.4	23.3	—	—
1875	4.7	—	—	—	—	39.3	2.3	32.8	20.9	—	—
1880	6.5	.2	—	—	—	39.2	2.4	32.2	19.6	—	—
1885	2.0	.2	—	—	—	39.8	3.0	30.5	22.6	—	—
1890	2.8	1.7	—	—	—	34.5	3.2	28.7	29.0	—	—
1895	1.8	6.7	.1	—	1.4	31.6	3.1	24.9	30.4	—	—
1900	3.7	20.1	—	—	6.3	25.5	2.7	19.6	22.1	—	—
1905	4.2	68.3	.1	—	5.1	10.6	1.1	7.9	6.6	—	—
1910	.1	80.8	.5	—	5.9	5.6	.6	2.8	3.8	—	—
1915	—	80.1	3.6	.2	7.8	4.7	.3	1.4	2.0	—	—
1920	—	73.5	8.7	.5	11.1	3.9	.2	.9	1.2	—	—
1925	—	77.9	11.7	1.4	6.8	1.7	—	.3	.3	—	—
1930	—	69.4	18.9	3.6	6.5	1.4	—	.1	.2	—	—
1935	—	70.5	17.9	2.6	7.4	1.5	—	.1	.1	—	—
1940	—	68.7	20.0	2.8	7.1	1.3	—	—	.1	—	—
1945	—	70.2	20.0	2.5	6.5	.7	—	—	—	—	—
1950	—	67.7	21.4	3.4	6.8	.6	—	—	—	—	—
1955	—	70.3	19.8	3.6	5.8	.5	—	—	—	—	—
1960	—	67.5	21.2	3.7	7.2	.4	—	—	—	—	—
1965	—	61.7	23.9	2.7	11.1	.5	—	—	—	.1	.1
1970	—	63.2	21.2	2.1	11.9	.5	—	—	—	1.0	.2
1975	—	64.2	12.0	2.5	16.4	.7	—	—	—	2.5	1.6
1980	—	69.8	15.2	2.2	6.6	.8	—	—	—	3.4	2.0

a. Percentages computed from British Thermal Unit values of total physical quantities of each energy source produced.
b. Includes natural gasoline and other natural gas liquids.
c. Includes sail and windmill power. Windmill power nears .2 percent of total wind power only in 1900.
Source: *Supra,* Table A–2.

APPENDIX B California Energy Resource Consumption Statistics

Table B–1. California Energy Consumption by Resource in Physical Quantities and Equivalents, 1850–1980[a]

Year	Coal and Coke	Oil	Natural Gas	NGL[b]	Hydro-electricity	Fuel Wood	Hydraulic Power	Wind Power[c]	Animal Power	Nuclear Power	Geothermal Power
1850	50	—	—	—	—	443	2	294	173	—	—
1855	86	—	—	—	—	786	22	735	640	—	—
1860	78	—	—	—	—	1,129	42	703	1,142	—	—
1865	150	—	—	—	—	1,403	58	698	784	—	—
1870	343	—	—	—	—	1,677	74	970	768	—	—
1875	562	—	—	—	—	1,735	81	1,158	738	—	—
1880	712	41	—	—	—	1,793	87	1,179	716	—	—
1885	1,044	325	—	—	—	1,742	106	1,069	791	—	—
1890	1,263	397	41	—	—	1,691	125	1,125	1,137	—	—
1895	1,710	1,183	110	—	17	1,633	126	1,031	1,257	—	—
1900	1,965	4,104	41	—	86	1,574	127	934	1,049	—	—
1905	1,160	30,848	148	—	300	1,588	123	883	730	—	—
1910	925	66,544	2,765	—	879	1,602	125	588	814	—	—
1915	699	90,946	21,891	13	1,885	1,610	72	358	492	—	—
1920	1,144	113,961	66,041	46	2,568	1,619	57	290	368	—	—
1925	460	162,745	187,789	273	5,171	1,501	25	169	205	—	—
1930	295	159,670	334,789	710	6,804	1,388	16	91	108	—	—
1935	200	145,482	284,109	442	7,751	1,296	10	53	65	—	—
1940	147	156,717	351,950	528	8,781	1,225	4	21	38	—	—
1945	607	261,186	502,442	651	11,619	1,040	4	12	26	—	—
1950	1,113	229,325	683,924	892	14,809	900	3	7	19	—	—
1955	1,520	283,850	1,020,395	710	14,555	715	—	—	—	—	—
1960	1,635	259,549	1,311,253	517	17,444	500	—	—	—	—	—
1965	1,863	398,793	1,799,000	603	30,054	773	—	—	—	300	300
1970	2,510	479,483	2,198,000	812	43,782	928	—	—	—	3,100	600
1975	2,624	522,759	1,892,000	849	59,829	1,031	—	—	—	6,120	3,956
1980	7,871	691,724	1,911,628	975	17,531	1,134	—	—	—	8,905	5,287

a. Physical quantities: Coal and Coke = 000 tons; Oil = 000 bbls; Natural Gas = 000,000 cubic feet; NGL = 000,000 gallons; Hydroelectricity = 000,000 kWh; Fuelwood = 000 cords; Waterpower, Wind power, and Animal power = 000 tons of coal; Nuclear and Geothermal = 000,000 kWh.

b. Includes natural gasoline and other natural gas liquids.

c. Includes sail and windmill power.

Sources: U.S. Minerals Yearbook; annual reports of the California State Mineralogist; Federal Power Commission data; California, *Statistical Abstract,* 1970, 1986, and 1990; and infra, Tables C–2, C–3, C–4, C–5, C–6, C–8, and C–14. Coal equivalents for waterpower, wind power, and animal power based on Sam H. Schurr and Bruce C. Netschert, *Energy in the American Economy,* 1850–1975 (Baltimore: Johns Hopkins University Press, 1960), 485–487.

Table B–2. California Energy Consumption by Resource in Trillions of BTUS, 1850–1980[a]

Year	Coal and Coke	Oil	Natural Gas	NGL[b]	Hydro-electricity	Fuel Wood	Hydraulic Power	Wind Power[c]	Animal Power	Nuclear Power	Geothermal Power
1850	1.3	—	—	—	—	9.3	—	7.7	4.5	—	—
1855	2.3	—	—	—	—	16.5	.6	19.5	16.8	—	—
1860	2.0	—	—	—	—	23.7	1.1	18.4	29.9	—	—
1865	3.9	—	—	—	—	29.4	1.5	18.3	20.5	—	—
1870	9.0	—	—	—	—	35.1	1.9	25.4	20.1	—	—
1875	14.7	—	—	—	—	36.4	2.1	30.3	19.3	—	—
1880	18.7	.2	—	—	—	37.6	2.3	30.9	18.8	—	—
1885	27.3	1.9	—	—	—	36.5	2.8	28.0	20.7	—	—
1890	33.1	1.8	—	—	—	35.4	3.3	29.5	29.8	—	—
1895	44.8	6.9	.1	—	1.6	34.2	3.3	27.0	32.9	—	—
1900	51.4	24.8	—	—	7.9	31.7	3.3	24.5	27.5	—	—
1905	30.3	178.9	.2	—	14.8	30.8	3.2	23.1	19.1	—	—
1910	24.2	386.0	3.0	—	32.6	31.1	3.3	15.4	21.3	—	—
1915	18.2	527.5	23.5	1.0	51.5	31.2	1.9	9.4	12.9	—	—
1920	29.8	661.0	71.0	3.7	90.1	31.4	1.5	7.6	9.6	—	—
1925	11.9	943.9	201.9	22.2	117.1	29.1	.7	4.4	5.4	—	—
1930	7.6	926.1	359.9	57.7	124.0	26.9	.4	2.4	2.8	—	—
1935	5.2	843.8	305.4	35.9	126.3	25.2	.3	1.4	1.7	—	—
1940	3.8	909.0	378.3	42.9	134.6	23.8	.1	.5	1.0	—	—
1945	15.8	1,514.9	540.1	52.9	174.9	20.2	.1	.3	.7	—	—
1950	29.0	1,330.1	735.2	72.5	192.2	17.5	.1	.2	.5	—	—
1955	39.7	1,646.3	1,096.9	57.7	170.8	13.9	—	—	—	—	—
1960	42.6	1,505.4	1,409.6	42.0	187.9	9.7	—	—	—	—	—
1965	49.0	2,313.0	1,933.9	49.0	324.8	15.0	—	—	—	3.2	3.2
1970	66.0	2,781.0	2,362.9	66.0	472.0	18.0	—	—	—	33.0	6.4
1975	69.0	3,032.0	2,033.9	69.0	645.0	20.0	—	—	—	66.0	42.6
1980	207.0	4,012.0	2,055.0	79.2	189.0	22.0	—	—	—	96.0	57.0

a. British Thermal Unit values computed from physical quantities of each energy source consumed.
b. Includes natural gasoline and other natural gas liquids.
c. Includes sail and windmill power.
Source: Supra, Table B–1. BTU conversions based largely upon Schurr and Netschert, *Energy in the American Economy.*

Table B–3. California Energy Consumption by Resource as a Percentage of Total State Consumption, 1850–1980[a]

Year	Coal and Coke	Oil	Natural Gas	NGL[b]	Hydro-electricity	Fuel Wood	Hydraulic Power	Wind Power[c]	Animal Power	Nuclear Power	Geothermal Power
1850	5.7	—	—	—	—	40.6	.2	33.7	19.8	—	—
1855	4.1	—	—	—	—	29.8	1.0	34.8	30.3	—	—
1860	2.7	—	—	—	—	31.5	1.5	24.5	39.8	—	—
1865	5.3	—	—	—	—	39.9	2.1	24.8	27.9	—	—
1870	9.8	—	—	—	—	38.4	2.1	27.8	22.0	—	—
1875	14.3	—	—	—	—	35.4	2.1	29.5	18.8	—	—
1880	17.2	.2	—	—	—	34.7	2.1	28.5	17.3	—	—
1885	23.3	1.6	—	—	—	31.1	2.4	23.9	17.7	—	—
1890	24.9	1.3	—	—	—	26.7	2.5	22.2	22.4	—	—
1895	30.0	4.6	.1	—	1.0	22.7	2.2	17.9	21.8	—	—
1900	30.2	14.0	—	—	4.6	18.6	2.0	14.4	16.1	—	—
1905	10.1	59.6	.1	—	4.9	10.3	1.1	7.7	6.4	—	—
1910	4.7	74.7	.6	—	6.3	6.0	.6	3.0	4.1	—	—
1915	2.7	77.9	3.5	.2	7.6	4.6	.3	1.4	1.9	—	—
1920	3.3	73.0	7.8	.4	9.9	3.5	.2	.8	1.1	—	—
1925	.9	70.6	15.1	1.7	8.8	2.2	—	.3	.4	—	—
1930	.5	61.4	23.9	3.8	8.2	1.8	—	.2	.2	—	—
1935	.4	62.7	22.7	2.7	9.4	1.9	—	.1	.1	—	—
1940	.3	60.8	25.3	2.9	9.0	1.6	—	—	.1	—	—
1945	.7	65.3	23.3	2.3	7.5	.9	—	—	—	—	—
1950	1.2	55.9	30.9	3.1	8.1	.7	—	—	—	—	—
1955	1.3	54.4	36.3	1.9	5.6	.5	—	—	—	—	—
1960	1.3	47.1	44.1	1.3	5.9	.3	—	—	—	—	—
1965	1.0	49.3	41.2	1.0	6.9	.3	—	—	—	.1	.1
1970	1.1	47.9	40.7	1.1	8.1	.3	—	—	—	.6	.1
1975	1.2	50.7	34.0	1.2	10.8	.3	—	—	—	1.1	.7
1980	3.1	69.7	30.6	1.2	2.8	.3	—	—	—	1.4	.8

a. Percentages computed from British Thermal Unit values of total physical quantities of each energy source consumed.
b. Includes natural gasoline and other natural gas liquids.
c. Includes sail and windmill power. Windmill power nears .2 percent of total wind power only in 1900.
Source: *Supra,* Table B–2.

APPENDIX C Other Statistical Data

Table C–1. California and United States Farm and Common Laborers Comparative Monthly Wages, 1850–1954

	Farm Laborers			*Common Laborers*		
Year	*U.S.*	*California*	*Percent Above*	*U.S.*	*California*	*Percent Above*
1850	$10.85	$60.00	453	$22.62	$130.00	475
1860	13.66	33.28	124	27.56	68.12	147
1870	16.57	29.19	76	40.30	60.06	49
1880	11.70	25.67	119	31.98	45.50	42
1890	13.93	22.40	61	37.96	50.96	34
1899	4.56	25.64	76	—	—	—
1909	21.30	34.17	60	—	—	—
1919	41.52	65.30	57	85.28	91.52	7
1929	40.40	62.50	55	—	—	—
1940	28.05	46.50	66	106.08	122.72	16
1948	91.00	161.00	76	—	—	—
1950	83.20	137.60	65	—	—	—
1954	116.80	163.20	40	—	—	—

Source: Stanley Lebergott, *Manpower in Economic Growth: The American Record since 1800* (New York: McGraw-Hill Book Company, 1964), 539–541. Common laborer monthly wages computed from daily wage figures multiplied by 26.

Table C–2. Work Animals in California, 1850–1950

Year	*Total Farms*	*Oxen*	*Horses*	*Mules*	*Total Animals*	*Number Per Farm*
1850	872	4,780	15,203	1,166	21,149	24.2
1855	9,000*	26,579	63,816*	1,872*	80,688	10.3
1860	18,716	26,000	112,427	2,577	141,004	7.5
1865	21,000*	20,000*	123,509*	7,425*	150,934	7.2
1870	23,724	14,000*	134,591	12,273	160,864	6.8
1875	29,000*	8,000*	161,070	16,240	185,310	6.4
1880	35,934	2,290	166,397	19,840	188,200	5.2
1885	44,000*	1,000*	185,658	22,086	208,744	4.7
1890	52,894	500*	295,676	37,539	333,715	6.3
1895	62,000*	—	359,545	44,123	403,668	6.5
1900	72,542	—	294,905	59,341	354,246	4.9
1905	80,000*	—	254,337	46,453	300,790	3.8
1910	88,197	—	346,976	51,623	398,599	4.5
1915	103,000*	—	379,610	53,900	433,510	4.2
1920	117,670	—	317,902	50,101	368,003	3.1
1925	136,409	—	254,573	45,241	299,814	2.2
1930	135,676	—	189,811	34,363	224,174	1.7
1935	150,360	—	161,261	33,343	194,604	1.3
1940	132,650	—	137,475	21,123	158,598	1.2
1945	138,917	—	122,730	12,231	134,961	1.0
1950	137,168	—	105,797	6,150	111,947	.8

* Estimated from available data.

Sources: Computed from data from U.S. Department of Commerce, *U.S. Census of Agriculture, 1950,* 2:48, 385, 387. California State Board of Agriculture, *Annual Report of the Statistician* (Sacramento, 1915), 42, 44. Hubert Howe Bancroft, *History of California,* (San Francisco: The History Company, 1890), 7:56, n. 4. Paul W. Gates, *California Ranchos and Farms, 1846–1862* (Madison: State Historical Society of Wisconsin, 1967), 31. J.F. Dewhurst and Associates, *America's Needs and Resources: A New Survey* (New York: Twentieth Century Fund, 1955), 1108. Dewhurst suggests that 15.7 percent of the total U.S. work animals were off the farm in 1850, 16.2 percent in 1860, and 22.5 percent in 1870. The figures here reflect those estimates. To conform to U.S. census data indicating California exceeded the national average for horses housed in urban places at the turn of the century, a slightly higher percentage was used in computing farm work animals from 1880 to 1930.

Table C–3. California Coal and Fuelwood Consumption, 1850–1935

	Coal (000 tons)					Wood (000 cords)
Year	California	Britain Australia	Pacific Northwest	Other Imports	Total	Total
1850	—	—	—	—	50	443
1855	—	—	—	—	86	786
1860	—	14	15	48	78	1,129
1865	61	27	34	28	150	1,403
1870	142	115	48	38	343	1,677
1875	167	195	171	29	562	1,735
1880	237	127	328	21	712	1,793
1885	72	398	524	30	1,023	1,742
1890	111	232	857	46	1,245	1,691
1895	75	474	1,093	38	1,680	1,633
1900	177	233	1,453	60	1,923	1,574
1905	47	—	—	1,100	1,100	1,588
1910	11	175	306	373	865	1,602
1915	10	—	284	319	614	1,610
1920	2	—	—	1,032	1,034	1,619
1925	1	—	—	369	370	1,501
1930	11	—	—	204	215	1,388
1935	8	—	—	117	125	1,296

Sources: For coal, California Division of Mines, *California Mineral Production and Directory of Mineral Producers for 1941,* Bulletin No. 122 (San Francisco: State Printing Office, September 1942), 17. U.S., Bureau of Mines, *Mineral Resources of the United States,* annual issues from 1882 to 1935. For wood, R.V. Reynolds and A.H. Pierson, *Fuel Wood in the United States, 1630–1930,* U.S. Department of Agriculture, Forest Service Circular No. 641 (Washington, D.C.: Government Printing Office, February 1942).

Table C–4. Estimated California Windmill Power, 1855–1950

Year	Total Windmills	HP Hours Per Year (000)	Year	Total Windmills	HP Hours Per Year (000)
1855	200	100	1905	28,000	14,000
1860	1,000	500	1910	30,869	15,435
1865	2,520	1,260	1915	41,200	20,600
1870	2,847	1,424	1920	47,068	23,534
1875	4,350	2,175	1925	47,743	23,876
1880	7,187	3,594	1930	40,703	20,352
1885	11,000	5,500	1935	30,072	15,036
1890	13,246	6,623	1940	19,899	9,950
1895	18,600	9,300	1945	13,891	6,946
1900	21,763	10,882	1950	6,858	3,429

The only data for the total number of windmills in California is an estimate of 26,000 farm mills for the year 1924 (L.J. Fletcher and C.D. Kinsman, *The Tractor on California Farms,* University of California Agricultural Experiment Station, Bulletin 415 [Berkeley, December 1926], 20). At one per farm, this would mean nineteen percent of the state's farms had a mill. A higher estimate for 1924 is used here to insure inclusion of nonfarm residences with mills for domestic water supply. Other annual estimates are based on probable annual mill sales and a percentage of residences expected to have mills. The average windmill generated approximately 500 horsepower per year.

Table C–5. Sailing Ships in California and Energy Equivalent in Coal, 1850–1905

Year	Tonnage of Coastal/ Inland Vessels[a]	Tonnage of San Francisco Arrivals[b]	Wind Power in Tons of Coal
1850	—	450,000	924,000
1855	500[c]	403,500	755,067
1860	1,250[c]	400,500	702,429
1865	4,750[c]	488,000	696,500
1870	15,545	738,500	968,785
1875	30,371	1,070,100	1,155,884
1880	43,639	1,105,800	1,174,784
1885	69,238	1,168,300	1,062,577
1890	96,100	1,089,800	1,117,263
1895	116,784	1,016,900	1,020,480
1900	157,182	907,700	921,926
1905	218,255	1,056,400	870,864

a. Cumulative tonnage of ships built in California.

b. Calculations for tons of coal for ships arriving at San Francisco assumes they spent half their sea life in trade with California.

c. Estimated from available data.

Sources: Benjamin C. Wright, *San Francisco's Ocean Trade—Past and Future: A Story of the Deep Water Service of San Francisco, 1848–1911* (San Francisco: A. Carlisle and Co., 1911), 211–212; John Lyman, *The Sailing Vessels of the Pacific Coast and Their Builders, 1850–1905,* Bulletin No. 2, Maritime Research Society of San Diego in Americana, 35 (April 1941). Wind energy in tons of coal is based on calculations by J. Frederic Dewhurst and Associates, *America's Needs and Resources,* 1104, and coal efficiency factors developed by Schurr and Netschert, *Energy in the American Economy,* 484–485.

Table C–6. California Waterpower, 1850–1930

Year	Effective Rated Horsepower		Equivalent Horsepower Hours (000)[b]	Tons of Coal (000)[c]
	Manufacturing	Mining		
1850	200[a]	—	202	1785
1860	2,500[a]	2,500[a]	5,655	2,283
1870	6,877	4,090[a]	13,772	74,302
1880	4,850	10,000[a]	20,579	86,844
1890	5,122	12,063	33,511	125,166
1900[d]	5,164	15,000	39,320	127,200
1910[d]	7,670	21,483	56,848	125,351
1920[d]	7,704	13,303	40,964	56,735
1930[d]	10,122	1,881	22,431	16,487

a. Estimated from available data.

b. Effective horsepower hours based on 2,600 working hours per year and efficiency ratings developed in J. Frederick Dewhurst and Associates, *America's Needs and Resources,* 1104–1105.

c. Equivalent tons of coal calculated using method by Schurr and Nestchert, *Energy in the American Economy,* 481–487.

d. After 1900, mines and manufacturers with waterwheels switched rapidly from using direct hydraulic power to generating their own electric power.

Sources: Eleventh Census, 1890, *Mineral Industries,* 7:61. Twelfth Census, 1900, *Manufacturers, Pt. I,* 7:cccxx. U.S. Census, *Manufacturers, 1905, Pt. IV, Special Reports,* 630; Thirteenth Census, 1910, *Mines and Quarries, 1909,* 11:718; Fourteenth Census, 1920, *Mines and Quarries, 1919,* 11:81. Fifteenth Census, 1930, *Mines and Quarries, 1929,* 84. Treasury Department, *Statistics of Mines and Mining, 1870,* 462–469. Louis C. Hunter, *A History of Industrial Power in the United States, 1780–1930, Volume One: Waterpower in the Century of the Steam Engine* (Charlottesville, VA: University of Virginia Press, 1979), 400. John J. Powell, *The Golden State and Its Resources* (San Francisco: Bacon and Co., 1874), 117.

Table C–7. Power in California Mining Mills, 1896*

Power Source	Stamp Mills	Percent	Other Mills	Total Percent	Mills	Percent
Water	234	42	68	37	302	41
Water & Steam	14	3	1	—	15	2
Not Specified	166	30	70	38	236	32
Steam	132	24	36	19	168	23
Electricity	4	—	1	—	5	—
Gasoline	6	1	2	1	8	1
Animal	—	—	9	5	9	1
Total	556	100	187	100	743	100

*For 28 counties.

Compiled from: California State Mining Bureau, *Thirteenth Report (Third Biennial) of the State Mineralogist, Two Years ending September 15, 1896* (Sacramento, 1896), 65–503 passim.

Table C–8. California Oil and Natural Gas Production, 1880–1960

Year	*Oil (000 barrels)*	*Natural Gas (M. cubic feet)*	*NGL (000 barrels)*
1880	41	—	—
1885	325	—	—
1890	307	41	—
1895	1,245	110	—
1900	4,320	41	—
1905	34,276	148	—
1910	77,698	2,765	—
1915	91,147	21,891	306
1920	103,337	66,041	1,148
1925	232,492	187,789	7,219
1930	228,100	334,789	19,755
1935	207,832	284,109	12,962
1940	223,881	351,950	15,122
1945	326,482	502,442	21,232
1950	327,607	558,398	28,328
1955	354,812	538,178	30,727
1960	305,352	517,535	28,644
1965	316,428	660,384	23,688
1970	375,291	677,689	21,193
1975	296,035	299,535	19,962

Sources: U.S., Bureau of Mines, *Mineral Resources of the United States,* 1906 to the present. California, Department of Natural Resources, Division of Mines, Bulletin No. 156, *Mineral Commodities of California* (San Francisco, August 1950). California, Department of Finance, *Statistical Abstract, 1970,* 13, Table B–5; California, State Mining Bureau, Bulletin No. 69, *Petroleum Industry of California,* by R.P. McLaughlin and C.A. Waring (Sacramento, October 1914). California, Division of Mines, Bulletin No. 122. *California Mineral Production, 1941* (San Francisco, 1942).

Table C–9. Consumption of Fuel Oil in California, 1919

Class of use	*Average per month (barrels)*	*Total for year (barrels)*	*Percent of total excluding steamships*
Railways	954,400	11,453,000	39.66
Government and Municipal	101,869	1,222,000	4.23
Public Utilities	459,726	5,517,000	19.10
Heating	118,950	1,427,000	4.94
Agriculture	9,485	114,000	.40
Industrial	664,534	7,974,000	27.62
Miscellaneous	97,357	1,168,000	4.05
Total, exclusive of ships	2,406,321	28,875,000	100.00
Steamships	346,724	4,161,000	—
Grand Total	2,753,045	33,037,000	—

Source: U.S., Geological Survey, Water Supply Paper No. 493, *Hydroelectric Power Systems in California and Their Extensions into Oregon and Nevada*, by Frederick Hall Fowler (Washington, D.C., 1923), 870, Table 182.

Table C–10. Energy Resources Used in California and United States Manufacturing by Percentage, Selected Years

Year	*Coal and Coke*	*Fuel Oil*	*Natural Gas*	*Purchased Electricity*
		California		
1910	.8	99.1	.1	—
1915	.4	99.1	.5	—
1919	.6	98.1	1.3	—
1929	.3	92.6	4.8	2.3
1937	.9	19.7	67.4	12.0
1947	3.8	17.3	57.5	21.4
1954	8.5	9.2	61.8	20.5
1958	.6	6.6	59.1	33.7
1962	.5	6.6	59.9	33.0
		United States		
1919	11.5	87.9	.6	—
1929	38.3	8.3	9.6	3.8
1937	54.8	7.8	30.0	7.4
1947	40.3	8.4	37.4	13.9
1954	28.2	7.9	46.9	17.0
1958	25.9	10.0	34.8	29.3

Sources: Computed from fuel data for manufacturing establishments found in manufacturing reports of the U.S. Census, 1910–1962.

Table C–11. California Oil Consumption in Thousands of Barrels, 1915–1960

Year	Fuel Oils	Gasoline	Kerosene	Lube Oils	Road Oil	Asphalt
1915	23,588	2,002	556	174	—	—
1925	74,817	19,334	500*	900*	900*	2,500*
1930	83,049	31,799	450*	1,800*	1,732	3,463
1935	70,631	35,910	420	1,821	1,988	2,463
1940	71,516	45,904	1,089	3,338	2,641	4,163
1945	146,267	65,760	1,691	3,820	675	7,455
1950	97,609	91,776	2,871	4,118	2,644	9,131
1955	107,832	126,000	1,000*	5,100*	2,600*	9,500*
1960	105,471	143,253	1,015	5,818	1,560	9,753

*Estimated from available data.

Sources: U.S., Bureau of Mines, *Mineral Resources of the United States,* annual issues from 1915 to 1960. U.S., Federal Trade Commission, *Report of the Federal Trade Commission on the Pacific Coast Petroleum Industry,* 2 parts (Washington, D.C., 1921 and 1922), 2:234–255. California, Department of Finance, *Statistical Abstract, 1970,* 13, Table B–5.

Table C–12. Automobiles and Trucks on California Farms, 1900–1970

Year	Total Autos	Farm Autos	Farm Trucks	Percent Farms With Autos	Percent Farms With Trucks
1915	191,196	16,380*	—	—	—
1920	595,187	71,518	6,416	53.1	5.0
1925	1,475,852	104,180*	23,694*	61.0*	15.2*
1930	2,099,293	136,842	40,971	76.8	26.3
1935	2,254,828	143,688*	49,493*	70.0*	27.4*
1940	2,955,952	150,534	58,015	79.9	35.1
1945	3,118,840	160,252	85,696	82.5	44.6
1950	4,976,296	171,305	110,563	80.8	51.9
1955	6,649,765	173,590	129,520	86.2	63.0
1960	8,569,295	127,098	131,019	88.1	71.1
1965	11,191,19	9110,159	126,135	88.4	78.8
1970	14,000,00	69,493	115,739	61.8	74.4

*Estimated from available data. 1915 autos estimated at 10 percent total automobile registration in California.

Sources: U.S., Department of Commerce, Bureau of the Census, *U.S. Census of Agriculture, 1950,* 2:223, 226, 232. U.S., Department of Commerce, Bureau of the Census, *United States Census of Agriculture, 1959,* v. 1, pt. 48:7. U.S., Department of Commerce, Bureau of the Census, *1969 Census of Agriculture,* v. 1, pt. 4:4. California, State Board of Agriculture, *Annual Report of the Statistician* (Sacramento, 1917), 245.

Table C–13. Tractors and Horses and Mules on California Farms, 1900–1970

Year	*Farms*	*Total Tractors*	*Percent Farms With Tractors*	*Horses and Mules*	*Percent Farms With Horses & Mules*
1900	72,542	—	—	354,246	91.0[a]
1905	80,000[a]	—	—	300,790	—
1910	88,197	—	—	398,599	90.0[a]
1915	103,000[a]	—	—	433,510	—
1920	117,670	13,852	10.3	368,003	78.0[a]
1925	136,409	29,948	19.0	299,814	61.1
1930	135,676	44,437	27.6	224,174	49.9
1935	150,360	49,814[a]	27.1[a]	194,604	43.7
1940	132,650	55,191	33.1	158,598	38.0
1945	138,917	79,289	40.6	134,961	30.0[a]
1950	137,168	124,582	54.2	111,947	23.8
1955	123,075	149,384	62.6	82,260[b]	19.7
1960	99,274	160,999	71.3	77,326[b]	20.2
1965	80,852	142,825	72.0	—	—
1970	77,875	138,914	72.3	—	—

a. Estimated from available data.

b. No adjustments made (see *Supra,* Table C–2, source note).

Sources: U.S., Department of Commerce, Bureau of the Census, *U.S. Census of Agriculture, 1950,* 2:384, 386, 357. U.S., Department of Commerce, Bureau of the Census, *United States Census of Agriculture, 1959,* v. 1, pt. 48:7, 9. U.S., Department of Commerce, Bureau of the Census, *1969 Census of Agriculture,* v. 1, pt. 4:4. California, State Board of Agriculture, *Annual Report of the Statistician* (Sacramento, 1917), 245.

Table C–14. California Electric Power Generation, 1895–1975 (Millions of Kilowatt Hours)

Year	Water Power	Steam Power[a]	Nuclear	Geothermal and Other[b]	Total
1895	17[c]	—	—	—	—
1900	86[c]	98[d]	—	—	184
1905	300[c]	—	—	—	—
1910	879[c]	381[d]	—	—	1,260
1915	1,885	368	—	—	2,253
1920	2,568	1,168	—	—	3,736
1925	5,171	1,043	—	—	6,214
1930	6,804	2,144	—	—	8,948
1935	7,751	1,309	—	—	9,060
1940	8,781	1,068	—	—	9,789
1945	11,619	2,817	—	—	14,136
1950	14,809	10,027	—	—	24,836
1955	14,555	27,957	—	—	42,512
1960	17,444	46,390	[e]	[e]	63,834
1965	30,523	62,527	[e]	[e]	93,050
1970	37,867	84,147	[e]	[e]	122,014
1975	40,780	86,193	6,120	3,956	137,049

a. Includes internal combustion generation.

b. Includes wind and solar power generation.

c. Estimated based on 3 million kWh per kW capacity given in "Western Hydroelectric Transmission Developments," *JEPG,* 35 (June 5, 1915): 442.

d. Estimated from available data.

e. Included with steam power generation.

Sources: U.S., Federal Power Commission, *Report to the Federal Power Commission on the Water Powers of California,* by Frank E. Bonner (Washington, D.C., 1928), 24, 28–29. U.S., Federal Power Commission, *Federal Power Statistics, 1920–1940* (Washington, D.C., 1941), 28, Table 1. Consumption of Fuel for Production of Electrical Energy (Washington, D.C., 1945). California, Railroad Commission, *Biennial Report, 1945/46* (Sacramento, 1946). California, Department of Finance, *Statistical Abstract, 1970,* 241, Table K-1, *1972,* 110, Table K–1, and *1978,* 126, Table K–1.

Table C–15. California and United States Manufacturing Horsepower (000), Selected Years, 1870–1930

Year	Steam Engines	Waterpower	Gas Engines	Generated Electricity	Purchased Electricity
			California		
1870	18.5	6.9	—	—	—
1880	28.1	4.9	—	—	—
1890	64.9	5.1	.4	.4	—
1900	105.9	5.2	3.3	6.1	9.7
1905	153.2	7.3	6.3	10.2	39.4
1910	193.5	7.7	10.1	27.1	116.5
1915	207.7	7.5	15.6	32.6	258.7
1920	211.3	7.7	25.0	51.1	521.2
1923	219.9	10.7	15.8	88.1	724.9
1925	228.4	7.9	22.2	125.6	874.8
1927	261.1	10.5	26.1	145.6	1,041.9
1930	296.6	10.1	31.9	168.2	1,230.5
			United States		
1870	1,215.7	1,130.4	—	—	—
1880	2,185.5	1,225.4	—	—	—
1890	4,581.6	1,225.2	8.9	15.6	—
1900	8,742.4	1,727.3	143.9	311.0	183.7
1905	10,825.3	1,647.9	289.4	150.9	441.6
1910	14,199.3	1,822.9	751.2	3,068.1	1,749.0
1920	17,038.0	1,765.3	1,259.4	6,969.7	9,347.6
1923	16,700.4	1,803.3	1,224.2	8,821.4	13,364.3
1925	16,915.7	1,800.8	1,185.7	10,254.7	15,864.6
1927	16,923.9	1,598.7	1,170.8	11,220.0	19,132.3
1930	17,361.9	1,559.6	1,233.9	12,376.4	22,775.7

Source: U.S. Census, 1870–1930.

Table C–16. California Farms with Electricity, 1900–1950

Year	*Total Farms*	*Farms With Central Station Service*	*Farms With Home Plants*	*Total Farms With Electricity*	*Percent Farms With Electricity*
1900	72,542	—	—	—	5.0*
1910	80,000*	—	—	—	10.0*
1915	88,197	14,000	1,500*	15,500*	17.6*
1920	103,000*	27,467*	2,410	29,877*	29.0
1925	136,409	—	—	60,000*	44.0*
1930	135,676	80,068	5,873*	85,941	63.3
1940	132,650	107,904	2,110	110,014	82.9
1945	138,917	114,308*	1,740*	116,048	83.5
1950	137,168	123,311	1,257	124,568	90.9

*Estimated from available data.

Sources: J.H. Davidson and F.E. Boyd, "The Electric Tractor on the Farm," *Journal of Electricity*, 41 (October 1, 1918): 297. U.S. Department of Commerce, *U.S. Census of Agriculture, 1950*, 2:212–213. "Electric Service in the American Home," *Electrical West*, 75 (May 15, 1920): 1134.

Table C–17. California Irrigation Pumping, 1870-1950

Year	*Total Farms*	*Percent Farms With Any Irrigation*	*Total Pumps*	*Percent Electric Pumps*	*Total Horsepower*
1870	23,724	—	45	—	568
1880	35,934	—	107	—	2,091
1890	52,894	—	408	—	13,478
1895	62,000[a]	25.0[a]	697[a]	—	20,910[a]
1900	72,542	35.4	986	—	24,933
1905	80,000[a]	40.0[a]	2,139	10.0[a]	45,106
1910	88,197	44.6	9,297	20.0[a]	128,143
1915	103,000[a]	52.0[a]	12,333	40.0[a]	208,123
1920	117,670	57.3	21,561	60.0[a]	414,726[b]
1925	136,409	60.0[a]	4,145[a]	75.0[a]	607,781[a]
1930	135,676	63.2	46,729	83.0	862,515[b]
1950	137,168	66.2	79,936	92.0	1,358,912[a]

a. Estimated from available data.

b. Includes horsepower for drainage.

Sources: U.S., Department of Commerce, Bureau of the Census, *Twelfth Census of the United States, 1900: Agriculture, Part II*, 6:830. U.S., Department of Commerce, Bureau of the Census, *Fourteenth Census of the United States, 1920: Irrigation and Drainage*, 7:134, 139. U.S., Department of Commerce, Bureau of the Census, *Fifteenth Census of the United States, 1930: Irrigation of Agricultural Lands*, 41, 93. U.S., Department of Commerce, *U.S. Census of Agriculture, 1950*, 2:212, and 3:35, 65.

Table C–18. Percentage of Selected Electric Appliances in California Electrified Homes, 1925*

Appliance	*51,000 Farm Homes*	*1,000,000 Total Homes*	*Appliance*	*51,000 Farm Homes*	*1,000,000 Total Homes*
Flat irons	69.8	100.0	Space or air heaters	18.8	29.0
Washing machines	36.1	35.0	Percolators	18.5	43.0
Vacuum cleaners	35.7	66.0	**Water heaters**	13.5	.8
Toasters	34.2	62.0	Waffle irons	12.9	18.0
Curling irons	24.0	23.0	Sewing machines	12.8	14.0
Ranges	22.7	1.3	Table grills	11.0	24.0
Portable fans	20.0	19.0	Heating pads	10.7	20.0

*Bold type indicates those items for which farm homes lead all homes.

Sources: "Appliance Sales Possibilities of the West," *Journal of Electricity,* 52 (February 1, 1924): 92–95. "Electrical Usage on California Farms," *Electrical World,* 90 (August 6, 1927): 269–270. University of California, Agricultural Experiment Station, Circular 316, *Electrical Statistics for California Farms,* by B.D. Moses (Berkeley, October 1929).

Table C–19. Status of Farm Housing in California Compared with Iowa: 1934 and 1950

	Percent of Farm Homes With Characteristic			
	1934[a]		*1950*[b]	
Farm Home Characteristics	*California*	*Iowa*	*California*	*Iowa*
Lighting:				
Kerosene/gasoline	11.8	73.1	—	—
Acetylene	.4	2.6	—	—
Piped gas	2.8	.6	—	—
Electricity				
Home power plant	.8	11.7	.9	.6
Powerline service	86.8	15.3	90.0	90.3
Heating:				
Fireplace	16.7	.8	—	—
Stove	80.0	61.9	—	—
Circulating heater	11.3	19.7	—	—
Furnace	4.1	29.4	—	—
Cooking:				
Wood/coal	74.4	95.4	—	—
Kerosene/gasoline	24.7	49.2	—	—
Gas	16.5	1.9	—	—
Electric	20.0	.9	—	—
Refrigeration:				
Ice	35.0	15.1	—	—
Mechanical (electric)	12.6	2.5	64.5	41.5
Home Freezer (electric)	—	—	21.2	14.9

(table continues)

Table C–19. *(continued)*

Farm Home Characteristics	*Percent of Farm Homes With Characteristic* 1934[a] California	1934[a] Iowa	1950[b] California	1950[b] Iowa
Laundry:				
Power washer (electric)	48.0	—	78.3	84.0
Hand washing machine	4.0	—	—	—
Fixed tub	42.0	—	—	—
Laundry location:				
Launder outside	40.0	—	—	—
Launder in kitchen	13.0	—	—	—
Launder in basement	3.0	—	—	—
Launder in laundry room	38.0	—	—	—
Water:				
Piped hot water	56.0	—	—	—
Piped cold water	90.0	—	—	—
Hand pump in house	1.0	—	—	—
Electric water pump	—	—	62.5	57.5
Electric water heater	—	—	41.4	24.9
Farm Power:				
Electric chick brooder	—	—	16.2	28.5
Electric feed grinder	—	—	1.9	1.3
Electric milker	—	—	11.3	—

a. Based upon a survey of 14,185 farm homes with an average house size of 5.46 rooms and family size of 3.9 in Alameda, Fresno, Sacramento, Santa Clara, Sonoma, and Stanislaus Counties (approximately 10 percent of the California's farms), and a comparable Iowa survey, both part of a Federal Civil Works Administration project which covered 300 counties nationwide. Percentages in each major subdivision may add up to more than 100 because farm homes fit in more than one category.

b. Computed from 1950 Census data.

Source: [Mary Seacrest], *California Annual Narrative Report, Acting Extension Specialist in Home Management, December 1, 1933–November 30, 1934* (Berkeley: College of Agriculture, Extension Service, April 15, 1935), 7–9, and Exhibits 2 and 29. U.S. Bureau of the Census, *U.S. Census of Agriculture, 1952,* 2:212–213.

Table C–20. Sources of California Oil Consumption, 1940–1975 (000 Barrels)

Year	*California*	*Other States*	*Foreign*	*Year*	*California*	*Other States*	*Foreign*
1940	155,717	1,000	—	1960	300,000	17,000	64,000
1945	246,586	14,600	—	1965	313,000	25,000	74,000
1950	223,125	6,200	—	1970	379,000	74,000	55,000
1955	256,850	12,000	15,000	1975	296,000	60,000	189,000

Sources: Computed from U.S., Energy Research and Development Administration, *Distributed Energy Systems in California's Future, A Preliminary Report,* by M. Christensen, *et al.,* 2 vols. (Washington, D.C., September 1977), 1:43. U.S. Bureau of Mines, *Mineral Resources of the United States,* annual issues.

Table C–21. Sources of California Natural Gas Consumption, 1940–1975 (Millions of Cubic Feet)

Year	*California*	*Southwest*	*Canada*	*Year*	*California*	*Southwest*	*Canada*
1940	351,950	—	—	1960	473,253	838,000	—
1945	502,442	—	—	1965	644,000	1,004,000	151,000
1950	535,885	148,039	—	1970	642,000	1,262,000	294,000
1955	470,395	550,000	—	1975	368,000*	1,159,000	365,000

*After 1970, PG&E began drawing less gas from its fields in favor of imports.

Sources: Computed from U.S., Energy Research and Development Administration, *Distributed Energy Systems in California's Future, A Preliminary Report,* by M. Christensen, *et al.*, 2 vols. (Washington, D.C., September 1977), 1:45. U.S. Bureau of Mines, *Mineral Resources of the United States,* annual issues.

Table C–22. California Electric Power Generation, 1970–1990 (Millions of Kilowatt Hours)

Year	*Hydro*	*Steam*	*Nuclear*	*Geothermal*	*Private Supply*	*Cogeneration*	*Wind and Solar*
1970	37,867	84,147	—	2,975	—	—	—
1975	40,780	86,193	6,120	3,956	—	—	—
1980	40,417	102,594	5,212	5,073	800	814	—
1985	34,769*	79,787	18,911	10,117	2,820	5,565	686
1990	25,906*	67,177	36,610	14,507	9,356	32,251	3,097

*Drought years account for decline in hydroelectric production.

Sources: California, Department of Finance, *Statistical Abstract, 1978,* 126, Table K–1, *1986,* 173, Table, and *1991,* 136, Table J–11.

Table C–23. California Electric Power Production under PURPA, 1982–1990 (Millions of Kilowatt Hours)

Year	*Cogeneration*	*Geothermal*	*Wind Power*	*Solar*	*Hydroelectric*
1982	611	510	4	—	65
1983	1,298	679	49	2	105
1984	2,777	693	189	11	191
1985	4,790	835	655	31	258
1986	7,455	1,263	1,218	60	359
1987	10,286	1,662	1,710	186	409
1988	18,131	2,367	1,824	314	430
1989	25,181	4,818	2,139	469	645
1990	31,756	6,354	2,418	679	478

Source: California, Department of Finance, *Statistical Abstract, 1991,* 136, Table J–11.

Table C–24. California Population, 1850–1990

Year	*State*	*North*	*South**
1850	127,373	121,860	5,513
1860	370,994	346,243	24,751
1870	560,247	528,215	32,032
1880	864,694	800,323	64,371
1890	1,213,398	1,012,046	201,352
1900	1,485,053	1,180,842	304,211
1910	2,377,549	1,626,239	751,310
1920	3,426,861	2,079,811	1,347,050
1930	5,677,251	2,744,456	2,932,795
1940	6,907,387	3,235,024	3,672,363
1950	10,586,223	4,933,974	5,652,249
1960	15,171,204	6,145,510	9,025,694
1970	20,039,000	8,315,600	11,723,400
1980	23,782,000	9,966,900	13,815,100
1990	29,976,000	11,985,600	17,990,400

*Southern California is generally considered to be comprised of eight counties: Imperial, Los Angeles, Orange, Riverside, San Bernardino, San Diego, Santa Barbara, and Ventura.

Sources: California, Department of Finance, *Statistical Abstract, 1970*, 13, Table B–5, and 1993, 13, Table B–3. The Commonwealth Club, *The Population of California* (San Francisco, CA: The Commonwealth Club, 1946), Table 12, 23–25. Warren S. Thompson, *et al., Growth and Changes in California's Population* (Los Angeles: Haynes Foundation, 1955).

Endnotes

The following journal title abbreviations are used:

EW - *Electrical World*

JEPG - *Electrical West.* The title varies over the years: *The Electrical Journal* (July–August 1895), *The Journal of Electricity* (September 1895–June 1899), *The Journal of Electricity, Power and Gas* (July 1899–December 1916), *Journal of Electricity* (January 1917–January 1921), *Journal of Electricity and Western Industry* (February 1921–June 1923), *Journal of Electricity* (July 1923–December 1926), *Electrical West,* January 1927–

PRP - *Pacific Rural Press*

Introduction

1. M. King Hubbert, "Energy from Fossil Fuels," in Smithsonian Institution, *Annual Report of the Board of Regents for 1950* (Washington, D.C., 1951): 271; Hubbert, "The Energy Resources of the Earth," *Scientific American,* 224 (September 1971): 70. Eugene Ayres and Charles Scarlott, *Energy Sources: The Wealth of the World* (New York: McGraw-Hill, 1952). These and other studies are cited by Eugene S. Ferguson, "Historical Roots of the Energy Crisis: A Basic Bibliography of Energy History," *Science, Technology & Society: Curriculum Newsletter of the Lehigh University STS Program,* 33 (December 1982): 1–21. On the exponential growth of energy use, also see Albert A. Bartlett, "Forgotten Fundamentals of the Energy Crisis," *Journal of Geological Education,* 28 (1980): 4–35.

2. William Cronon, *Changes in the Land: Indians, Colonists, and the Ecology of New England* (New York: Hill and Wang, 1983); Cronon, *Nature's Metropolis: Chicago and the Great West* (New York: W.W. Norton, 1991); Theodore Steinberg, *Nature Incorporated: Industrialization and the Waters of New England* (Cambridge: Cambridge University Press, 1991). All discuss the process of identifying resources as commodities for exploitation under the capitalist system. Donald Worster weaves some of these ideas together in *The Wealth of Nature: Environmental History and the Ecological Imagination* (New York: Oxford University Press, 1993), chs. 1 and 4. William G. Robbins's discussion of the fundamental role of capitalism in European-American settlement of the American West, in *Colony and Empire: The Capitalist Transformation of the American West* (Lawrence: University of Kansas Press, 1994) applies to the nation as a whole.

3. Dolores Greenberg's "Energy, Power, and Perceptions of Social Change in the Early Nineteenth Century," *American Historical Review,* 95 (June 1990): 693–714, explains how our basic assumptions about energy and social change are "based on misconceptions first formulated in the early nineteenth century." George Basalla, "Energy and Civilization," in *Science, Technology and the Human Prospect,* Chauncey Starr, ed. (New York: Pergamon Press, 1980), 39–52, carefully questions long-held assumptions that the standard of living, as well as the quality of civilization, is proportional to the quantity of energy a society uses. Langdon Winner, "Energy Regimes and the Ideology of Efficiency," in *Energy and Transport: Historical Perspectives on Policy Issues,* George H. Daniels and Mark H. Rose, eds. (Beverly Hills, CA: Sage Publications, 1982), 261–277, provides a cogent analysis of energy and economic growth in American society.

4. My thinking about this began several years ago, when I discovered Gerald D. Nash's argument that, "historically, three forces have operated to shape the nature of California's economy: environment and climate, population, and technology." I am grateful to him. See Nash, "Stages of California's Economic Growth, 1870–1970: An Interpretation," *California Historical Quarterly,* 51 (Winter 1972): 315–329.

5. Samuel P. Hays, *Beauty, Health, and Permanence: Environmental Politics in the United States, 1955–1985* (New York: Cambridge University Press, 1987), 32–35, discusses the relationship between rising standards of living and environmental values.

6. George Basalla, "Some Persistent Energy Myths," in *Energy and Transport: Historical Perspectives on Policy Issues,* George H. Daniels and Mark H. Rose, eds. (Beverly Hills, CA: Sage Publications, 1982), 27–28. Basalla, "Energy and Civilization," 39–52. Also see Greenberg, "Energy, Power, and Perceptions."

7. Lewis Mumford, "Authoritarian and Democratic Technics," *Technology and Culture,* 5 (Winter 1964): 7. Carroll Pursell, "The American Ideal of Democratic Technology," in *The Technological Imagination: Theories and Fictions,* Teresa de Lauretis, Adreas Huyssen, and Kathleen Woodward, eds. (Madison, WI: Coda Press, 1980), 11–25, discusses this topic in detail.

8. T. Allan Comp, "The Best Arena: Industrial History at the Local Level," *History News,* 37 (May 1982): 10; Greenberg, "Energy, Power, and Perceptions," 694; Thomas Parke Hughes, "Technology as a Force for Change in History: The Effort to Form a Unified Electric Power System in Weimar Germany," in *Industrielles System und politische Entwicklung in der Weimarer Republik,* Hrsg. von Hans Mommsen, D. Petzina, and Bernd. Weisbrod, eds. (Dusseldorf, 1974), 166; David E. Nye, *Electrifying America: Social Meanings of a New Technology, 1880–1940* (Cambridge, MA: The M.I.T. Press, 1990), 182; Winner, "Energy Regimes and the Ideology of Efficiency," 261–277. American society's values as they relate to energy are discussed in Ian Barbour, *et al., Energy and American Values* (New York: Praeger, 1982).

9. Alvin M. Weinberg, "Reflections on the Energy Wars," *American Scientist,* 66 (March–April 1978): 154–155. Daniel T. Spreng and Alvin M. Weinberg, "Time and Decentralization," *Daedalus,* 109 (Winter 1980): 139. Barbour, *et al., Energy and American Values,* 73.

10. See, for example, Martin V. Melosi, *Coping With Abundance: Energy and Environment in Industrial America* (Philadelphia: Temple University Press, 1985); the brief, laudatory narrative by Joseph M. Dukert, *A Short Energy History of the United States (And Some Thoughts About the Future)* (Washington, D.C.: Edison Electric Institute, 1980); Sam H. Schurr and Bruce C. Netschert, *Energy in the American Economy, 1850–1975* (Baltimore: Johns Hopkins Press for Resources of the Future, 1960); Dolores Greenberg, "Energy Flow in a Changing Economy, 1815–1880," *An Emerging Independent American Economy, 1815–1870,* Joseph R. Frese and Jacob Judd, eds. (Tarrytown, NY: Sleepy Hollow Press, 1980), 29–58; John Opie, chs. 1–2 in Barbour, *et al., Energy and American Values.* California historians also ignore their own energy history; the only general state history that deals somewhat broadly with energy resources is W.H. Hutchinson, *California: The Golden Shore by the Sundown Sea,* 2nd ed. (Belmont, CA: Star Publishing Company, 1988).

11. For example, see Harold I. Sharlin, *The Making of the Electrical Age: From the Telegraph to Automation* (London: Abelard-Schuman, 1963).; D. Clayton Brown, *Electricity for Rural America: The Fight for the REA* (Westport, CT: Greenwood Press, 1980); and Nye, *Electrifying America.*

Thomas Parke Hughes, *Networks of Power: Electrification in Western Society, 1880–1930* (Baltimore: Johns Hopkins University Press, 1983), devotes a chapter to California's hydroelectric developments. Richard F. Hirsh, *Technology and Transformation in the American Electric Utility Industry* (Cambridge, MA: Cambridge University Press, 1989), uses California as a case study for changes in the electric power industry during the past thirty years.

12. Forest Hill, "The Shaping of California's Industrial Pattern," *Proceedings of the 30th Annual Conference of the Western Economic Association* (1955): 63–68, is a good introduction to California's conditioning factors. Also see Schurr and Netschert, *Energy in the American Economy,* 34.

13. Brooke Hindle, ed., *America's Wooden Age: Aspects of Its Early Technology* (Tarrytown, NY: Sleepy Hollow Restorations, 1975), 12.

14. Cronon, *Nature's Metropolis,* 55–56, 70, 72–73. Also see Donald Worster, "Transformations of the Earth: Toward an Agroecological Perspective in History," *Journal of American History,* 76 (March 1990): 1089.

15. Donald Worster, *Under Western Skies: Nature and History in the American West* (New York: Oxford University Press, 1992), 14, 71, 87. I am particularly grateful to Jeffrey K. Stine for guiding me into the literature exploring the relationship between technology and the environment. See Stine, "Engineering a Better Environment," in *[Proceedings of a] Conference on Critical Problems and Research Frontiers in History of Science and History of Technology, 30 October–3 November, 1991* (Madison, WI: [Society for the History of Technology, 1993]), 310–325; Stine and Joel A. Tarr, "Technology and the Environment: The Historians Challenge," *Environmental History Review,* 18 (Spring 1994): 1–8.

16. Worster, *Rivers of Empire: Water, Aridity, and the Growth of the American West* (New York: Oxford University Press, 1985), 50–51, explicitly makes the link to Mumford's concept of the "megamachine" (Lewis Mumford, *The Myth of the Machine: Technics and Human Development* [New York: Harcourt, Brace & World, 1967]).

17. The quoted phrase is from Worster, "Transformations of the Earth," 1089.

18. Worster, *Under Western Skies,* 251–252. An environmental history of California remains to be written, but a model for such an undertaking is William L. Preston, *Vanishing Landscapes: Land and Life in the Tulare Lake Basin* (Berkeley: University of California Press, 1981).

19. The general history of California is traced in Richard B. Rice, William A. Bullough, and Richard J. Orsi, *The Elusive Eden: A New History of California* (New York: Alfred A. Knopf, 1988), and Walton Bean and James Rawls, *California, An Interpretive History,* 6th ed. (New York: McGraw-Hill, 1993). For colonialism in California's economic past, see Gerald D. Nash, *The American West in the Twentieth Century: A Short History of an Urban Oasis* (Albuquerque: University of New Mexico Press, 1977), 9–41; Hill, "California's Industrial Pattern," 63–68.

20. On the transformation wrought by World War Two, see Gerald D. Nash's *The American West Transformed: The Impact of the Second World War* (Bloomington: Indiana University Press, 1985), and Nash, *World War II & the West: Reshaping the Economy* (Lincoln: University of Nebraska Press, 1990).

Chapter I/A Land Apart

1. Quoted in John McPhee, *Assembling California* (New York: Farrar, Straus and Giroux, 1993), 109.

2. James J. Parsons, "The Uniqueness of California," *American Quarterly,* 7 (Spring 1955): 45.

3. David Hornbeck, *California Patterns: A Geographical and Historical Atlas* (Palo Alto: Mayfield Publishing Co., 1983), 6–10. McPhee, *Assembling California, passim.* W.H. Hutchinson, *California: The Golden Shore by the Sundown Sea,* 2nd ed. (Belmont, CA.: Star Publishing Company, 1988), 5–21.

4. The following description is drawn principally from Richard B. Rice, William A. Bullough, and Richard J. Orsi, *The Elusive Eden: A New History of California* (New York: Alfred A. Knopf, 1988), 8–21; Raymond F. Dasmann, *California's Changing Environment* (San Francisco: Boyd and Fraser Publishing Co., 1981); Allan A. Schoenherr, *A Natural History of California* (Berkeley: Uni-

versity of California Press, 1992); and David N. Hartman, *California and Man* (Dubuque, IA: Wm. C. Brown Co., Inc., 1964). Also see Bern Kreissman, *California: An Evironmental Atlas & Guide* (Davis, CA: Bear Klaw Press, 1991).

5. For prehistoric Native American life, I have relied on Joseph L. Chartkoff and Kerry Kona Chartkoff, *The Archaeology of California* (Stanford: Stanford University Press, 1984), chs. 2–4.

6. Arthur F. McEvoy, *The Fisherman's Problem: Ecology and Law in the California Fisheries, 1850–1980* (Cambridge, MA: Cambridge University Press, 1986), 21; William Cronon, *Changes in the Land: Indians, Colonists, and the Ecology of New England* (New York: Hill and Wang, 1983), 166; Norris Hundley, Jr., *The Great Thirst: Californians and Water, 1770s–1990s* (Berkeley: University of California Press, 1992), 13–19; Preston, *Vanishing Landscapes: Land and Life in the Tulare Lake Basin* (Berkeley: University of California Press, 1981), 31–46.

7. Good general histories of California are Rice, Bullough, and Orsi, *The Elusive Eden;* Walton Bean and James Rawls, *California, An Interpretive History,* 6th ed. (New York: McGraw-Hill, 1993); and David Lavender, *California: Land of New Beginnings* (Lincoln: University of Nebraska Press, 1987). Except where particularly noted, my summary of California's history up to American conquest is drawn from these sources.

8. Raymond F. Dasmann, *California's Changing Environment,* (San Francsico: Boyd & Fraser Publishing Co., 1981), 25; Preston, *Vanishing Landscapes,* 59–60. Also see Gayle M. Groenendaal, "California's First Fuel Crisis and Eucalyptus Plantings" (MA thesis, California State University, Northridge, 1985), 23–25.

9. James C. Williams, "Civil Engineering on the Spanish Frontier: Alta California Water Systems," (paper presented at the Spanish Beginnings in California, 1542–1822, an International Symposium, Santa Barbara, CA, July 1991); Gary Kurutz and Jean Bruce Ward, "Some New Thoughts on an Old Mill," *California Historical Quarterly,* 53 (Summer 1974): 139–164; Edith Buckland Webb, *Indian Life at the Old Missions* (Los Angeles: W.F. Lewis, 1952), 64–65; James Dix Schuyler, *Reservoirs for Irrigation, Water-Power, and Domestic Water-Supply* (New York: J. Wiley and Sons, 1901), 125. Spanish and Mexican California's relationship to water is particularly well summarized in Hundley, *The Great Thirst,* 25–62.

10. Quoted in Walton Bean, *California: An Interpretive History,* 2nd ed. (New York: McGraw-Hill, 1973) 91.

Chapter II/Sweat, Toil, Invention

1. Kenneth Thompson and Richard A. Eigenheer, "The Agricultural Promise of the Sacramento Valley: Some Early Views," *Journal of the West,* 18 (October 1979): 37; Kenneth Thompson, "The Perception of the Agricultural Environment," *Agricultural History,* 49 (January 1975): 232–237; Kenneth Thompson, "Negative Perceptions of Early California," *The California Geographer,* 18 (1978): 1–15; Gilbert C. Fite, *The Farmers' Frontier, 1850–1900* (New York: Holt, Rinehart and Winston, 1966), 164–165.

Examples of works with a positive view of California published in the 1850s include: E. Gould Buffum, *Six Months in the Gold Mines,* John W. Caughey, ed. (Los Angeles: Ward Ritchie Press, 1959); Bayard Taylor, *El Dorado, or, Adventures in the Path of Empire* (Glorieta, NM: Rio Grande Press, 1967); and Frank Marryat, *Mountains and Molehills or Recollections of a Burnt Journal,* intro. by Robin Winks (Philadelphia: Lippincott, 1962).

2. James S. Kemp, "The Foot-hills of the Sierra," *The Resources of California* (June 1883): 9.

3. California State Mining Bureau, *Fourth Annual Report of the State Mineralogist for the Year Ending May 15, 1884* (Sacramento, 1884), 36; Titus Fey Cronise, *Natural Wealth of California* (San Francisco: H.H. Bancroft and Co., 1868), 598, 386. On the perceived abundance of energy resources, see Cronise, *Natural Wealth,* 94–224, 511–599 *passim.* Cronise, *The Agricultural and Other Resources of California* (San Francisco: A. Roman, 1870), 27. Also see: John J. Powell, *The Golden State and Its Resources* (San Francisco: Bacon and Co., 1874), 118, 186; W.W. Morrow, "Our Material Resources," *The Resources of California* (October 1881): 5. For U.S. energy use, see Sam H. Schurr and Bruce C. Netschert, *Energy in the American Economy, 1850–1975* (Baltimore: Johns Hopkins Press for Resources of the Future, 1960), 49–67.

4. Albert L. Hurtado, "California Indians and the Workaday West: Labor, Assimilation, and Survival," *California History,* 69 (Spring 1990): 2–11; Hurtado, *Indian Survival on the California Frontier* (New Haven: Yale University Press, 1988). A number of sources concerning immigrant and minority labor in California are found in Doyce B. Nunis Jr. and Gloria Ricci Lathrop, eds., *A Guide to the History of California* (New York: Greenwood Press, 1989). I have made no attempt to deal with the exploitation of workers performing human labor, although the availability of inexpensive labor has played a part in decisions to use or not to use other energy resources in work.

Human labor is estimated at 3,000 hours of work per year per person; 205,940 work-hours is equivalent to burning one ton of coal (J. Frederic Dewhurst and Associates, *America's Needs and Resources, A New Survey* [New York: Twentieth Century Fund, 1955], 1102). For a good general discussion of human labor, see Louis C. Hunter and Lynwood Bryant, *A History of Industrial Power in the United States, 1780–1930, Volume 3: The Transmission of Power* (Cambridge, MA: The M.I.T. Press, 1991), 3–5, 15–28.

5. Hill, "California's Industrial Pattern," 64; Hubert Howe Bancroft, *History of California,* 7 vols. (San Francisco: The History Company, 1884–1890), 7:71–72.

6. Marryat, *Mountains and Molehills,* 169. U.S. Department of Agriculture, *Clearing New Land,* Farmers' Bulletin No. 150, by Franklin Williams, Jr. (Washington, D.C., 1902), 21; Bancroft, California, 7:58; California State Board of Agriculture, *Annual Report of the Statistician* (Sacramento, 1915), 36; Reynold M. Wik, "Some Interpretations of the Mechanization of Agriculture in the Far West," *Agricultural History,* 49 (January 1975): 78; John Qunicy Adams Warren, *California Ranchos and Farms, 1846–1862,* edited with and introduction by Paul W. Gates (Madison: State Historical Society of Wisconsin, 1967), 31; Sampson Wright to W.S.M. Wright, November 27, 1852, Wright Family Papers, in possession of author; Allen P. Lillie, "Logging Redwoods in Early Gilroy" and Juanita Swenor, "Notes on Logging Redwoods," *Sketches of Gilroy,* James C. Williams, ed. (Gilroy, CA: Gilroy Historical Society, 1980), 16, 20; Bert Hurt, "A Sawmill History of the Sierra National Forest, 1852–1940" (unpublished MS, Sierra National Forest Headquarters, Fresno, CA, February 1941), 1–31 *passim.* Oxen were not replaced in logging until the 1880s, when steam donkeys and logging railroads were introduced.

7. Sampson Wright to W.S.M. Wright, November 27, 1852; Ernest Q. Wiltsee, *The Pioneer Miner and the Pack Mule Express* (San Francisco: California Historical Society, 1931); John S. Hittell, *The Resources of California,* 7th ed. (San Francisco: A.L. Bancroft and Co., 1879), 19; Bancroft, California, 6:461; Hunter, *Transmission of Power,* 47–51.

8. Bancroft, *California,* 7:153, n. 49; Sampson Wright to W.S.M. Wright, November 27, 1852; W.H. Hutchinson, *California: The Golden Shore by the Sundown Sea,* 2nd ed. (Belmont, CA: Star Publishing Company, 1988), 176, indicates that horse snowshoes were developed for teams and pack strings working in the Sierra Nevada during the winter months.

9. John S. Hittell, *The Commerce and Industries of the Pacific Coast of North America,* 2nd ed., (San Francisco: A.L. Bancroft and Co., 1882), 616; U.S. Census Office, Eighth Census, 1860, *Manufactures of the United States in 1860* (Washington, D.C., 1865), 3:23–26; Bancroft, *California,* 7:79–80.

10. James C. Williams, "The Trolley: Technology and Values in Retrospect," *San José Studies,* 3 (November 1977): 75–76; U.S. Census Office, Eleventh Census, 1890, *Report on the Transportation Business in the United States, Pt. 1, Statistics of Transportation* (Washington, D.C., 1895), 14:689, 704, and *Report on the Statistics of Agriculture in the United States* (Washington, D.C., 1895), pt. 1, 5:580–587.

11. Bancroft, *California,* 6:454; Cronise, *Natural Wealth,* 657; U.S. Census Office, Tenth Census, 1880, *Report on the Social Statistics of Cities; Pt. II, The Southern and Western States* (Washington, D.C., 1887), 19:780, 795, 805–806, 819. Julia M. Carpenter, "My Journey West," 1882, manuscripts collection, Bancroft Library, Berkeley, CA; American Public Works Association, *History of Public Works in the United States, 1776–1976* (Chicago: American Public Works Association, 1976), 434. On the environmental impact of animal-based urban transportation and other matters of waste, see the *History of Public Works,* 434–438; Clay McShane, "Transforming the Use of Urban Space: A Look at the Revolution in Street Pavements, 1880–1924," *Journal of Urban History,* 5

(May 1979): 296–298; Joel A. Tarr and Francis Clay McMichael, "Decisions about Wastewater Technology, 1850–1932," *Journal of the Water Resources Planning and Management Division,* ASCE, 103 (May 1977): 47–61; Martin V. Melosi, ed., *Pollution and Reform in American Cities, 1870–1930* (Austin: University of Texas Press, 1980).

12. Hittell, *Resources,* 19, 228; Powell, *The Golden State,* 53; *PRP,* 39 (December 27, 1890): 552, and various other nineteenth and early twentieth century issues.

13. University of California, Agricultural Experiment Station, *Cost of Work Horses on California Farms,* Bulletin No. 401, by R.L. Adams (Berkeley, April 1926); University of California, Agricultural Experiment Station, *The Tractor on California Farms,* Bulletin No. 415, by L.J. Fletcher and C.D. Kinsman (Berkeley, December 1926), 26; Robert E. Ankli and Alan L. Olmstead, "The Adoption of the Gasoline Tractor in California," *Agricultural History,* 55 (July 1981): 224–225.

14. Reynold M. Wik, *Steam Power on the American Farm* (Philadelphia: University of Pennsylvania Press, 1953), 16–17. Bancroft, California, 7:84; "Smyth's Tree Feller," *The Engineer of the Pacific,* 2 (March 1879): 3; Williams, *Clearing New Land,* 10–11; U.S. Department of Agriculture, *Insect and Fungous Enemies of the Grape East of the Rocky Mountains,* Farmers' Bulletin 284, by A.L. Quaintance and C.L. Shear (Washington, D.C., 1907), 41–42. *PRP,* 36 (July 7, 1888): 18–20, and v. 41 (March 7, 1891): 227.

15. San Joaquin Valley Agricultural Society, *Transactions and Second Annual Report* (1861): 131. "Horse-power Centrifugal Pumps," *Industry,* 1 (July 1884): 188; A.J. Bowie, Jr., "Proposed Electric Pumping for Stock Wells," *JEPG,* 12 (June 1902): 111; *PRP,* 39 (December 27, 1899): 549.

16. E.W. Hilgard, "The Agricultural Soils of California," in U.S. Department of Agriculture, *Annual Report,* (Washington, D.C., 1878), 476; *State Mineralogist, 1884,* 47; Warren, *California Ranchos,* 62. Earle D. Ross, "Retardation in Farm Technology Before the Power Age," *Agricultural History,* 30 (January 1956): 12–13; Wik, "Some Interpretations of the Mechanization of Agriculture," 77–82; Leo Rogin, *The Introduction of Farm Machinery in Its Relation to the Productivity of Labor in the Agriculture of the United States During the Nineteenth Century* (Berkeley: University of California Press, 1931). Also see Claude B. Hutchison, ed., *California Agriculture* (Berkeley: University of California Press, 1946) and, for a good synthesis of California agricultural development, Lawrence J. Jelinek, *Harvest Empire: A History of California Agriculture* (San Francisco: Boyd and Fraser Publishing Company, 1979). In general, concludes Vaclav Smil, *Energy in World History* (Boulder, CO: Westview Press, 1994), 86, horses provided the labor equivalent of thirteen to fifteen men in periods of heavy field work.

17. Wik, "Some Interpretations of the Mechanization of Agriculture," 78; Rogin, *Introduction of Farm Machinery,* 42–44; Hittell, *Resources,* 228; Cronise, *Agricultural and Other Resources,* 58; W.S.M. Wright, Record Book "For Hired Labor Exclusively," 1876–1878, Wright Family Papers.

18. Cronise, *Agricultural and Other Resources,* 58; Rogin, *Introduction of Farm Machinery,* 175, 186. Wright, Record Book "For Hired Labor." W.S.M. Wright, Ranch Ledger, 1871–1880, Wright Family Papers, p. 16.

19. James F. Shepard, "The Development of Wheat Production in the Pacific Northwest," *Agricultural History,* 49 (January 1975): 266; Wik, "Some Interpretations of the Mechanization of Agriculture," 79; Roy Bainer, "Science and Technology in Western Agriculture," *Agricultural History,* 49 (January 1975): 56.

20. Bainer, "Science and Technology," 56–57; F. Hal Higgins, "John M. Horner and the Development of the Combined Harvester," *Agricultural History,* 32 (January 1958): 14–24; Rogin, *Introduction of Farm Machinery,* 120.

21. Quoted in Higgins, "John M. Horner," 21.

22. Quoted in Wik, "Some Interpretations of the Mechanization of Agriculture," 79.

23. Bainer, "Science and Technology," 57; Cronise, *Natural Wealth,* 149; E.J. Wickson, *Rural California* (New York: The McMillan Co., 1923), 65; Higgins, "John M. Horner," 21–24; Wik, "Some Interpretations of the Mechanization of Agriculture," 79–82; *PRP,* 59 (January 27, 1900): 60; Rogin, *Introduction of Farm Machinery,* 121–125, 213–241 *passim;* Smil, *Energy in World History,* 71.

24. U.S. Department of Agriculture, Farmers' Bulletin, *The Agricultural Outlook* (Washington,

D.C., February 7, 1914), 21; Gordon True, "Raising Horses for Farm and Market," *PRP,* 91 (January 29, 1916): 121, 143.

25. Wayne Dinsmore, "Farmers Should Breed Draft Horses," *PRP,* 93 (January 27, 1917): 120, and "Heavy Horses Have Best Market Prospects," *PRP,* 94 (October 20, 1917): 405. For the 1920s, see Fletcher and Kinsman, *The Tractor,* 20; L.J. Fletcher, "Electricity in California Agriculture," *CREA Bulletin,* 2 (December 28, 1925): 8. E.J. Wickson, *Rural California,* 246, also wrote, in 1923, that horses still rendered chief service on farms and were unlikely to be displaced by autos, trucks, and tractors.

26. Smil, *Energy in World History,* 80.

Chapter III/Fuelwood and Coal

1. California State Mining Bureau, *Fourth Annual Report of the State Mineralogist for the Year Ending May 15, 1884* (Sacramento, 1884), 33; *PRP,* 24 (December 23, 1882): 489; Dolores Greenberg, "Energy Flow in a Changing Economy, 1815–1880," *An Emerging Independent American Economy, 1815–1870,* ed. Joseph A. Frese and Jacob Judd (Tarrytown, NY: Sleepy Hollow Press, 1980), 42. For America's experience with wood, see Brooke Hindle, ed., *America's Wooden Age* (Tarrytown, NY: Sleepy Hollow Restorations, 1975). Vaclav Smil, *Energy in World History of North America,* 2nd ed. (San Francisco: A.L. Bancroft and Co., 1882), 208, notes that, before fossil fuels, urban growth was limited by the fact that cities required "croplands and woodlands at least forty, and commonly about one hundred times, larger than the size of the settlement itself."

2. California wood use data developed from U.S. Census Office, Tenth Census, 1880, *Report on the Forests of North America* (Washington, D.C., 1884), 9:489; U.S. Department of Agriculture, *Fuel Wood Used in the United States, 1630–1930,* Forest Service Circular No. 641, by R.V. Reynolds and A.H. Pierson (Washington, D.C., February 1942), 6–10, 17–18; U.S. Department of Agriculture, Forest Service Circular No. 181, *Consumption of Firewood in the United States,* by Albert H. Pierson (Washington, D.C., September 23, 1910). Also see "Cheap Country Home," *PRP,* 1 (January 28, 1871): 49; *State Mineralogist, 1884,* 49.

3. Delphine Ferrier, "Farm House Furnishings and Kitchen Efficiency," *The U.C. Journal of Agriculture,* 3 (February 1916): 183. Also see Ken Butti and John Perlin, "Solar Water Heaters in California, 1891–1930," *The Co-Evolution Quarterly,* 15 (Fall 1977): 5–6; Harry C. Ramsower, *Equipment for the Farm and Farmstead* (Boston: Ginn and Co., 1917), 222–224; Arthur H. Cole, "The Mystery of Fuel Wood Marketing in the United States," *Business History Review,* 44 (Autumn 1970): 343–346.

4. Pierson, *Consumption of Firewood,* 1–2; Greenberg, "Energy Flow in a Changing Economy," 41; W.S.M. Wright Ranch Ledger, 1871–1880, p. 12, Wright Family Papers; *PRP,* 1 (February 18, 1871): 109, and v. 24 (December 23, 1882): 489; Louis C. Hunter, *A History of Industrial Power in the United States, 1780–1930, Volume One: Waterpower in the Century of the Steam Engine* (Charlottesville: University of Virginia Press, 1979), 399; W.H. Hutchinson, *California: The Golden Shore by the Sundown Sea,* 2nd ed. (Belmont, CA: Star Publishing Company, 1988), 170.

5. Hubert Howe Bancroft, *History of California,* 7 vols. (San Francisco: The History Company, 1884–1890), 7:77; John S. Hittell, *The Commerce and Industries of the Pacific Coast of North America,* 2nd ed. (San Francisco: A.L. Bancroft and Co., 1882), 725–726; Titus Fey Cronise, *Natural Wealth of California* (San Francisco: H.H. Bancroft and Co., 1868), 171; Randy Milliken, Dan Foster, and Leslie Schupp-Wessel, *Cultural Resource Survey of the U.S.F.S.-Plumas Eureka State Park Land Exchange,* ARR No. 05–11–97 (CY81) (Plumas National Forest, CA, 1981), 11, 34. State Mineralogist, 1884, 234–237; *The Resources of California* (July 1881): 7; Robert G. Elston and Donald Hardesty, Intermountain Research, *Archeological Investigations on the Hopkins Land Exchange,* 2 vols., U.S. Forest Service Contract No. 53–91V9–0–80077 (Tahoe National Forest, Nevada City, CA, April 1981), 96–97.

6. San Joaquin Valley Agricultural Society, *Transactions and Second Annual Report* (1861): 102, 115. Cronise, *Natural Wealth,* 131, 135; The *Stockton City, San Joaquin, Stanislaus, Calaveras, Toulumne and Contra Costa Counties Directory . . . 1884–5* (San Francisco, 1884), 612. For a general

understanding of the difficulties in gathering data and analyzing wood fuel production and distribution, see Cole, "Fuel Wood Marketing," 346–347.

7. Hutchinson, *California: The Golden Shore,* 113–119; W.H. Hutchinson, *California Heritage: A History of Northern California Lumbering,* rev. ed. (Santa Cruz, CA: The Forest History Society, 1974), n.p.; Bert Hurt, "A Sawmill History of the Sierra National Forest, 1852–1940," mimeographed, [1941]; Hittell, *Commerce and Industries,* 419–421, 584; Eliot Lord, *Comstock Mining and Miners* (1883, reprint ed., Berkeley: Howell-North, 1959), 256–258.

8. Cronise, *Natural Wealth,* 179.

9. Hank Johnson, *Thunder in the Mountains: The Life and Times of Madera Sugar Pine* (Los Angeles: Trans-Anglo Books, 1968), 14–16; U.S. Treasury Department, *Report on the Internal Commerce of the United States, 1890* (Washington, D.C., 1891), 186–187; California, Railroad Commission, *Annual Report* and *Biennial Report* (Sacramento, 1889–1912), *passim.* The Railroad Commission heard several complaints from wood cutters and dealers between 1911 and 1914 and almost always substantially reduced the railroad rates.

10. *State Mineralogist, 1884,* 33; Report of T. Butler King to John M. Clayton, U.S. Secretary of State, March 22, 1850, in Bayard Taylor, *El Dorado, or, Adventures in the Path of Empire* (Glorieta, NM: Rio Grande Press, 1967), 232; Bancroft, *History of California,* 6:193, n. 31; "Firewood," *California Culturist,* 3 (July 1860): 19.

11. Gayle Groenendaal, "California's First Fuel Crisis and Eucalyptus Plantings" (MA thesis, California State University, Northridge, 1985), 110–112. Groenendaal's thesis suggesting a statewide fuel crisis is misleading; the rapid timber depletion in and near gold fields and cities is better termed a great environmental misfortune.

12. *Ibid.,* 113, 161–163; "Firewood," *California Culturist,* 20; Bancroft, *History of California,* 7:51, n. 9; *Resources of California* (November 1884): 12; Jaqueline Anna Lanson, "Eucalyptus in California: Its Distribution, History, and Economic Value" (M.A. thesis, University of California, Berkeley, 1952), 35–36.

13. *PRP,* 24 (December 9, 1882): 444, and v. 37 (January 26, 1889): 83; Lanson, "Eucalyptus," 51, 87; C.H. Sellers, *Eucalyptus: Its History, Growth, and Utilization* (Sacramento: A.J. Johnson, 1910), 47–51; University of California, Agricultural Experiment Station, *Eucalyptus in California,* Bulletin No. 196, by Norman D. Ingham (Sacramento, 1908), 31; Groenendaal, "California's First Fuel Crisis," ch. 8, offers a synthesis of efforts to commercialize eucalyptus.

14. *Internal Commerce, 1890,* 186–187; Pierson, "Consumption of Firewood," 6–7.

15. Watson A. Goodyear, *The Coal Mines of the Western Coast of the United States* (San Francisco: A.L. Bancroft and Co., 1877), 153. Prices are from Bancroft, *History of California,* 7:111, n. 17.

16. Quoted in Gerald T. White, *Scientists in Conflict: The Beginnings of the Oil Industry in California* (San Marino, CA: The Huntington Library, 1968), 25.

17. California, Assembly, Report of the Sub-Committee on Mines and Mining Interests of the Assembly concerning the State Geological Survey, *App. Journal, 14th Sess.,* Doc. No. 24, 1863; Frank Marryat, *Mountains and Molehills or Recollections of a Burnt Journal,* intro. by Robin Winks (Philadelphia: Lippincott, 1962). Also see Michael L. Smith, *Pacific Visions: California Scientists and the Environment, 1815–1915* (New Haven, CT: Yale University Press, 1987).

18. Margaret (Ballard) Thompson, "History of the Coal Mining Industry in the Mount Diablo Region, 1859–1885" (unpublished M.A. thesis, University of California, Berkeley, 1931), 14–22, 55. *State Mineralogist, 1884,* 268; Cronise, *Natural Wealth,* 158–160; Dan L. Mosier, *Harrisville and the Livermore Coal Mines* (San Leandro, CA: Mine Road Books, 1978), 27. Total Diablo production compiled from *State Mineralogist, 1884,* 268, and U.S. Bureau of Mines, *Mineral Resources of the United States* (Washington, D.C., 1882), 97–98, and succeeding issues through 1900, hereafter referred to as *Minerals Yearbook.*

19. Mosier, Harrisville, 35. Dan L. Mosier, *California Coal Towns, Coaling Stations & Landings* (San Leandro, CA: Mines Road Books, 1978).

20. "Our Material Resources," *The Resources of California* (October 1881): 5.

21. Thompson, "History of Coal Mining," 14–17, 55–57; Mosier, *Harrisville,* 12, 91–92; U.S.

Census Office, Eleventh Census, 1890, *Report on Mineral Industries in the United States* (Washington, D.C., 1892), 7:359.

22. California, Division of Mines, *California Mineral Production and Directory of Mineral Producers for 1941* (Sacramento, 1942), 17, provides accurate total California coal production figures. Total California coal consumption data compiled from *Minerals Yearbook, 1882,* 97–98, and succeeding issues through 1900; Goodyear, *Coal Mines,* 135; *Internal Commerce, 1885,* 316.

23. Hittell, *Commerce and Industries,* 309; John L. Howard, "Fuel Conditions in California," *JEPG,* 12 (November 1902): 213; John D. Issacs, "Recent Improvements in Coal-Handling Machinery," *Transactions of the Technical Society of the Pacific Coast,* 12 (July 1895–December 1896): 54–66. Goodyear, *Coal Mines,* 86–152 *passim; State Mineralogist,* 1884, 270–272.

24. Goodyear, *Coal Mines,* 153. Mosier, *Harrisville,* 88, 114; Thompson, "History of Coal Mining," 86–87; *Minerals Yearbook, 1887,* 209; *State Mineralogist, 1887,* 164; Bancroft, *History of California,* 6:777, n. 47 and 7:26. Hutchinson, *California: The Golden Shore,* 163, indicates the price of the best imported coal in San José, only fifty miles from San Francisco, reached as much as $53 per ton and "a whopping $110 in Marysville" during the nineteenth century.

25. Lawrence J. Jelinek, *Harvest Empire: A History of California Agriculture* (San Francisco: Boyd and Fraser Publishing Company, 1979), 39–43; Margery Holburne Saunders, "California Wheat, 1867–1910: Influence of Transportation on the Export Trade and Location of Producing Areas" (unpublished M.A. thesis, University of California, Berkeley, 1960); Morton Rothstein, "West Coast Farmers and the Tyranny of Distance: Agriculture on the Fringes of the World Market," *Agricultural History,* 49 (January 1975): 275.

26. Saunders, "California Wheat," 52–54; W.A. Starr, "Abraham Dubois Starr: Pioneer California Miller and Wheat Exporter," *California Historical Society Quarterly,* 27 (September 1948): 197.

27. *Internal Commerce, 1885,* 39–41; *State Mineralogist, 1884,* 37; *Minerals Yearbook, 1882,* 97–98, and succeeding issues through 1900.

28. *Internal Commerce, 1885,* 39.

29. Rodman Paul, "The Wheat Trade Between California and the United Kingdom," *Mississippi Valley Historical Review,* 45 (December 1958): 392–412; *Minerals Yearbook, 1883–1884,* 19; Horace Davis, "California Breadstuffs," *Journal of Political Economy,* 2 (September 1894): 532–534; Howard, "Fuel Conditions," 213.

30. Dan L. Mosier, *Corral Hollow Coal Mining District* (Fremont, CA: Mine Road Books, 1983), 57; M.E. Dittmar, "Northern California's Resources," *California Mines and Minerals* (San Francisco: L. Roesch Co., 1899), 389; *California Mineral Production, 1941,* 146; V. Covert Martin, *et al., Stockton Album, Through the Years* (Stockton, CA, 1959), 117.

31. *Minerals Yearbook, 1907,* 223, and *1908,* 213–215.

32. Robert Schorr, "The Briquetting Plant at Stockton, California," *The Engineering and Mining Journal,* 78 (August 18, 1904): 262; *Minerals Yearbook, 1904,* 458.

33. Edward W. Parker, "Condition of the Coal-Briquetting Industry in the United States," in U.S. Geological Survey, *Contributions to Economic Geology, 1906, Part II—Coal, Lignite and Peat,* Bulletin No. 316, (Washington, D.C., 1907), 476, 473–477; "Briquette Plants in California," *The Engineering and Mining Journal,* 77 (April 28, 1904): 682; Robert Schorr, "The Schorr Briquette Press," *The Engineering and Mining Journal,* 80 (October 7, 1905): 627; *Minerals Yearbook, 1905,* 555.

34. *The Engineering and Mining Journal,* 79 (March 2, 1905): 425; *Minerals Yearbook, 1906,* 653, *1909,* 203, and *1913,* pt. 2:383–392; Parker, "Coal-Briquetting Industry," 477; "Report on Fuel Resources," *Transactions of the Commonwealth Club of California,* 7 (June 1912): 202; Jefferson B. Rodgers, "Thermal and Physical Properties of Fuel Briquettes Made from Agricultural and Other Waste Products," *Agricultural Engineering,* 17 (May 1936): 199–204.

35. "Discussion on Lamp Black Briquettes," *JEPG,* 21 (November 14, 1908): 319; [W.B. Cline], "Pacific Coast Gas Association: President's Address," *JEPG,* 25 (September 24, 1910): 266; *Minerals Yearbook, 1911,* 106, 271–273, *1940,* 929, and annual issues from 1907 for national and California production and consumption statistics.

36. Goodyear, *Coal Mines,* 153.

Chapter IV/Candles, Kerosene, and Gas

1. Walter S. Tower, *A History of the American Whale Fishery* (Philadelphia: University of Pennsylvania, 1907), 58–64; Sam H. Schurr and Bruce C. Netschert, *Energy in the American Economy, 1850–1975* (Baltimore: Johns Hopkins Press for Resources of the Future, 1960), 96; Hittell, *Commerce and Industries*, 717; Harold F. Williamson, *et al.*, *The American Petroleum Industry*, 2 vols. (Evanston, IL: Northwestern University Press, 1959), 1:36–37; Hubert Howe Bancroft, *History of California*, 7 vols. (San Francisco: The History Company, 1884–1890), 7:93; Arthur F. McEvoy, *The Fisherman's Problem: Ecology and Law in the California Fisheries, 1850–1980* (Cambridge, MA: Cambridge University Press, 1986), 74–75.

2. California State Mining Bureau, *Fourth Annual Report of the State Mineralogist for the Year Ending May 15, 1884* (Sacramento, 1884), 49; Hittell, *Commerce and Industries*, 720–721; *The [Weekly] Price Current*, various issues, March 19, 1883 to September 28, 1885; W.S.M. Wright Notebook, 1865–1867, Wright Family Papers; U.S. Treasury Department, *Report on the Internal Commerce of the United States, 1890* (Washington, D.C., 1891), 287; Bancroft, *History of California*, 7:103, n.1, and 114.

3. J.A. Graves, *Seventy Years in California, 1857–1927* (Los Angeles: Times-Mirror Press, 1927), 28; Bancroft, California, 7:93; Titus Fey Cronise, *Natural Wealth of California* (San Francisco: H.H. Bancroft and Co., 1868), 623; Hittell, *Commerce and Industry*, 720.

4. W.H. Hutchinson, *Oil, Land and Politics: The California Career of Thomas Robert Bard*, 2 vols. (Norman: University of Oklahoma Press, 1965), 1:53; W.H. Hutchinson, *California: The Golden Shore by the Sundown Sea*, 2nd ed. (Belmont, CA: Star Publishing Company, 1988), 121–122; Gerald T. White, *Scientists in Conflict: The Beginnings of the Oil Industry in California* (San Marino, CA: The Huntington Library, 1968); Bancroft, *History of California*, 7:77; Hittell, *Commerce and Industry*, 715–716; Cronise, *Natural Wealth*, 294, 642; *State Mineralogist, 1884*, 49; Williamson, *et al.*, *American Petroleum Industry*, 1:36.

5. *Ibid.*, 1:43–60.

6. *Ibid.*, 320, 337; "Hale's Patent Kerosene Burner," *California Culturist*, 2 (August 1859): 85–86; Gerald T. White, *Formative Years in the Far West: A History of the Standard Oil Company of California and Predecessors through 1919* (New York: Appleton-Century Crofts, 1962), 4, 14; Hutchinson, *Oil, Land and Politics*, 1:53.

7. Quoted in White, *Scientists in Conflict*, 61–62; *ibid.*, 8–10, 24, 43–44, 60; Hutchinson, *Oil, Land and Politics*, 1:53–64.

8. Quoted in California, Division of Mines, *California Mineral Production and Directory of Mineral Producers for 1941* (Sacramento, 1942), 22.

9. White, *Scientists in Conflict*, 92, 129; White, *Formative Years*, 8; Cronise, *Natural Wealth*, 675–676; "The Colusa Oil Excitement," *California Rural Home Journal*, 1 (April 1, 1865): 31, and issues for April 15 and May 1, 1865.

10. Williamson, *et al.*, *American Petroleum Industry*, 1:57–58, 311–312; White, *Formative Years*, 14–16, 19; Hutchinson, *Oil, Land and Politics*, 1:107, 118, 140; *State Mineralogist, 1884*, 297; Graves, *Seventy Years*, 29; *The [Weekly] Price Current*, January 15, 1883; Saratoga Springs Cash Book, November 1, 1886–October 31, 1889, 27 and 69, Wright Family Papers.

11. D. Clayton Brown, "Farm Life: Before and After Electrification," in *Proceedings of the Annual Meeting, June 16–18, 1974, University of California, Davis [of] the Association for Living Historical Farms and Agricultural Museums* (Washington, D.C., 1975), 1; Graves, *Seventy Years*, 29; "How Oils Explode," *PRP*, 1 (January 7, 1871): 13.

12. Harry C. Ramsower, *Equipment for the Farm and Farmstead* (Boston: Ginn and Co., 1917), 101–106; *PRP*, 94 (November 24, 1917): 538.

13. Hardison quoted in *Petroleum In California: A Concise and Reliable History of the Oil Industry of the State* (Los Angeles:, 1900), 98. See also Hutchinson, *Oil, Land, and Politics*, 2:53–54; White, *Formative Years*, 171; Advertisement, *PRP*, 36 (July 14, 1888): 35; U.S. Department of Agriculture, *Modern Conveniences for the Farm*, Farmers' Bulletin No. 270, by Elmina T. Wilson (Washington, D.C., 1906), 12.

14. U.S. Bureau of Mines, *Minerals Yearbook 1911*, 106; [Mary Seacrest], *California Annual Nar-*

rative Report, Acting Extension Specialist in Home Management, December 1, 1933–November 30, 1934 (Berkeley: College of Agriculture, Extension Service, April 15, 1935), 7–9, and Exhibits 2 and 29; Glenn A. Kennedy, "It Happened in Stockton, 1900–1925" (mimeographed, Haggin Gallery and Pioneer Museum, Stockton, CA, 1967), 197; *PRP,* 80 (April 2, 1910): 277, v. 94 (July 7, 1917): 23 and (November 17, 1917): 517; White, *Formative Years,* 171, 488–489, 513.

15. Julia A. Carpenter, "My Journey West," 1882, manuscripts collection, Bancroft Library, Berkeley, CA. *Oil, Paint and Drug Reporter* correspondent quoted in Williamson, *et al. American Petroleum Industry,* 2:175. The price of poor quality coal in San Francisco ran around $6.00 per ton.

16. California, Railroad Commission, *Biennial Report for of the Board of Railroad Commissioners of the State of California, for the Years of 1895 and 1896* (Sacramento, 1896), 12–13.

17. *Internal Commerce, 1888,* 135, and *1890,* 176.

18. Railroad Commission, *Biennial Report . . . 1895 and 1896,* 11–12.

19. *Ibid.,* 10.

20. White, *Formative Years,* 97, 513; *State Mineralogist, 1884,* 308; *Internal Commerce,* 1888, 135, and *1890,* 176; U.S. Federal Trade Commission, *Report of the Federal Trade Commission on the Pacific Coast Petroleum Industry,* 2 parts (Washington, D.C., 1921 and 1922), 2:250–251; *Minerals Yearbook, 1937,* 1045. Data in *Internal Commerce* is given in pounds. Conversion to gallons is based upon 7.4805 U.S. liquid gallons equalling one cubic foot and forty cubic feet equalling one U.S. shipping ton (two thousand pounds).

21. Williamson, *et al., American Petroleum Industry,* 1:33–40; California, State Mining Bureau, *Production and Use of Petroleum in California,* by P.W. Prutzman, Bulletin No. 32 (Sacramento, 1904), 162.

22. James C. Williams, "Cultural Conflict: The Origins of American Santa Barbara," *The Southern California Quarterly,* 60 (Winter 1978): 355; Bancroft, *History of California,* 6:775; Charles M. Coleman, *PG&E of California: The Centennial Story of Pacific Gas and Electric Company, 1852–1952* (New York: McGraw-Hill, 1952), 11.

23. E.C. Jones, "The History of Gas Lighting in San Francisco," *JEPG,* 24 (June 11, 1910): 542–543; Bancroft, *History of California,* 6:779–780, n. 54; Coleman, *PG&E,* 11–15.

24. Coleman, *PG&E,* 34–45; Bancroft, *History of California,* 7:100; U.S. Census Office, Ninth Census, 1870, *Statistics of the Wealth and Industry of the United States* (Washington, D.C., 1872), 3:497, Table 9B, and Eleventh Census, 1890, *Report on the Manufacturing Industries in the United States, Pt. III, Selected Industries* (Washington, D.C., 1895), 6:708–725, Tables 1 and 2. For Los Angeles's experience, see Douglas R. Littlefield and Tanis C. Thorne, *The Spirit of Enterprise: The History of Pacific Enterprises from 1886 to 1889* (Los Angeles: Pacific Enterprises, 1990), 12–15.

25. Coleman, *PG&E,* 29–34; Littlefield and Thorne, *Spirit of Enterprise,* 15–16. Public utilities did not come under comprehensive regulation until 1911, when the State Railroad Commission was expanded to oversee gas, water, and electric companies as well as railroad firms. See chapter 11.

26. *Ibid.,* 38; *State Mineralogist, 1896,* 567–569; *Stockton and Its Resources, 1893: A Souvenir of the Evening Mail* (Stockton, CA, [1893]), 58, 72 (at the Haggin Gallery, Stockton).

27. *Ibid.,* 24–43; Hittell, *Commerce and Industry,* 734; *State Mineralogist, 1888,* 548; Littlefield and Thorne, *Spirit of Enterprise,* 17; Hutchinson, *Oil, Land and Politics,* 1:241; White, *Formative Years,* 60–61.

28. E.C. Jones, "The Manufacture of Water Gas," *JEPG,* 30 (June 21, 1913): 569–572; White, *Formative Years,* 61.

29. White, *Formative Years,* 60–63.

30. E.C. Jones, "The Development of Oil-Gas in California," *JEPG,* 24 (May 28, 1910): 491–494, and "The Manufacture of Oil Gas—II," *JEPG,* 30 (May 10, 1913): 432–433; Coleman, *PG&E,* 49–50; Littlefield and Thorne, *Spirit of Enterprise,* 26, 30.

31. Williamson, *et al., American Petroleum Industry,* 1:236–238.

32. White, *Formative Years,* 24; Hutchinson, *Oil, Land and Politics,* 1:235–236.

33. White, *Formative Years,* 25.

34. Ramsower, *Equipment for the Farm,* 106–111.

35. "Illuminating Gas Plant," *Industry,* 3 (July 1891): 196; *PRP,* 80 (January 8, 1910): 32, and (April 2, 1910): 277.

36. T.A. Ferguson, "Commercial Value of Acetylene Gas as an Illuminant," *Engineering News,* 35 (May 14, 1896): 317; A. Lipschutz, "The Use of Acetylene in Railway Station and Train Lighting," *Journal of the Association of Engineering Societies,* 24 (June 1900): 355–369; Dr. W.H. Birchmore, *Acetylene Gas: Its History and Utilization* (New York: J.B. Colt Company, 1897), pamphlet, "Energy Products—Oil, Gas, etc." box, Romaine Trade Catalog Collection, University of California, Santa Barbara.

37. *Engineering News,* 35 (April 2, 1896): 216; "Dangers of Acetylene Gas," *The Pacific Electrician,* 18 (June 1897): 64.

38. *JEPG,* 4 (August 1897): 97; "Properties and Dangers of Acetylene," *JEPG,* 6 (August 1989): 46; Vivian B. Lewis, "Acetylene [four parts]", *JEPG,* 6 (December 1898): 124, and v. 7 (January 1899): 10–11, (February 1899): 32–33; and (March 1899): 47–50.

39. Ramsower, *Equipment for the Farm,* 112–120; U.S. Department of Agriculture, *Experiment Station Work, LXXII,* Farmers' Bulletin 517 (Washington, D.C., 1912), 21–23; *Extension Specialist Narrative Report, December 1, 1933–November 30, 1934.*

Chapter V/Wind, Tide, and Sun

1. Quoted in Louis C. Hunter and Lynwood Bryant, *A History of Industrial Power in the United States, 1780–1930, Volume Three: The Transmission of Power* (Cambridge, MA: The M.I.T. Press, 1991), 89, n. 76.

2. Hubert Howe Bancroft, *History of California,* 7 vols. (San Francisco: The History Company, 1884–1890), 6:18, 7:96; *The California Agriculturist,* 4 (March 1, 1874): 57; "Tustin's Self-Regulating Wind Wheel," *California Culturist,* 2 (September 1859): 138; T. Lindsay Baker, *A Field Guide to American Windmills* (Norman: University of Oklahoma Press, 1985), 25, 6–13, and 430, n. 2; "James B. Johnson's Improved Patent Wind Wheel," *California Culturist,* 1 (July 1858): 79–80; *California Culturist,* 3 (July 1860): 40, plus unpaginated advertisement.

3. *The California Farmer,* 13 (June 8, 1860): 127, and v. 15 (May 17, 1861): 93; "Philips' Self-Regulating Wind Power," *California Culturist,* 2 (April 1860): 451.

4. San Joaquin Valley Agricultural Society, *Transactions and First Annual Report* (1860), 55, *Second Annual Report* (1861), 79–88, 164, and *Third Annual Report* (1863), 148; *California Culturist,* 3 (July 1860): 40.

5. California State Mining Bureau, *Fourth Annual Report of the State Mineralogist for the Year Ending May 15, 1884* (Sacramento, 1884), 47–48; Roger S. Manning, "The Windmill in California," *Journal of the West,* 14 (July 1975): 34; Quincy Adams Warren, *California Ranchos and Farms, 1846–1862,* edited with and introduction by Paul W. Gates (Madison: State Historical Society of Wisconsin, 1967), 79.

6. U.S. Department of Agriculture, *Irrigation in Humid Climates,* Farmers' Bulletin No. 46, by F.H. King (Washington, D.C., 1896), 18–20; U.S. Census Office, Eleventh Census, 1890, *Report on Agriculture by Irrigation in the Western Part of the United States* (Washington, D.C., 1894) 5:59–82; *The Resources of California* (July 1881): 2–3. Also see U.S. Geological Survey, *Pumping Water for Irrigation,* Water Supply and Irrigation Paper No. 1, by Herbert Wilson (Washington, D.C., 1896), 25–35.

7. *The Resources of California* (August 1881): 2, and (April 1883): 3; Hittell, *The Commerce and Industries of the Pacific Coast of North America,* 2nd ed. (San Francisco: A.L. Bancroft and Co., 1882), 679, 786; U.S. Department of Agriculture, *The Use of Windmills in Irrigation in the Semiarid West,* Farmers' Bulletin 394, by P.E. Fuller (Washington, D.C., 1910), 41, 43.

8. Hittell, *Commerce and Industries,* 679; John S. Hittell, *The Resources of California,* 7 ed. (San Francisco: A.L. Bancroft and Co., 1879), 70; San Joaquin Valley Agricultural Society, *Second Annual Report* (1861), 109; *The Resources of California* (February 1881): 2; *An Illustrated History of San Joaquin County, California* (Chicago: Lewis Publishing Co., 1890), 116, 565. Also see Bob Turner, "Windmills: A Doer of Work" (unpublished MS, San Joaquin County Historical Museum, Lodi, CA), 6–8.

9. *Illustrated History of San Joaquin County,* 564–565. *The Resources of California* (September 1884): 1–2; *Stockton Weekly Mail,* March 22, 1890, clipping, Haggin Museum, Stockton, California. R. Coke Wood and Leonard Covello, *Stockton Memories: A Pictorial History of Stockton, California* (Fresno, CA, 1977), 89. *Stockton City . . . Directory, 1884–5,* 612. Advertisments in *PRP,* 41 (June 13, 1891): 588 and (June 20, 1891): 601, v. 42 (July 25, 1891): 83, and v. 44 (July 9, 1892): 40.

10. U.S. Census Office, Eleventh Census, 1890, *Report on the Manufacturing Industries of the United States, Pt. I, Totals for States and Industries* (Washington, D.C., 1895), 6:324, and pt. 2:303, 414; "The Peerless 'IXL' Windmill," *Second Special Booster Edition of the Byron Times,* 1910, Haggin Museum, Stockton, California. Advertisements in *PRP,* 24 (August 5, 1882): 90 and 24 (December 23, 1882): 506, v. 36 (July 14, 1888): 37, v. 39 (January 27, 1889): 77, v. 39 (November 30, 1889): 513, v. 42 (July 18, 1891): 63, and v. 43 (March 12, 1892): 249; *Engineer of the Pacific,* 1 (December 1877): 7, and v. 3 (March 1880): ii, iv.

11. *PRP,* 41 (April 4, 1891): 336, and v. 44 (December 31, 1892): 559; U.S. Bureau of the Census, *Manufacturers, 1905; Pt. II, States and Territories* (Washington, D.C., 1907) 70, 79, and *Census of Manufactures, 1914, Vol. I, Reports by States with Statistics for Principal Cities and Metropolitan Districts* (Washington, D.C., 1918), 127.

12. George P. Low, "Gas and Power Engineering Development in Monterey County, Cal.," *JEPG,* 14 (September 1904): 360; *State Mineralogist, 1888,* 227; Volta Torrey, *Wind-Catchers: American Windmills of Yesterday and Today* (Brattleboro, VT: S. Greene Press, 1976), 107.

13. "Wind Power," *Industry,* 1 (July 1889): 188; "The Old Dutch Windmill at Golden Gate Park, San Francisco," *JEPG,* 15 (March 1905): 71–77.

14. McKinnon's Challenge Double Header Geared Windmill system still stands on the E.E. Harden Ranch in Salinas. According to T. Lindsay Baker, Curator of Science and Technology at the Panhandle-Plains Historical Museum, Canyon, Texas, there may be no other surviving mills of this type in the United States (personal correspondence, November 21, 1982). Description of the mill is derived from personal inspection by the author, informed by U.S. Geological Survey, *The Windmill: Its Efficiency and Economic Use,* Water Supply Paper No. 42, Part II, by Edward Charles Murphy (Washington, D.C., 1901), 83–101; Betty Farrell Doty, "Salinas Valley Wind Kept the Wheels Turning at Old McKinnon Mill," *Salinas Californian,* April 26, 1975, 4A–5A; David Doty, "The McKinnon Windmill" (unpublished research paper, Gavilan College, Gilroy, CA, 1979), in possession of author; Baker, Field Guide, 162. and, courtesy of T. Lindsay Baker, *Illustrated Catalogue, Challenge Mill Company, Batavia, Illinois* (Batavia, IL, [ca. 1881]). *Wind Power and How It May Be Utilized* (Batavia, IL, Challenge Wind Mill and Feed Mill Company, [ca. 1885], reprint by Batavia Historical Society, 1975); *Challenge Wind Mill and Feed Mill Company, Manufacturers of Dandy Steel Wind Mills [Catalog]* (Chicago, IL, [ca. 1895]).

15. U.S. Geological Survey, *Development and Application of Water Near San Bernardino, Colton and Riverside, Cal.,* Water Supply Paper No. 60, Part II, by Joseph Barlow Lippincott (Washington, D.C., 1902), 115–134; U.S. Department of Agriculture, *Modern Conveniences for the Farm,* Farmers' Bulletin No. 270, by Elmina T. Wilson (Washington, D.C., 1906), 8–9; Doty, "The McKinnon Windmill," 9.

16. "Modern Plumbing in Country Home," *PRP,* 94 (September 8, 1917): 237. In a windshield survey made while traveling some two thousand miles in northern and central California during August 1982, the author counted 155 Aeromotor type windmills in working condition.

17. Benjamin C. Wright, *San Francisco's Ocean Trade, Past and Future: A Story of the Deep Water Service of San Francisco, 1848–1911* (San Francisco: A. Carlisle and Co., 1911), 211–212. Wind energy consumption is based on calculations by S. Frederic Dewhurst and Assoc., *America's Needs and Resources, A New Survey* (New York: Twentieth Century Fund, 1955), 1104, and coal efficiency factors developed by Sam H. Schurr and Bruce C. Netschert, *Energy in the American Economy, 1850–1975* (Baltimore: Johns Hopkins Press for Resources of the Future, 1960), 484–485.

18. William Armstrong Fairburn, *Merchant Sail,* 6 vols. (Center Lovell, ME: Fairburn Marine Educational Foundation, 1945–1955), 3:1902, 1905–1906.

19. *Ibid.,* 3:1906–1907, 1913, 1918–1919.

20. *Ibid.,* 4:2629; Wright, *San Francisco's Ocean Trade,* 132, 211–212.

21. Roger Olmsted, *Scow Schooners of San Francisco Bay*, ed. Nancy Olmsted (Cupertino, CA: California History Center Foundation, 1987).

22. Quoted in Warren, *California Ranchos*, 55; U.S. Census Office, Tenth Census, 1880, *Report on the Agencies of Transportation in the United States* (Washington, D.C., 1883), 4:22–25; James Dixon, *Dixon's Marine Guide to the Port of San Francisco* (New York, [1888]), 103; California State Board of Agriculture, *Annual Report of the Statistician* (Sacramento, 1915), 235–236.

23. Olmsted, *Scow Schooners*, ch. 4.

24. "Wave Motors," *Industry*, 2 (September 1889): 21; "Wave Motor Experiments," *JEPG*, 2 (January 1896): 1–2.

25. *The Pacific Electrician*, 18 (July 1897): 76; Albert W. Stahl, "The Utilization of the Power of Ocean Waves," *Transactions of the American Society of Mechanical Engineers*, 13 (1891–1892): 494; "Wave Motor Experiments," 2; *Engineering News*, 35 (April 2, 1896): 216.

26. Stahl, "Power of Ocean Waves," 492, 497–498.

27. *Ibid.*, 499–501.

28. Charles C. Moore and Co., Engineers, Inc., *Electricity in the "City of the Holy Cross,"* [reprint from *JEPG*, May 1905] (San Francisco, 1905): 14; Robert C. Catren, "A History of the Generation, Transmission, and Distribution of Electrical Energy in Southern California" (Ph.D. diss., University of Southern California, 1951), 145; "Wave-Motor," *JEPG*, 26 (February 11, 1911): 144.

29. Bancroft, *History of California*, 7:33, n. 9; Titus Fey Cronise, *Natural Wealth of California* (San Francisco: H.H. Bancroft and Co., 1868), 394; *State Mineralogist, 1884*, 52.

30. University of California, Agricultural Experiment Station, *Fruit Dehydration. I. Principles and Equipment*, Bulletin No. 698, by R.L. Perry, *et al.* (Berkeley, December 1946), 3; University of California, Agricultural Experiment Station, *The Principles and Practice of Sun-Drying Fruit*, Bulletin No. 388, by A.W. Christie and L.C. Barnard (Berkeley, May 1925), 3; R.C. Griffin, "Dehydration of Fruit in California," *JEPG*, 51 (November 1, 1923): 333. University of California, Agricultural Experiment Station, *The Dehydration of Prunes*, Bulletin No. 404, by A. W. Christie (Berkeley, August 1926), 3–17; University of California, Agricultural Experiment Station, *Walnut Dehydraters: Characteristics, Heat Sources, and Relative Costs*, Bulletin 531, by P.F. Nichols, B.D. Moses, and D.S. Glenn (Berkeley, June 1932), 3–4.

31. Stella Haverland Rouse, "Solar Furnace Built Here," *Santa Barbara News Press*, February 4, 1979, clipping, "Solar and Steam Heating" box, Romaine Trade Catalog Collection, University of California, Santa Barbara; Wilson Clark, *Energy for Survival* (Garden City, NY: Anchor Press, 1974), 365–366; C.G. Abbot, *The Sun and the Welfare of Man* (New York: Smithsonian Institution, 1929), 204–208; Ethan B. Kapstein, "The Transition to Solar Energy: An Historical Approach," in *Energy Transitions: Long-Term Perspectives*, ed. Lewis J. Perelman, August W. Giebelhaus, and Michael D. York (Boulder, CO: Westview Press, Inc., 1981), 111–112.

32. The story of solar energy in California is detailed in Butti and Perlin, "Solar Water Heaters in California, 1891–1930," The Co-Evolution Quarterly, 15 (Fall 1977): 4–13. Also see Butti and Perlin, *A Golden Thread: 2000 Years of Solar Architecture and Technology* (Palo Alto, CA: Cheshire Books, 1980), particularly 114–155; Kapstein, "Transition to Solar Energy," 109–123.

33. Quoted in Butti and Perlin, "Solar Water Heaters," 9. Also see *PRP*, 54 (November 13, 1897): 313.

34. Butti and Perlin, "Solar Water Heaters," 9–10; "Solar Heater," *JEPG*, 25 (August 13, 1910): 156.

35. Mrs. Susan Swaysgood, "Home Improvement, No. 11: Sun Heated Water for the Farm," *PRP*, 85 (February 22, 1913): 249.

36. University of California, College of Agriculture Experiment Station, *Solar Energy and Its Uses for Heating Water in California*, Bulletin 602, by F.A. Brooks (Berkeley, November 1936), 3, 23; University of California, College of Agriculture Experiment, *The Solar Heater*, Station Bulletin 469, by A.W. Farrell (Berkeley, June 1929); *CREA Newsletter*, No. 7 (August 31, 1929): 22; *CREA Bulletin*, 7 (November 1931): 45; *California Annual Narrative Report, Acting Extension Specialist in Home Management, December 1, 1924–November 30, 1925* (Berkeley, College of Agriculture Exten-

sion Service, December 20, 1925), 32, *December 1, 1927–November 30, 1928* (July 19 1929), 3, *December 1, 1931–November 30, 1932* (December 20, 1932), 29, and *December 1, 1932–November 30, 1933* (December 18, 1933), 47.

Chapter VI/Waterpower and Steam Engines

1. Mark Twain, *A Connecticut Yankee in King Arthur's Court* (New York: New American Library, 1963), 15.

2. Titus Fey Cronise, *Natural Wealth of California,* (San Francisco: H.H. Bancroft and Co., 1868), 48; Hubert Howe Bancroft, *History of California,* 7 vols. (San Francisco: The History Company, 1884–1890), 6:15–21.

3. Bert Hurt, "A Sawmill History of the Sierra National Forest, 1852–1940" (unpublished MS, Sierra National Forest Headquarters, Fresno, CA, February 1941), 1–31 *passim; The Alta Californian,* November 10, 1879; Heckendorn and Wilson, *Miners and Businessmen's Directory . . . of the Citizens of Tuolumne, and Portions of Calaveras, Stanislaus and San Joaquin Counties. . . .* (Columbia, CA: The Clipper Office, 1856), 100–101; Bancroft, *History of California,* 6:183, 525, 765; Cronise, *Natural Wealth,* 124.

4. *Sacramento Bee,* February 10, 1894, 1; Bancroft, *History of California,* 6:525; Cronise, *Natural Wealth,* 124–162 *passim;* "Means and Methods on Jersey Farm," *PRP,* 36 (April 7, 1888): 298–299.

5. "Extensive Flouring Mills," *The California Farmer,* 15 (June 28, 1861): 140; *Alta Californian,* June 27, 1863, 1; James Dix Schuyler, *Reservoirs for Irrigation, Water-Power and Domestic Water-Supply* (New York: J. Wiley and Sons, 1901), 179–180; Charles M. Coleman, *PG&E of California: The Centennial Story of Pacific Gas and Electric Company, 1852–1952* (New York: McGraw-Hill, 1952), 117–120. On waterpower and the transformation of New England rivers, see Theodore Steinberg, *Nature Incorporated: Industrialization and the Waters of New England* (Cambridge, MA: Cambridge University Press, 1991).

6. John J. Powell, *The Golden State and Its Resources* (San Francisco: Bacon and Co., 1874), 186, and assessor records cited on 117; Cronise, *Natural Wealth,* 133, 140, 597–599; U.S. Bureau of the Census, *Manufacturers, 1905, Pt. IV, Special Reports on Selected Industries* (Washington, D.C., 1908), 630, Table 8.

7. Coleman, *PG&E,* 92; Louis C. Hunter, *A History of Industrial Power in the United States, 1780–1930, Volume One: Waterpower in the Century of the Steam Engine* (Charlottesville: University of Virginia Press, 1979), 398, 401. Also see the section about California gold mining in Louis C. Hunter and Lynwood Bryant, *A History of Industrial Power in the United States, 1780–1930, Volume Three: The Transmission of Power* (Cambridge, MA: The M.I.T. Press, 1991), 393–413.

8. Quoted in W. Parker Ireland, *Origin, Growth and Development of the Hydro-electric and Irrigation Systems in Placer County, California,* a report prepared for the Placer County Water Users' Association, December 1, 1914, p. 2, in "Flumes, Ditches, Reservoirs" Historical File, Tahoe National Forest Headquarters, Nevada City, CA. For data on mining ditches, see U.S. Treasury Department, *Statistics of Mines and Mining in the States and Territories West of the Rocky Mountains for the Year Ending 1869* [2nd annual report] (Washington, D.C., 1870), 15, and *Year Ending 1871* [4th annual report] (1873), 46; California State Mining Bureau, *Tenth Annual Report of the State Minerologist for the Year Ending October 15, 1890* (Sacramento, 1890), 123–125, and *1896,* 526–566; "Ditches and Flumes," typed synopsis, Historical Statistics File, Tahoe National Forest Headquarters, Nevada City, CA, n.d.; Bancroft, *History of California,* 6:413, n. 9; J.W. Johnson, "Early Engineering in California," *California Historical Quarterly,* 29 (September 1950): 193–194; Helen Rocca Goss, "The Golden Rock Water Ditch," *The Historical Society of Southern California Quarterly,* 43 (March 1961): 1–16; Hunter and Bryant, *Transmission of Power,* 409–411. Also see Hunter, *Waterpower,* 411; Hamilton Smith, Jr., "Water Power with High Pressures and Wrought-Iron Pipe," *Transactions of the American Society of Civil Engineers,* 13 (February 1884): 15.

9. Goss, "Golden Rock," 2–3; Johnson, "Early Engineering," 194–196; *State Mineralogist, 1890,* 123.

10. Goss, "Golden Rock," 3–6; Johnson, "Early Engineering," 196–197; Smith, "Water Power," 35.

11. J.D. Galloway, "The Design of Rock-fill Dams," *Transactions of the American Society of Civil Engineers,* 104 (1939): 2–5; Johnson, "Early Engineering," 197–199; *State Mineralogist, 1890,* 123; Marsden Manson, "Rainfall on the Pacific Coast of North and South America and the Factors of Water Supply in California," *Journal of the Association of Engineering Societies,* 30 (March 1903): 113–114.

12. Hunter, *Waterpower,* 398. Hunter and Bryant, *Transmission of Power,* 404–407; Arthur F. McEvoy, *The Fisherman's Problem: Ecology and Law in the California Fisheries, 1850–1980* (Cambridge, MA: Cambridge University Press, 1986), 47–48, 84. For the essential account of hydraulic mining's rise, fall, and environmental impact, see Robert L. Kelley, *Gold vs. Grain: The Hydraulic Mining Controversy in California's Sacramento Valley* (Glendale, CA: Arthur H. Clark Co., 1959).

13. Hunter, *Waterpower,* 397–400.

14. Smith, "Water Power," 38; Hunter, *Waterpower,* 400–401; and *State Mineralogist, 1888,* 785, 791.

15. Hunter and Bryant, *Transmission of Power,* 412–413; Smith, "Water Power," 32–37.

16. Hunter, *Waterpower,* 405.

17. *Ibid.,* 407–408. For in-depth contemporary discussions concerning the development and charactersitics of the tangential waterwheel, see W.A. Doble, "The Tangential Water-Wheel," *Transactions of the American Institute of Mining Engineers,* 29 (1899): 852–894, and "Water Wheels of Impulse Type," *Transactions of the International Engineering Congress, 1915* (San Francisco, 1916): 7:503–559. Also see *State Mineralogist, 1888,* 790–802; and Coleman, *PG&E,* 112–114.

18. "The Mineralogical Report for 1888," *Industry,* 1 (March 1889): 116; U.S. Treasury Department, *Statistics of Mines and Mining in the States and Territories West of the Rocky Mountains for the Year ending 1870 [3d annual report]* (Washington, D.C., 1872), 462–469; U.S. Census Office, Eleventh Census, 1890, *Report on Mineral Industries in the United States* (Washington, D.C., 1892), 7:61; *State Mineralogist, 1884,* 148, *1886/87,* 23, 45–46, and *1890,* 25; Arthur DeWint Foote, "A Water-Power and Compressed Air Transmission Plant for the North Star Mining Company, Grass Valley, Cal.," in *State Mineralogist, 1896,* 706–720; J.H. Collier, Jr., "Deep Mining at the Utica, Angels, Calaveras County, California," *California Mines and Minerals* (San Francisco: L. Roesch and Co., 1899), 111–112.

19. Hunter, *Waterpower,* 415. U.S. Bureau of the Census, Thirteenth Census, 1910, *Mines and Quarries, 1909* (Washington, D.C., 1913), 11:718. U.S. Bureau of the Census, Fourteenth Census, 1920, *Mines and Quarries, 1919* (Washington, D.C., 1922), 11:81.

20. W.A. Starr, "Abraham Dubois Starr: Pioneer Miller and Wheat Exporter," *California Historical Society Quarterly,* 27 (September 1948): 195; *State Mineralogist, 1884,* 37.

21. Cronise, *Natural Wealth,* 597–599; Powell, *The Golden State,* 118–119; "Our Material Resources," *The Resources of California* (October 1881): 5; Smith, "Water Power," 15.

22. "Small Pelton Water-Motors," *PRP,* 38 (September 7, 1889): 214; U.S. Office of Experiment Stations, *Current Wheels: Their Use in Lifting Water for Irrigation,* Bulletin No. 146, (Washington, D.C., 1904), 4, 17; John S. Hittell, *The Commerce and Industries of the Pacific Coast of North America,* 2nd ed. (San Francisco: A.L. Bancroft and Co., 1882), 404; *PRP,* 43 (April 9, 1892): 328, and v. 37 (June 15, 1889): 575. For a discussion of waterpower drawing from water supply systems, see Hunter and Bryant, *Transmission of Power,* 155–161.

23. Thomas S. Ashton, *The Industrial Revolution, 1760–1830* (London: Oxford University Press, 1948), 58; Jeremy Atack, Fred Bateman, and Thomas Weiss, "The Regional Diffusion and Adoption of the Steam Engine in American Manufacturing," *Journal of Economic History,* 40 (June 1980): 281–308. Sam H. Schurr and Bruce C. Netschert, *Energy in the American Economy, 1850–1975* (Baltimore: Johns Hopkins Press for Resources of the Future, 1960), 61.

24. Bancroft, *History of California,* 6:20, 76, 415, 462, 507, 782; Hittell, *Commerce and Industries,* 585; U.S. Census Office, Ninth Census, 1870, *Statistics of the Wealth and Industry of the United States* (Washington, D.C., 1872), 3:497, Table 9; Tenth Census, 1880, *Report on the Manufactures of the United States* (Washington, D.C., 1883), 2:511, Table 3; Eleventh Census, 1890, *Report on the Manufacturing Industries in the United States, Pt. III, Selected Industries,* (Washington, D.C., 1895), pt. 1:758, Table 3.

25. Bancroft, *History of California*, 6:782, n. 60; *State Mineralogist, 1884*, 142; Hittell, *Commerce and Industries*, 542. Tenth Census, 1880, *Manufactures*, 2:511, Table 3; U.S. Census Office, *Report on the Manufacturing Industries of the United States, Pt. II, Statistics of Cities* (Washington, D.C., 1895), 6:12 and 15, Table 2.

26. *State Mineralogist, 1892*, 335. Watson A. Goodyear, *The Coal Mines of the Western Coast of the United States* (San Francisco: A.L. Bancroft and Co., 1877), 49; "Condenser, Filterer, and Feed Water Heater Combined," *Engineer of the Pacific*, 2 (April 1879): 3; "Surface Condensing in San Francisco: The Wheeler Condenser," *Industry*, 1 (July 1888): 185; Hittell, *Commerce and Industries*, 682.

27. Warren, *California Ranchos*, 43, 80; Reynold M. Wik, *Steam Power on the American Farm* (Philadelphia: University of Pennsylvania Press, 1953), 18, 29–30, 57; Leo Rogin, *The Introduction of Farm Machinery in Its Relation to the Productivity of Labor in the Agriculture of the United States During the Nineteenth Century* (Berkeley: University of California Press, 1931), 175, 186–189.

28. Goodyear, *Coal Mines*, 153; Cronise, *Natural Wealth*, 375.

29. Wik, *Steam Power on the American Farm*, 57; Hittell, *Commerce and Industries*, 432, 675–678.

30. Wik, *Steam Power on the American Farm*, 54–56, and Wik, "Some Interpretations of the Mechanization of Agriculture in the Far West," *Agricultural History*, 49 (January 1975): 82; Rogin, *Introduction of Farm Machinery*, 187–189.

31. Hittell, *Commerce and Industries*, 408; "Steam Power Irrigation," *PRP*, 43 (January 2, 1892): 9–10; U.S. Geological Survey, *Irrigation Near Bakersfield, California*, Water Supply and Irrigation Paper No. 17, by Carl Ewald Grunsky (Washington, D.C., 1898), 94; U.S. Census Office, Eleventh Census, 1890, *Report on Agriculture by Irrigation in the Western Part of the United States* (Washington, D.C., 1894) 5:89; Twelfth Census, 1900, *Agriculture, Part II, Crops and Irrigation* (Washington, D.C., 1902), 6:831; "Land Reclamation," *PRP*, 36 (April 14, 1888): 325.

32. "Farmers' Steam Generator and Engine," *PRP*, 36 (May 26, 1888): 469; University of California, Agricultural Experiment Station, *The Tractor on California Farms*, Bulletin No. 415, by L.J. Fletcher and C. D. Kinsman (Berkeley, December 1926), 20; Hunter and Bryant, *Transmission of Power*, 55–59 and 222, n. 69.

33. Bancroft, *History of California*, 6:127–129, 136, 138; U.S. Treasury Department, *Report on the Internal Commerce of the United States, 1890* (Washington, D.C., 1891), 124; Buffum, *Six Months in the Gold Mines*, 101; U.S. Bureau of Mines, *Minerals Yearbook 1882*, 22, and *1887*, 209; U.S. Census Office, Tenth Census, 1880, *Report on the Agencies of Transportation in the United States* (Washington, D.C., 1883), 4:22–25 and 716, Table 4.

34. *Ibid.*, 4:702, Table 1, and 705, Table 5; *Internal Commerce, 1885*, 43–45.

35. California State Board of Agriculture, *Annual Report of the Statistician* (Sacramento, 1911), 251, *1915*, 234–236, and *1918*, 262–266.

36. *Internal Commerce, 1885*, 45–46; W.H. Hutchinson, *California: The Golden Shore*, 177–179, 182.

37. Martin V. Melosi, *Coping With Abundance: Energy and Environment in Industrial America* (Philadelphia: Temple University Press, 1985), 21; Neill C. Wilson and Frank J. Taylor, *Southern Pacific: The Roaring Story of a Fighting Railroad* (New York: McGraw-Hill, 1952), 219; Dan L. Mosier, *Harrisville and the Livermore Coal Mines* (San Leandro, CA: Mine Road Books, 1978), 115–120; *State Mineralogist, 1884*, 264, 271–272; *Minerals Yearbook, 1883/84*, 100, *1886*, 362–365, and *1891*, 336–338; Hittell, *Commerce and Industries*, 308; California, Railroad Commission, *Annual Report* and *Biennial Report*, 1888–1901 (Sacramento, 1888–1901), *passim*.

38. William Cronon, *Nature's Metropolis: Chicago and the Great West* (New York: W.W. Norton, 1991), 73. On the railroads and the technological sublime, see John Kasson, *Civilizing the Machine: Technology and Republican Values in America, 1776–1900* (New York: Grossman Publishers, 1976), 172–180.

39. Mansel G. Blackford, *The Politics of Business in California, 1890–1920* (Columbus, OH: Ohio State University Press, 1977), 5–12.

40. Powell, *The Golden State*, 94.

41. Earle E. Williams, *Carrell of Corral Hollow* ([Stockton, CA], 1980), 194.

42. James C. Williams, "The Trolley: Technology and Values in Retrospect," *San José Studies*, 3 (November 1977): 77–78; Hilton, "Transport, Technology and the Urban Pattern," *Journal of Contemporary History*, 4 (July 1969): 125; U.S. Census Office, Eleventh Census, 1890, *Report on the Transportation Business in the United States, Pt. 1, Statistics of Transportation* (Washington, D.C., 1895), 14:689, 791–792. George W. Hilton's *The Cable Car in America*, rev. ed. (La Jolla, CA: Howell-North Books, 1982) is the definitive work on cable traction.

43. W.H. Hutchinson, *California Heritage: A History of Northern California Lumbering*, rev. ed. (Santa Cruz, CA: The Forest History Society, 1974), n.p; Bancroft, History of California, 7:7, 70; Wik, *Steam Power*, 63, 70; F. Hal Higgins, "The Celebrated Plow of Philander H. Standish," *American West*, 3 (Spring 1966): 26; R. Douglas Hurt, "Steam Plowing," *Journal of the West*, 21 (January 1982): 107; *California Culturist*, 3 (September 1860): 141.

44. Higgins, "Celebrated Plow," 25–27, 91.

45. Wik, *Steam Power*, 67–68. *PRP*, 1 (January 14, 1871): 17, and (February 18, 1871): 105.

46. Wik, *Steam Power*, 75–85, 94–97, 229; Roy Bainer, "Science and Technology in Western Agriculture," *Agricultural History*, 49 (January 1975): 57; "Jacob Price's New Field Locomotive," *PRP*, 40 (August 23, 1890): 165.

47. *PRP*, 39 (February 8, 1890): 134, 149, and (April 26, 1890): 434; *PRP*, 40 (August 16, 1890): 139, (September 27, 1890): 273–274, (November 22, 1890): 443, and (December 27, 1890): 550; Robert I. Slyter, "Towle Brothers Railroad and Mills, Washington Mining District," undated photocopied MS, Historical File, Tahoe National Forest Headquarters, Nevada City, CA, [pp. 3 and 11]; W. Turrentine Jackson, *et al.*, *History of Tahoe National Forest: 1840–1940*, U.S. Forest Service Contract #43–9A63–1–1745 (Davis, CA: Jackson Research Projects, May 1982), 105; Wik, "Some Interpretations of the Mechanization of Agriculture," 82.

48. Thomas H. Means, "Plowing With Engines," *PRP*, 80 (August 13, 1910): 128; "Traction Engines," *Industry*, 2 (October 1889): 35; Wik, *Steam Power*, 210–211; Rogin, *Introduction of Farm Machinery*, 147–153.

Chapter VII/Oil and Natural Gas

1. Federal Trade Commission, *Report of the Federal Trade Commission on the Pacific Coast Petroleum Industry*, 2 parts (Washington, D.C., 1921 and 1922), 1:18.

2. John J. Hittell, *The Commerce and Industries of the Pacific Coast of North America* (San Francisco: H.H. Bancroft and Co., 1882), 317.

3. Titus Fey Cronise, *Natural Wealth of California* (San Francisco: H.H. Bancroft and Co., 1868), 598, 626, 675; Gerald T. White, *Formative Years in the Far West: A History of the Standard Oil Company of California and Predecessors through 1919* (New York: Appleton-Century Crofts, 1962), 16–18; Gerald T. White, "California's Other Mineral," *Pacific Historical Review*, 34 (May 1970): 140; W.H. Hutchinson, *Oil, Land and Politics: The California Career of Thomas Robert Bard*, 2 vols. (Norman: University of Oklahoma Press, 1965), 1:140–143.

4. White, *Formative Years*, 59–89; Hutchinson, *Oil, Land and Politics*, 1:330–339; "Our Oil Interest," *The Resources of California* (October 1883): 7.

5. *The [Weekly] Price Current* (San Francisco, February 26, 1883): 1; "Los Angeles Oil" and "The Oil Trade," *The Resources of California* (November 1883): 7, and (November 1884): 16.

6. California State Mining Bureau, *Fourth Annual Report of the State Mineralogist for the Year Ending May 15, 1884* (Sacramento, 1884), 280, and *1887*, 76; White, *Formative Years*, 134; "The Oil Trade," *The Resources of California* (November 1884): 16. Petroleum viscosity is measured on the *Baume* (B.) degree scale; 12 to 15 degree B. is used for asphalt, 18 degree B. for fuel oil, and above 18 degree B. is refined into kerosene, gasoline, and other napthas. Two-thirds of California's oil, as late as 1909, was under 19 degree B. See Mansel G. Blackford, *The Politics of Business in Califonia, 1890–1920* (Columbus, OH: Ohio State University Press, 1977), 41, 185, n. 3.

7. R.G. Paddock, "Liquid Fuel—Its Application, Past and Present," *Journal of the Association of Engineering Societies*, 28 (April 1902): 233–246; *State Mineralogist, 1884*, 107, 284–285, 303, and *1887*, 76; White, *Formative Years*, 43, 85; Hutchinson, *Oil, Land and Politics*, 1:327–329, and 2:13, 72.

8. California State Mining Bureau, Bulletin No. 32, *Production and Use of Petroleum in California* by P.W. Prutzman (Sacramento, 1904), 100; Hutchinson, *Oil, Land and Politics,* 2:13; White, *Formative Years,* 616, n. 59; W.C. Watts, "Oil as Fuel in Los Angeles County," *State Mineralogist, 1896,* 662; Admiral Selwyn, "The Existing State of the Fluid Fuel Question," *Transactions of the Technical Society of the Pacific Coast,* 10 (September 1893): 187–198.

9. White, *Formative Years,* 233–234, and "California's Other Mineral," 142; Prutzman, *Production and Use,* 43, 45; Lionel V. Redpath, *Petroleum in California: A Concise and Reliable History of the Oil Industry of the State* (Los Angeles: R.V. Redpath, 1900), 97–99; *Engineering and Mining Journal,* 78 (July 7, 1904): 2; Neill C. Wilson and Frank J. Taylor, *Southern Pacific: The Roaring Story of a Fighting Railroad* (New York: McGraw-Hill, 1952), 219, 221; Ralph Andreano, "The Structure of the California Petroleum Industry, 1895–1911," *Pacific Historical Review,* 34 (May 1970): 174–181; Blackford, *Politics of Business,* 8–9.

10. Quoted in White, *Formative Years,* 136. *State Mineralogist, 1887,* 23–30.

11. White, *Formative Years,* 135, 137; Hutchinson, *Oil, Land and Politics,* 2:8–19.

12. Prutzman, *Production and Use,* 114–115, 124–125; "California Petroleum as Fuel for Steamers," *Engineering and Mining Journal,* 77 (April 28, 1904): 678; Paddock, "Liquid Fuel," 247; U.S. Department of the Navy, *Report of the U.S. Naval 'Liquid Fuel' Board* (Washington, D.C., 1904), 396; J.H. Hopps, "Marine Use of Fuel Oil," *JEPG,* 26 (March 18, 1911): 245–247; Harold F. Williamson, *et al., The American Petroleum Industry,* 2 vols. (Evanston, IL: Northwestern University Press, 1959), 2:182; White, "California's Other Mineral," 140.

13. Redpath, *Petroleum in California,* 99; "Petroleum Fuel on the Pacific Coast," *Industry,* 1 (May 1889): 151. Hutchinson, *Oil, Land and Politics,* 2:47, 64, 73; Richard Charles Schwarzman, *The Pinal Dome Oil Company: An Adventure in Business, 1901–1917* (New York: Arno Press, 1976), 47–49; Watts, "Oil as Fuel," 662–663.

14. White, *Formative Years,* 155–157, 239; Hutchinson, *Oil, Land and Politics,* 2:82–83; "A New California Pipeline," *Engineering and Mining Journal,* 78 (November 3, 1904): 712.

15. White, *Formative Years,* 239–241.

16. *Ibid.,* 241, 259; *Report of the Federal Trade Commission,* 1:22, 155–157; Williamson, *et al., American Petroleum Industry,* 2:72.

17. J.A. Graves, *Seventy Years in California, 1857–1927* (Los Angeles: Times-Mirror Press, 1927), 381; Hutchinson, *Oil, Land and Politics,* 2:80–81; Gerald D. Nash, "Oil in the West: Reflections on the Historiography of an Unexplored Field," *Pacific Historical Review,* 34 (May 1970): 199; U.S. Bureau of Mines, *Minerals Yearbook 1900,* 354, *1901,* 350, and *1910,* 89; Arthur M. Johnson, "California and the National Oil Industry," *Pacific Historical Review,* 34 (May 1970): 157.

18. *Report of the Federal Trade Commission,* 1:51–52, 68–70.

19. David T. Day and Ralph Arnold (U.S.G.S. petroleum experts) to George Otis Smith (U.S.G.S. Director), November 11, 1908, as quoted in California, State Council of Defense, *Report of the Committee on Petroleum* (Sacramento, July 7, 1917), 38.

20. *Report of the Committee on Petroleum,* 42–46, 51. J.N. Gillett (California oil industry lobbyist) to J.A. Elston (U.S. House of Representatives), May 9, 1917, in State Council of Defense, Committee on Petroleum, Petroleum Investigation Correspondence, File 3415–1, California State Archives, Sacramento, CA, hereafter SCD/CSA.

21. "Fuel Resources of California," *Transactions of the Commonwealth Club,* 197. See also State of California, Conservation Commission, *Report of the Conservation Commission of the State of California, January 1, 1913* (Sacramento, 1912).

22. Gillett to Elston, May 9, 1917, File 3415–1, SCD/CSA; "Conserving the Oil Resources," *Transactions of the Commonwealth Club of California,* 7 (June 1912): 208–209, and "Remarks by A.L. Weil," 211.

23. Gillett to Elston, May 9, 1917, and *The Daily Midway Driller,* May 7, 1917, 2, newspaper clipping, File 3415–1, SCD/CSA; "Shortage of Fuel on Coast is Feared," *San Francisco Examiner,* May 27, 1917, newspaper clipping, File 3415–7, Newspaper Clippings, SCD/CSA.

24. R.P. McLaughlin to E.L. Doheny, May 12, 1917, and William D. Stephens to Max Thelen, May 9, 1917, File 3415–1, SCD/CSA.

25. Max Thelen to William D. Stephens, January 21, 1918, File 3415–9, SCD/CSA, discusses the Committee on Petroleum's successful work. The committee's study of fuel substitution is detailed in Max Thelen to H.H. Jones (President, San Diego Gas and Electric Company), May 18, 1917, and similar letters to John A. Britton (Vice President and General Manager, Pacific Gas and Electric Company), other utility and railroad company officials, File 3415–1, SCD/CSA, and *Report of the Committee on Petroleum,* 151–169, 182–183. Henley C. Booth (Southern Pacific Company lawyer) to Max Thelen, June 4, 1917, outlines general problems inherent in reconverting locomotives to coal, File 3415–1, SDC/CSA. William Sproule (President, Southern Pacific Company) to D.M. Folsom, May 9, 1918, in United States Fuel Administration, Oil Division, Office of the Federal Oil Director for the Pacific Coast, Records of D.M. Folsom, Federal Oil Director for the Pacific Coast, February–December 1918, Record Group 67, Box 2624, Folder "Coal vs. Oil," National Archives, San Bruno, CA. Also see D.M. Folsom, "The Fuel Oil Situation on the Pacific Coast," *JEPG,* 45 (December 1, 1920): 523.

26. *Report of the Federal Trade Commission,* 1:41, 51, 168–174; U.S. Fuel Administration, *Final Report of the United States Fuel Administrator,* 1917–1919 (Washington, D.C.: Government Printing Office, 1921), 268–271, 279; White, *Formative Years,* 477, 549, 559–560; Joe S. Bain, *The Economics of the Pacific Coast Petroleum Industry,* 3 vols. (Berkeley: University of California Press, 1944–1947), 1:36; Johnson, "California and the National Oil Industry," 164; *Minerals Yearbook, 1919,* pt. 2:521; Arthur F.L. Bell, "Present and Future Supply of Petroleum as Fuel on the Pacific Coast," *JEPG,* 26 (March 18, 1911): 237–239; David M. Folsom, "The War- and After-the-War Value of Fuel Oil," *JEPG,* 41 (October 15, 1918): 367–368; "Fuel Problems of the Pacific Coast," *Mechanical Engineering,* 41 (February 1919): 264–269 *passim;* D.M. Folsom, "The Fuel Oil Situation," *Transactions of the Commonwealth Club of California,* 15 (December 1920): 367. Hydroelectricity as an alternative energy resource to petroleum during World War I is discussed *infra,* Chapter 12.

27. *Minerals Yearbook, 1934,* pt. 1:663. Bain, *Pacific Coast Petroleum,* 1:15–26 and 2:8, 17; Johnson, "California and the National Oil Industry," 165; F.H. Rosetti, "Pinching Back the Oil Wells," *California Journal of Development,* 13 (July 1923): 6; Kenny A. Franks and Paul F. Lambert, *Early California Oil: A Photographic History, 1865–1940* (College Station: Texas A. & M. University Press, 1985), 131–213 *passim;* Blackford, *Politics of Business,* 41, 54–59.

28. *State Mineralogist, 1887,* 55, 75, 181–183, also *1888,* 560–561, and *1896,* 349–352, 567–569; *Minerals Yearbook, 1883/84,* 238–242; Louis Stotz and Alexander Jamison, *History of the Gas Industry* (New York: Stettiner Brothers, 1938), 79–80.

29. Pacific Gas and Electric Co., *Eighth Annual Report of the Pacific Gas and Electric Company for the Fiscal Year Ended December 31, 1913* (San Francisco, 1914), 22, and *Sixteenth Annual Report, 1921* (1922), 6; Kempster B. Miller, "Oakland's Gas System," *JEPG,* 24 (June 25, 1910): 585; L.H. Newbert, "Suburban Gas Distribution," *JEPG,* 25 (October 8, 1910): 322–325; E.C. Jones, "Distribution of Gas," *Pacific Service Magazine,* 5 (July 1913): 49–54.

30. *Ibid.,* 50–57; Leon B. Jones, "Welding of High Pressure Pipe Lines," *JEPG,* 24 (September 21, 1912): 257–260; Newbert, "Suburban Gas," 322–325; E.C. Jones, "A Recent High Pressure Installation," *JEPG* 19 (September 28, 1907): 271–274; Guy R. Kinsley, "System of Heating and Lighting by Gas—P.P.I.E.," *JEPG,* 32 (March 21, 1914): 241–243.

31. C.S.S. Forney, "High Pressure Gas Distribution," *JEPG,* 27 (October 7, 1911): 326–327; "Transmission Lines in the San Diego District," *EW,* 60 (November 23, 1912): 1099; Rudolph Van Norden, "Central California Gas Company's System," *JEPG,* 31 (November 29, 1913): 477–491; G.L. Bayley, "Discussion of Papers at San Francisco Meeting," *Journal of the American Society of Mechanical Engineers,* 37 (December 1915): 698.

32. Pacific Gas and Electric Company, *Nineteenth Annual Report of the Pacific Gas and Electric Company for the Fiscal Year Ended December 31, 1923* (San Francisco, 1924), 26; Douglas R. Littlefield and Tanis C. Thorne, *The Spirit of Enterprise: The History of Pacific Enterprises from 1886 to 1889* (Los Angeles: Pacific Enterprises, 1990), 72–95. Development of California's electric power transmission system is discussed *infra,* Chapters 9 and 12.

33. Wilson is quoted in Littlefield and Thorne, *Spirit of Enterprise,* 52.

34. Walter G. Fitzgerald, "Why Gilroy Ceased to Operate Its Municipal Gas Plant," *Pacific Municipalities,* 21 (December 1909): 116, 120.

35. *Ibid.,* 120–121.

36. James C. Williams, "Engineering California Cities," in *Science-Technology Relationships: Relations Science-Technique,* ed. Alexandre Herléa (San Francisco: San Francisco Press, 1993), 396–397.

37. "Ukiah, California, Origin of the Council-Manager Plan," Institute for Governmental Studies, University of California, *Public Affairs Report* (March 1989): 11; H.W. Burkhart, "Municipal Gas in Long Beach Proving a Success," *Pacific Municipalities,* 39 (April 1925): 107–108. The issue of municipal ownership of electric utilities is explored *infra,* Chapter 12.

38. Martin V. Melosi, *Coping With Abundance: Energy and Environment in Industrial America* (Philadelphia: Temple University Press, 1985), 155–156; Claude C. Brown, "California's Natural Gas Resources," *California Journal of Development,* 21 (September 1931): 42–45.

39. *Ibid.,* 45; Mark Requa, "The Fuel Resources of California," *Transactions of the Commonwealth Club of California,* 7 (June 1912): 198.

40. Quoted in White, *Formative Years,* 420, and see 356–360.

41. *Ibid.,* 411, 419; California, Railroad Commission, *Report of the Railroad Commission of California from July 1, 1919 to June 30, 1920* (Sacramento, 1920), 100, and *July 1, 1922 to June 30, 1923* (1924), 86; California State Board of Agriculture, *Annual Report of the Statistician* (Sacramento, 1912), 223, and (1913), 191; *Minerals Yearbook, 1912,* 337, and *1915,* 983–984; Littlefield and Thorne, *Spirit of Enterprise,* 45–47; Charles T. Hutchinson, "Compressor Plant of the Southern California Gas Company," *Western Engineering,* 4 (May 1914): 344; U.S. Geological Survey, *Hydroelectric Power Systems of California and Their Extensions into Oregon and Nevada,* Water Supply Paper 493, by Frederick Hall Fowler (Washington, D.C., 1923), 874; "California's Fuel Outlook," Transactions of the Commonwealth Club of California, *The Commonwealth,* 24 (June 21, 1948): 57–58; Elizabeth M. Sanders, *The Regulation of Natural Gas: Policy and Politics, 1938–1978* (Philadelphia: Temple University Press, 1981), 26.

42. "High Pressure Gas Transmission," *JEPG,* 31 (August 16, 1913): 158; California, Department of Natural Resources, Division of Mines, *Mineral Commodities of California,* Bulletin No. 156 (San Francisco, August 1950), 72; *Minerals Yearbook, 1938,* 938; Sanders, *Regulation of Natural Gas,* 24–25; "Western Power and Fuel Outlook—2. Natural Gas," *Federal Reserve Bank of San Francisco, Monthly Review* (May 1949): 58–59; Littlefield and Thorne, *Spirit of Enterprise,* 72–89 *passim.*

43. "Western Power and Fuel Outlook—2. Natural Gas," 53–56; White, *Formative Years,* 145–148, 244, 250–251. Also see Blackford, *Politics of Business,* 54–55.

44. *Ibid.,* 420–421, 477–478, 559–560. *Minerals Yearbook, 1919,* pt. 2:521, *1921,* pt. 2:243; Kenyon L. Reynolds, "The Natural Gasoline Industry," *California Engineer,* 3 (October 1924): 38–39.

45. *Report of the Federal Trade Commission,* 1:168–174; Williamson, *et al., American Petroleum Industry,* 2:154–159, 391–393; Reynolds, "Natural Gasoline," 38.

46. *Minerals Yearbook, 1930,* pt. 2:452, and *1934,* 742; E.F. English, "The Small Town Gas Company," *California Journal of Development,* 21 (September 1931): 12–13, 50; "Gas Competition," *JEPG,* 65 (August 1, 1930): 58–61; Kendall Beaton, *Enterprise in Oil: A History of Shell in the United States* (New York: Appleton-Century Crofts, 1957), 408, 502.

47. *Ibid.,* 337–338.

48. *Ibid.,* 387; Brown, "California's Natural Gas Resources," 45–46; Johnson, "California and the National Oil Industry," 165–166; "Western Power and Fuel Outlook—2. Natural Gas," 54.

49. Alan McEwen, "The Sharkey Oil Control Act," *California Journal of Development,* 22 (April 1932): 10, 45–46; Beaton, *Enterprise in Oil,* 387.

Chapter VIII/Fueling California

1. Hardison, quoted in Lionel V. Redpath, *Petroleum in California: A Concise and Reliable History of the Oil Industry of the State* (Los Angeles: R.V. Redpath, 1900), 98.

2. Warren S. Thompson, *et al., Growth and Changes in California's Population* (Los Angeles:

Haynes Foundation, 1955), ch. 2; California, Department of Finance, *Statistical Abstract, 1970,* 13, Table B–5. Southern California's early history is told by Robert G. Cleland, *The Cattle on a Thousand Hills* (San Marino: The Huntington Library, 1951).

3. Walton Bean and James Rawls, *California, An Interpretive History,* 6th ed. (New York: McGraw-Hill, 1993), 191–193; Richard B. Rice, William A. Bullough, and Richard J. Orsi, *The Elusive Eden: A New History of California* (New York: Alfred A. Knopf, 1988), 311–319.

4. Fred W. Veihe, "Black Gold Suburbs: The Influence of the Extractive Industry on the Suburbanization of Los Angeles, 1890–1930," *Journal of Urban History,* 8 (November 1981): 3. Among the most thorough works dealing with transportation and the growth of Los Angeles is Scott Bottles, *Los Angeles and the Automobile: The Making of the Modern City* (Berkeley: University of California Press, 1987), and he deals with the distinction between "dispersal" and "decentralization," p. 259, n. 21. Also see Spencer Crump, *Ride the Big Red Cars: How Trolleys Helped Build Southern California* (Los Angeles: Crest Publications, 1962); Robert M. Fogelson, *The Fragmented Metropolis: Los Angeles, 1850–1930* (Cambridge, MA: Harvard University Press, 1967); Reyner Banham, *Los Angeles: The Architecture of Four Ecologies* (Los Angeles: Harper and Row, 1972); Sam Bass Warner, Jr., *The Urban Wilderness: A History of the America City* (New York: Harper and Row, 1973), 132–149; and Mark S. Foster, "The Model-T, The Hard Sell, and Los Angeles Urban Growth: The Decentralization of Los Angeles during the 1920s," *Pacific Historical Review,* 44 (November 1975): 459–485.

5. Veihe, "Black Gold Suburbs," 3, 8, 14–19. Veihe further develops his position on the importance of oil to the evolution of the Los Angeles Basin in "The Social-Spatial Distribution of the Black Gold Suburbs of Los Angeles, 1900–1930," *Southern California Quarterly,* 78 (Spring 1991): 33–54. In other areas of southern California such as Bakersfield, oil also had a profound community development influence.

6. U.S. Bureau of Mines, *Survey of Fuel Oil and Kerosene Distribution in District Five [annual reports 1924–1949]* (San Francisco, 1924–1949), mimeographed. Natural gas consumption data in U.S. Bureau of Mines, *Minerals Yearbooks,* 1920–present; *supra,* Table C–15.

7. *PRP,* 40 (December 27, 1890): 550, v. 41 (March 7, 1891): 227 and (June 18, 1891): 588, v. 42 (July 18, 1891): 64, and v. 44 (November 12, 1892): 409; *Engineering News,* 28 (August 4, 1892): 106; Roger Olmsted, *Scow Schooners of San Francisco Bay,* ed. Nancy Olmsted (Cupertino, CA: California History Center Foundation, 1987), ch. 4.

8. "Gasoline Pumping Plant," *PRP,* 54 (September 11, 1897): 161; Chas. H.F. Lubcke, "Recent Uses of the Gas Engine," *JEPG,* 7 (February 1899): 33–34; L.K. Sherman, "The Condition of Water and Power Development in Southern California," *Journal of the Society of Engineers,* 5 (October 1900): 345–346; California, State Board of Trade, *Ninth Annual Report for 1898* (March 14, 1899), 11–21; A.F. Bridge and Ralph Reynolds, "Costs of Pumping Water at Isleton," *JEPG,* 30 (February 25, 1913): 131–132.

9. Howard C. Kegley, "A Profitable Pumping Plant," *Farm Engineering,* 1 (June 1914): 5; J.V. Tuttle, "A Gas Engine and 13,000 Hens," *Farm Engineering,* 1 (April 1914): 22; B.D. Moses and W.P. Duruz, "Electric Power for Orchard Spraying," *JEPG,* 54 (February 15, 1925): 129–131; L.S. Wing, "Rural Electrification from an Economic and Engineering Standpoing—I," *JEPG,* 57 (September 1, 1926): 164; University of California, Agricultural Experiment Station, *The Tractor on California Farms,* Bulletin No. 415, by L.J. Fletcher and C.D. Kinsman (Berkeley, December 1926), 20. On the tempermental character of gasoline engines, see "Measuring the Speed of Gas Engines," *Farm Engineering,* 3 (October 1915): 69; University of California, Agricultural Experiment Station, *The Selection and Cost of a Small Pumping Plant,* Circular No. 117, by B.A. Etcheverry (Berkeley, April 1914), 21.

10. U.S. Census Office, Eleventh Census, 1890, *Report on the Transportation Business in the United States, Pt. 1, Statistics of Transportation* (Washington, D.C., 1895), 14:689; U.S. Bureau of the Census, *Census of Electric Industries, 1917: Electric Railways* (Washington, D.C., 1920), 123, Table 141; Rudolph Van Norden, "Pacific Electric Railway Interurban System," *JEPG,* 26 (January 7, 1911): 1–31; U.S. Geological Survey, *Hydroelectric Power Systems of California and Their Extensions into Oregon and Nevada,* Water Supply Paper 493, by Frederick Hall Fowler (Washington,

D.C., 1923), 23; William A. Myers, *Iron Men and Copper Wires: A Centennial History of the Southern California Edison Company* (Glendale, CA: Trans-Anglo Books, 1983), 52–67; Bottles, *Los Angeles and the Automobile,* 29–33. Also see George H. Hilton and John F. Due, *The Electric Interurban Railways in America* (Palo Alto, CA: Stanford University Press, 1960); Spencer Crump, *Ride the Big Red Cars;* David E. Nye, *Electrifying America: Social Meanings of a New Technology, 1880–1940* (Cambridge, MA: The M.I.T. Press, 1990), 85–137.

11. In general, see Sam Bass Warner, Jr., *Streetcar Suburbs: The Process of Growth in Boston, 1870–1900* (Cambridge, MA: Harvard University Press, 1962). Specifically, see Bottles, *Los Angeles and the Automobile,* 29–45; James C. Williams, "The Trolley: Technology and Values in Retrospect," *San José Studies,* 3 (November 1977): 82–83.

12. Bottles, *Los Angeles and the Automobile,* 35–48.

13. Bottles, *Los Angeles and the Automobile,* 48; Warren J. Belasco, "Cars Versus Trains, 1980 and 1910," in *Energy and Transport: Historical Perspectives on Policy Issues,* ed. by George H. Daniels and Mark H. Rose (Beverly Hills, CA: Sage Publications, 1982), 41, 42–50. Also see Belasco, *Americans on the Road: From Autocamp to Motel, 1910–1945* (Cambridge, MA: The M.I.T. Press, 1979); James J. Flink, *America Adopts the Automobile, 1895–1910* (Cambridge, MA: The M.I.T. Press, 1970).

14. Ashleigh E. Brilliant, "Some Aspects of Mass Motorization in Southern California, 1919–1929," *Southern California Quarterly,* 47 (June 1965): 191; Foster, "The Model-T," 466, n. 20; Bottles, *Los Angeles and the Automobile,* 49–51, 58, 93. In other American cities, people also called for adoption of the automobile as a solution to the sanitation problem caused by thousands of horses pulling delivery wagons and buggies (Mark S. Foster, "The Automobile and the City," *Michigan Quarterly Review,* 19/20 [Fall 1980/Winter 1981], 465).

15. *Ibid.,* 53–54, 93, 98–99; California State Board of Agriculture, *Annual Report of the Statistician* (Sacramento, 1911), 249 and 261, *1917,* 245, and *1918,* 272; California, *Statistical Abstract, 1970,* 234.

16. Bottles, *Los Angeles and the Automobile,* 97–115, 123, 158–174; David Brodsly, *L.A. Freeway: An Appreciative Essay* (Berkeley: University of California Press, 1981), 82–93.

17. Quoted in Brilliant, "Mass Motorization," 205. Also see David L. Lewis, "Sex and the Automobile: From Rumble Seats to Rockin' Vans," *Michigan Quarterly Review,* 19/20 (Fall 1980/Winter 1981): 518–528.

18. Brilliant, "Mass Motorization," 194–203.

19. *Statistician Report, 1911,* 260–261, *1915,* 247–251, *1917,* 243–247, *1918,* 268–272, and *1921,* 281.

20. Harold F. Williamson, *et al., The American Petroleum Industry,* 2 vols. (Evanston, IL: Northwestern University Press, 1959), 2:238.

21. Gerald T. White, *Formative Years in the Far West: A History of the Standard Oil Company of California and Predecessors through 1919* (New York: Appleton-Century Crofts, 1962), 446; Richard Charles Schwarzman, *Pinal Dome Oil Company: An Adventure in Business, 1901–1917* (New York: Arno Press, 1976), 207–209, 226.

22. Joe S. Bain, *The Economics of the Pacific Coast Petroleum Industry,* 3 vols. (Berkeley: University of California Press, 1944–1947), 1:191, and 2:53–55, 242–243; Schwarzman, *Pinal Dome,* 235; Williamson, *American Petroleum Industry,* 2:271, 274, 486; White, *Formative Years,* 506–511; Brilliant, "Mass Motorization," 203–204.

23. *Ibid.,* 205. Foster, "Model-T", 471–480; Williams, "The Trolley," 74–90; Folke T. Kihlstedt, "The Automobile and the Transformation of the American House," *Michigan Quarterly Review,* 19/20 (Fall 1980/Winter 1981): 555–570.

24. Bottles, *Los Angeles and the Automobile,* 175–210; Brodsly, *L.A. Freeway,* 10.

25. Bottles, *Los Angeles and the Automobile,* 211–232. The auto club proposal is quoted on 216.

26. John Thompson, "From Waterways to Roadways in the Sacramento Delta," *California History,* 59 (Summer 1980): 144–169. The concept of nature-alienation stems from William L. Preston's suggestion that the federal land survey and other federal and state land disposal practices were, in effect, "land-alienation policies." See his *Vanishing Landscapes: Land and Life in the Tulare Lake Basin* (Berkeley: University of California Press, 1981), 76–80, 97–98, 168.

27. H. Marshall Goodwin, Jr., "Right-of-Way Controversies in Recent California Highway-Freeway Construction," *Southern California Quarterly,* 56 (Spring 1974): 61–62; Paul B. Smith, "Highway Planning in California's Mother Lode: The Changing Townscape of Auburn and Nevada City," *California History,* 59 (Fall 1980): 204–221; Brodsly, *L.A. Freeway,* 97–137 *passim;* Bottles, *Los Angeles and the Automobile,* 233–234.

28. Reynold M. Wik, *Henry Ford and Grass Roots America* (Ann Arbor: University of Michigan Press, 1972), 24–30; Flink, *America Adopts the Automobile,* 112, 69–70, 108–111.

29. Wik, *Henry Ford,* 14–24; Flink, *America Adopts the Automobile,* 66–68; "The Automobile in Agriculture," *PRP,* 76 (September 19, 1908): 177.

30. Edw. O. Amundsen, "Farm Uses for Autos," *The University of California Journal of Agriculture,* 2 (December 1914): 138; "Novel Way of Clearing Land," *Farm Engineering,* 5 (January 1917): 9; "On Converting the Automobile," *Farm Engineering,* 3 (October 1915): 70; Wik, *Henry Ford,* 30–33; California, Department of Public Works, Division of Engineering and Irrigation, Bulletin No. 21, *Irrigation Districts in California,* by Frank Adams (Sacramento, 1929), 240; Bain, *Pacific Coast Petroleum,* 2:93.

31. Wik, *Henry Ford,* 84–85, as well as Wik, *Steam Power on the American Farm* (Philadelphia: University of Pennsylvania Press, 1953), 202–203, and Wik, "Some Interpretations of the Mechanization of Agriculture in the Far West," *Agricultural History,* 49 (January 1975), 358–360. Also see "California's Thirst for Tractors," *PRP,* 92 (August 19, 1916): 186; Robert C. Williams, *Fordson, Farmall, and Poppin' Johnny: A History of the Farm Tractor and Its Impact on America* (Champaign: University of Illinois Press, 1987).

32. Wik, *Henry Ford,* 92–93, and Wik,"Some Interpretations of the Mechanization of the American Farm," 360.

33. Robert E. Ankli and Alan L. Olmstead, "The Adoption of the Gasoline Tractor in California," *Agricultural History,* 55 (July 1981): 214–215; "California's Thirst for Tractors," 186; G.M. Walker, "The Modern Power Farmer," *The University of California Journal of Agriculture,* 7 (May 1921): 11; Wayne D. Rasmussen, "The Impact of Technological Change on American Agriculture, 1862–1962," *Journal of Economic History,* 22 (December 1962): 579.

34. "A New Tractor," *Farm Engineering,* 3 (March 1916): 209; Albert Marple, "Smallest Tractor Ever Built," *Farm Engineering,* 4 (December 1916): 96.

35. For a judicious study of Holt's Caterpillar tractor, see Reynold M. Wik, "Benjamin Holt and the Invention of the Track-Type Tractor," *Technology and Culture,* 20 (January 1979): 90–107. Also see Pliny E. Holt, "The Development of the Caterpillar Tractor and Its Application to Industry," *Mechanical Engineering,* 48 (June 1948): 657–661; R.P. O'Neill, "The Caterpillar Tractor," *California Engineer,* 5 (February 1927): 160–161; Wik, "Some Interpretations of the Mechanization of Agriculture," 81; Roy Bainer, "Science and Technology in Western Agriculture," *Agricultural History,* 49 (January 1975): 66.

36. W.H. Hutchinson, *California: The Golden Shore by the Sundown Sea,* 2nd ed. (Belmont, CA: Star Publishing Company, 1988), 170, notes that horse snowshoes developed for use in winter months in the Sierra Nevada "were also used on horses harvesting asparagus on the peat lands of the Delta until the 'caterpiller' tractor was perfected."

37. For a thorough discussion of California farm adoption of the tractor, see Ankli and Olmstead, "Adoption of the Gasoline Tractor," 213–230.

38. L.J. Fletcher, "Orchard Tractors," *The California Countryman,* 8 (April 1922): 3–4; Fletcher and Kinsman, *The Tractor,* 10; Ankli and Olmstead, "Adoption of the Gasoline Tractor," 213, 227–230.

39. Fletcher and Kinsman, *The Tractor,* 3–16; John E. Pickett, "You Can't Fool an Old Horse-fly," *PRP,* 110 (August 29, 1925): 199.

40. Wik, *Henry Ford,* 97.

41. U.S. Department of Commerce, Bureau of the Census, *United States Census of Agriculture,* 1959, pt. 48, 1:8; U.S. Department of Commerce, Bureau of the Census, *1969 Census of Agriculture,* pt 4., 1:4.

Chapter IX/Hydroelectricity

1. Dr. Wellington Adams, "Evolution of the Electric Railway: Its Commercial and Scientific Aspect," *Journal of the Association of Engineering Societies,* 3 (November 1883): 259; Joseph Wetzler, "The Electric Railway of Today," *Scribner's Magazine,* 7 (April 1890): 443; "The Electrical Exposition," *Harper's Weekly,* 40 (May 30, 1896): 547; R.B. Owens, "Electricity as a Factor in Modern Development," *Cassier's Magazine,* 16 (June 1899): 212. Electricity has become a topic of sustained interest among historians of technology. Concerning the rise of electrical technology, the development of power systems, and the history of the electric power industry, see Thomas P. Hughes, *Networks of Power: Electrification in Western Society, 1880–1930* (Baltimore: Johns Hopkins University Press, 1983); Harold Passer, *The Electrical Manufacturers, 1897–1900* (Cambridge, MA: Harvard University Press, 1953); Louis C. Hunter and Lynwood Bryant, *A History of Industrial Power in the United States, 1780–1930, Volume Three: The Transmission of Power* (Cambridge, MA: The M.I.T. Press, 1991), Pt. 2; Harold L. Platt, *The Electric City: Energy and the Growth of the Chicago Area, 1880–1930* (Chicago: University of Chicago Press, 1991); and Harold I. Sharlin, *The Making of the Electrical Age: From the Telegraph to Automation* (London: Abelard-Schuman, 1963). For discussion of the social impact of electricity in America, see David E. Nye, *Electrifying America: Social Meanings of a New Technology, 1880–1940* (Cambridge, MA: The M.I.T. Press, 1990).

2. Charles M. Coleman, *PG&E of California: The Centennial Story of Pacific Gas and Electric Company, 1852–1952* (New York: McGraw-Hill, 1952), 51–61, is a key source for the development of early California community lighting systems; "Illumination by Electric Light," *Engineer of the Pacific,* 1 (July 1878): 1; U.S. Bureau of Census, *Central Electric Light and Power Stations, 1902* (Washington, D.C., 1905), 90. For a general discussion of the rise and importance of electricity in California and the West, see David E. Nye, "Electrifying the West, 1880–1940," *The American West As Seen By Europeans and Americans,* ed. Rob Kroes (Amsterdam: Free University Press, 1989), 183–202. The bulk of this article is woven into his *Electrifying America.*

3. Robert C. Catren, "A History of the Generation, Transmission, and Distribution of Electrical Energy in Southern California" (Ph.D. diss., University of Southern California, 1951), 40–43; Douglas R. Littlefield and Tanis C. Thorne, *The Spirit of Enterprise: The History of Pacific Enterprises from 1886 to 1889* (Los Angeles: Pacific Enterprises, 1990), 22; Coleman, *PG&E,* 63–77; Sidney Harold Elman, "Some Aspects of the California Midwinter International Exposition of 1894" (unpublished M.A. thesis, University of Southern California, 1950), 107–108; U.S. Census Office, Eleventh Census, 1890, *Report on the Transportation Business in the United States, Pt. 1, Statistics of Transportation* (Washington, D.C., 1895), 14:689, 704; "Eight-Horse-Power Electric Motor," *Industry,* 3 (October 1890): 41. A description of the San José light tower in the Paris journal *La Lumière Electrique, Journal d'Electricité* that also bemoaned street lighting in Paris (Coleman, *PG&E,* 73), plus local lore in San José, hints that perhaps later French tower proposals (Nye, *Electrifying America,* 29), which eventually resulted in Eiffel's famous tower, stemmed from the San José effort.

4. Hunter and Bryant, *Transmission of Power,* 411.

5. A.H. Markwart, "Power in California," *Journal of the Franklin Institute,* 204 (August 1927): 147–148, 153–156; Coleman, *PG&E,* 92–106; Louis C. Hunter, *A History of Industrial Power in the United States, 1780–1930, Volume One: Waterpower in the Century of the Steam Engine* (Charlottesville: University of Virginia Press, 1979), 398, 401, 413; A.M. Hunt and Wynn Meredith, "Electric Power: Its Generation and Utilization for Mining Work on the Pacific Coast," *California Mines and Minerals* (San Francisco: L. Roesch Co., 1899), 75. For examples of mining ditches being taken over by hydroelectric companies, see Rudolph W. Van Norden, "History of the Bear River Ditch" and W. E. Meservey, "Early History of the Ditches Owned by 'Pacific Service' in Nevada County," *Pacific Service Magazine,* 5 (July 1913): 58–61, 62–63; James C. Williams, "Centerville-De Sabla Project Historical Report and Project Significance and Recommendations," *Cultural Resources Inventory and Management Plan for the Proposed Improvements to the De Sabla-Centerville Hydroelectric System, Butte County, California, FERC No. 803* (Sacramento: Public Anthropological Research, 1985), B1–B27; Historic American Engineering Record, *The Battle Creek Hy-*

droelectric System: An Historical Study, by Terry S. Reynolds and Charles Scott (Washington, D.C., [1980]), 7–18.

6. Coleman, *PG&E,* 56, 78, 217; U.S. Department of Interior, Bureau of Land Management, *Report: Determination of Eligibility of the Excelsior Ditch, Nevada Co., California, for inclusion into the National Register of Historic Places,* by Robert E. Gray (Sacramento, September 1980), 31; California State Mining Bureau, *Sixth Annual Report of the State Mineralogist for the Year Ending May 15, 1886* (Sacramento, 1886), 25, and *1888,* 76–77, 798; Nye, *Electrifying America,* 195–196. For developments of electricity in mining during the 1890s, see Thomas Haight Leggett, "Electric Power Transmission Plants and the Use of Electricity in Mining Operations" in *State Mineralogist, 1894,* 419–451; W.F.C. Hasson, "Electric-Power Transmission Plants in California" in *State Mineralogist, 1896,* 673–678; Hunt and Meredith, "Electric Power," 73–87.

7. Hunt and Meredith, "Electric Power," quoting the State Mineralogist on 74; also see 75–76, 80–87.

8. E.J. Molera, "On the Recent Progress in Electricity," *Transactions of the Technical Society of the Pacific Coast,* 2 (February 1885): 70; Markwart, "Power in California," 153; U.S. Geological Survey, *Hydroelectric Power Systems of California and Their Extensions into Oregon and Nevada,* Water Supply Paper 493, by Frederick Hall Fowler (Washington, D.C., 1923), 30–31. Haupt is quoted in Hunter and Bryant, *Transmission of Power,* 184, n. 137.

9. M.L. Holman, "Presidential Address, 1908," *Transactions of the American Society of Mechanical Engineers,* 30 (1908): 602, as quoted in Hunter and Bryant, *Transmission of Power,* 255; also see 180–181, 254.

10. Tesla and others who developed AC technology discovered that voltage (pressure) could be raised through a transformer which caused amperes (current) to be proportionally reduced. Since wire resistance increases with current and causes energy loss through heat, reduction in current meant AC power incurred less resistance and could be transmitted over longer distances than DC power using smaller copper wire. A single-phase system required two wires, one from the generating point and one back from the load source. Current transmitted in a two-phase system is sent over two circuits which are out of phase by 90 degrees, but four wires are needed, two for each circuit. In a three-phase system, since the currents of the three circuits are always 120 degrees apart and their sum is always zero, proper connection of the circuits required only three wires. Thus, the three-phase system resulted in the most economical multipurpose power by reducing the cost of copper wire 25 percent over a two-phase system. See Hunter and Bryant, *Transmission of Power,* 244–249; Donald C. Jackson, "Theory and Practice in the Development of a Technological Style: California's Early 3-Phase AC Power Systems" (Paper presented at the annual meeting of the Society for the History of Technology, Philadelphia, October 1982), 5. Also see Jackson, *Building the Ultimate Dam: John S. Eastwood and Control of Water in the West* (Lawrence: University of Kansas Press, 1995).

11. Much of this story is drawn from Jackson, "Theory and Practice," 6–9, as well as Hunter and Bryant, *Transmission of Power,* 246–247; Fowler, *Hydroelectric Power Systems,* 1; Markwart, "Power in California," 148–149; G.O. Wilson, "History of Electrical Transmission," *JEPG,* 33 (July 25–August 15, 1914): 71–74; Thomas Parke Hughes, "The Science-Technology Interaction: The Case of High-Voltage Power Transmission Systems," Technology and Culture, 17 (October 1976): 647; Hughes, *Networks of Power,* 111–117, 282–283; William A. Myers, *Iron Men and Copper Wires: A Centennial History of the Southern California Edison Company* (Glendale, CA: Trans-Anglo Books, 1983), 22–27; Robert McF. Doble, "Hydro-Electric Power Development and Transmission in California," *Journal of the Association of Engineering Societies,* 34 (March 1905): 8–9; Catren, "Electrical Energy in Southern California," 45–48, 71–72; Coleman, *PG&E,* 107.

12. "The Folsom-Sacramento Power Transmission," *JEPG,* 1 (September 1895): 55–69; Archie Rice, "The Story of the Folsom Power Plant," *JEPG* (February 26, 1910): 179–185; Hasson, "Electric-Power Transmission," 673–674; Fowler, *Hydroelectric Power Systems,* 110; Jackson, "Theory and Practice," 9–10.

13. C.E. Grunsky, "The Industrial Problem of the Pacific Coast: I. Address of the Retiring President," *Journal of the Association of Engineering Societies,* 14 (May 1895): 378; Hunter and Bryant,

Transmission of Power, 255, 257. Most historians also have generally accepted this positon on Niagara Falls, among them Joseph M. Dukert, *A Short Energy History of the United States (And Some Thoughts About the Future)* (Washington, D.C.: Edison Electric Institute, 1980); Hughes, *Networks of Power;* Martin V. Melosi, *Coping With Abundance: Energy and Environment in Industrial America* (Philadelphia: Temple University Press, 1985); John Opie in Barbour, *et. al., Energy and American Values* (New York: Praeger, 1982); Passer, *The Electrical Manufacturers;* Sharlin, *Making of the Electrical Age;* Sam H. Schurr and Bruce C. Netschert, *Energy in the American Economy, 1850–1975* (Baltimore: Johns Hopkins Press for the Resources of the Future, 1960); Nye, *Electrifying America,* 196, gives some recognition to Folsom for showing that AC power transmission had the ability to make factory location flexible.

14. Reynold M. Wik, "Some Interpretations of the Mechanization of Agriculture in the Far West," *Agricultural History,* 49 (January 1975): 78.

15. Quoted in "Popular Reflections," *JEPG,* 1 (July 1895): 27.

16. C.W. Whitney, "Hydro-Electric Power Plants in California," *California Journal of Technology* (February 1906): 6–7. Eastwood was later noted for his work with multiple arch dams and on the Pacific Light and Power Corporation's Big Creek hydroelectric project. See Jackson, "Theory and Practice," 10–11; Jackson, *Building the Ultimate Dam,* ch. 4.

17. Williams, "Centerville-De Sabla Project," B1–B27. Reynolds and Scott, *Battle Creek Hydroelectric System,* 21–22, describes similar beginnings for a power company near Red Bluff and Redding. Also see the brief histories of California companies in Fowler, *Hydroelectric Power Systems.*

18. James C. Williams, "De Sabla Historical Report," *An Archaeological and Historical Investigation of Site CA-But-868H at the De Sabla Powerhouse, Butte County, California* (Paradise, CA: Professional Archaeological Services, April 1988), 99–144. The Nevada County powerhouse was subsequently renamed "Rome" in honor of Romulus Colgate. Coleman, *PG&E,* chs. 12–14, are relied on for much of the following sketch of the electric power activities of de Sabla, Tregidgo, and Martin. See also "Chronological Diagram showing the Origin and Development of 'Pacific Service'" (1920), California State Archives, Bin 6059.

19. W.F.C. Hasson, "Electric Transmission of Power Long Distances," *Transactions of the Technical Society of the Pacific Coast,* 10 (May 1893): 49–72; F.A.C. Perrine, "Some Data of a 30,000 Volt Line," *JEPG,* 5 (May 1898): 162–163; Paul M. Downing, "The Developed High Tension Net-Work of a General Power System," *Transactions of the American Institute of Electrical Engineers,* 29, Pt. 1 (1910): 714–715. As well as belonging to the PCETA, both de Sabla and Martin presented papers. See E.J. de Sabla, Jr., "Lightning in Nevada County" and John Martin, "Insulators for Transmission Lines" in the proceedings of the PCETA published in *JEPG,* 5 (November 1897): 39–40, 41–42.

20. John D. Galloway, "Hydro-Electric Developments on the Pacific Coast," *Transactions of the American Society of Civil Engineers,* 86 (1923): 803–804; W.G. Vincent, Jr., "The Interconnected Transmission System of California," *JEPG,* 54 (June 15, 1925): 567–568; "World's Largest Transmission System," *EW,* 59 (June 1, 1912): 1198.

21. John D. Galloway, "Hydro-electric Power Station, California Gas & Electric Corporation, De Sabla, Cal.," *Engineering News,* 54 (August 10, 1905): 131–135; Samuel Storrow, *Report Upon the Hydraulic Plants of the California Gas & Electric Company* (San Francisco, Cal, April 24, 1905; edition of September 1909), n.p., typescript MS at the PG&E Corporate Library.

22. James C. Williams, "Pre-1940 Hydroelectric Developments Historic Evaluation Survey," Pacific Gas and Electric Company, San Francisco, Land Department Contract 17–83, October 27, 1983. Hughes, *Networks of Power,* 278. Acquisition of the Great Western Power Company by PG&E gave it additional access into San Francisco through four submarine transmission cables which Great Western had installed between 1911 and 1924. See "The Transbay Cable of the Great Western Power Company," *JEPG,* 28 (March 16, 1912): 239–244; John A. Koontz, Jr., "Laying the Largest Submarine Power Cable," *JEPG,* 33 (July 4, 1914): 1–2; "Eight Mile Cable Under San Francisco Bay," *EW,* 81 (February 17, 1923): 399–400; James B. Black, "Laying Bay Cable No. 4," *California Engineer,* 2 (April 1924): 6–7.

23. Williams, "Pre-1940 Hydroelectric Developments," n.p. "Salt Springs Rock-Fill Dam,"

Western Construction News, 4 (April 10, 1929): 183–184; "The Big Meadows Dam," *JEPG,* 27 (September 30, 1911): 287–289; E. Court Eaton, "The Melones Dam on the Stanislaus River, California," *Western Construction News,* 2 (May 25, 1927): 34–41; California, Department of Finance, *Statistical Abstract, 1970* 175–177, Table G–10.

24. Catren, "Electrical Energy in Southern California," 71–81; Littlefield and Thorne, *Spirit of Enterprise,* 34–56 *passim.* For a general history of Southern California Edison, see Myers, *Iron Men and Copper Wires.* For a recent detailed account of the Santa Ana River system, see Mark T. Swanson and David De Vries, *The Santa Ana River Hydroelectric System,* Report Prepared for U.S. Army Corps of Engineers, Los Angeles District, Statistical Research Technical Series No. 47 (Tuscon, AZ: Statistical Research, Inc., 1994).

25. Littlefield and Thorne, *Spirit of Enterprise,* 36; Laurence D. Shoup, *The Hardest Working Water in the World: A History and Significance Evaluation of the Big Creek Hydroelectric System* (Fair Oaks, CA: Theodoratus Cultural Research, October 1988, for Southern Califoria Edison Company), 30–31; Catren, "Electrical Energy in Southern California," 79–80; Hank Johnston, *The Railroad that Lighted Southern California* (Los Angeles: Trans-Anglo Books, 1965), 10.

26. Shoup, *Hardest Working Water,* 29–30.

27. *Ibid.,* 32–33. Rudolph Van Norden, "Pacific Electric Railway Interurban System," *JEPG,* 26 (January 7, 1911): 1–31; Fowler, *Hydroelectric Power Systems,* 23; Myers, *Iron Men and Copper Wires,* 52–67; Vincent, "Interconnected Transmission System," 567; "The Transmission of Electrical Energy in Southern California," *EW,* 54 (October 28, 1909): 1037–1039; Catren, "Electrical Energy in Southern California," 91–97.

28. Shoup, *Hardest Working Water,* 25–29, 33–42. To work out financing for the Big Creek project, Huntington created a bonded indebtedness for Pacific Light and Power Corporation by assessing stockholders a certain amount per share owned. To pay his assessments, Eastwood, who had few personal funds, was forced to sell off all his stock rather cheaply to Huntington. In the end, he had nothing for his efforts.

29. *Ibid.,* 43–80 *passim.* Also see David H. Redinger, *The Story of Big Creek* (Los Angeles: Angelus Press, 1949). Redinger worked on Big Creek from its inception through 1947. Engineering journals across the nation watched the Big Creek project carefully, and Stone and Webster, the principal construction firm on the project, published reports in its own journal. See A.J. Farnsworth, "The Big Creek Power Development," *Stone and Webster Public Service Journal,* 11 (September 1912): 170–175; Robert Sibley, "The Pacific Light and Power System," *JEPG,* 28 (February 24, 1912): 151–174; J.H. Anderthon, "Electrical Features of the Big Creek Development," *Stone and Webster Public Service Journal,* 13 (November 1913): 326–336; "The 150,000-Volt Big Creek Development," *EW,* 63 (January 3, 1914): 33–38, continued in (January 10, 1914): 85–88; Walter E. Jessup, "A High Head Hydro-Electric Power Development in the Sierra Nevada Mountains, California," *The Wisconsin Engineer,* 18 (February 1914): 193–211; H.C. Hoyt, "The Big Creek Development of the Pacific Light and Power Company," *General Electric Review,* 17 (August 1914): 828–840.

30. "Merger of California Hydroelectric Systems" and "Power Companies Accept Offer of Los Angeles," *EW,* 68 (December 9, 1916): 1134–1135, and 1136–1137; Shoup, *Hardest Working Water,* 86–155; Redinger, *Big Creek, passim;* Southern California Edison Company, "Construction of the Florence Lake Tunnel and General Information Concerning the Big Creek Hydro-electric Development of the Southern California Edison Company," June 30, 1925, on file at the Water Resources Center Archives, University of California, Berkeley; Harry W. Dennis and H.A. Barre, "Growth of the Use of Electric Power in Southern California and Probabilities of its Future Growth with Reference to Sources of Hydraulic Power," *Transactions of the American Society of Civil Engineers,* 86 (1923): 822–828. A number of articles appear in industry journals concerning the work at Big Creek between 1919 and 1929. For examples, see Malcolm B. Arthur and Carl F. Meyer, "The Big Creek Hydroelectric Project of the Southern California Edison Company," *The Journal of the Worcester Polytechnic Institute,* 27 (April 1924): 1–94; "Big Creek-San Joaquin Hydro-Electric Project of the Southern California Edison Company," *Western Construction News,*

2 (June 25, 1927): 28–36; "High Head and Large Units at Big Creek 2-A," *EW*, 92 (November 17, 1928): 989–992; "Edison Adds 237,000 More Horse-Power in 1928," *JEPG*, 61 (November 1, 1928): 252–266. The Southern Sierras Power Company also undertook important construction work in the eastern Sierra Nevada, notably involving the use of multiple-arch dams. See James C. Williams, *Evaluation of the Historic Resources of the Lee Vining Creek (FERC Project Number 1388) and Rush Creek (FERC Project Number 1389) Hydroelectric Systems, Mono County, California* (Fair Oaks, CA: Theodoratus Cultural Resources, Inc., July 1989, for Southern California Edison Company), *passim*.

31. Carroll W. Pursell, Jr., "The Technical Society of the Pacific Coast, 1884–1914," *Technology and Culture*, 17 (October 1976): 702–717; J. Richard, "The High Pressure Hydraulic System of Distributing Power in Cities, with Some Remarks on Other Methods," *Transactions of the Technical Society of the Pacific Coast*, 3 (November 1886): 87–100; N.S. Keith, "The Transmission of Power by Means of Electricity," *Transactions*, 4 (April 1887): 87–104; W.F.C. Hasson, "Electric Transmission of Power Long Distances," *Transactions*, 10 (May 1893): 49–72.

32. "Engineering Associations in the West," *JEPG*, 9 (January 1900): 69; "The Transmission Convention at Santa Cruz," *JEPG*, 4 (August 1897): 101; "The Tamalpias Meeting of the Transmission Association," *JEPG*, 10 (July 1900): 3–4; Jackson, "Theory and Practice," 11–12.

33. "Editorial," *JEPG*, 8 (December 1899): 131.

34. C.L. Cory, "Biography," *JEPG*, 10 (July 1900): 26; F.A.C. Perrine, "Personal," *JEPG*, 5 (April): 1898, 134; Hughes, *Networks of Power*, 268.

35. "Proceedings of the Pacific Coast Electric Transmission Association," *JEPG*, 6 (July 1898): 10. At the Association's October 1897 meeting, four papers dealt with lightning, reviewing the general problems and experiences of various California power companies (*JEPG*, 5 [November 1897]: 34–42). The papers and proceedings of the Pacific Coast Electric Transmission Association appear in various issues of *JEPG*, vols. 5–14, August 1897–July 1904.

36. Quoted in Hughes, *Networks of Power*, 268.

37. F.A.C. Perrine, "Some Data of a 30,000 Volt Line," *JEPG*, 5 (May 1898): 162–163. Also see "Western Hydroelectric Transmission Developments," *JEPG*, 34 (June 5, 1915): 443; Paul M. Downing, "The Developed High Tension Net-Work of A General Power System," *Transactions of the American Institute of Electrical Engineers*, 29, Pt. 1 (1910): 714–715.

38. F.A.C. Perrine, "Tests and Calculations for a Forty-Mile Aluminium Wire Transmission Line," *JEPG*, 8 (August 1899): 41–42; "Editorial," *JEPG* 8 (August 1899): 47; "PCETA Proceedings," *JEPG* 8 (August 1899): 50, 53; F.A.C. Perrine, "Elements of Design Particularly Pertaining to Long Distance Transmission," *Transactions of the American Institute of Electrical Engineers*, 18 (1901): 831–869.

39. "Editorial," *JEPG*, 10 (August 1900): 40–41.

40. "Wanted: An Equation of Success," *JEPG*, 12 (July 1902): 123.

41. "Engineering Associations in the West," 69.

42. "It is Coming Westward," *JEPG*, 13 (May 1903): 230; "Editorial," *JEPG*, 15 (February, 1905): 53; Pursell, "Technical Society of the Pacific Coast," 716.

43. Andrew R. Boone, "Future Electric Transmission: An Interview with Dr. Harris J. Ryan," *JEPG*, 55 (September 1, 1925): 168; George H. Rowe, "Alternating Current Laboratory Testing," *JEPG*, 15 (January, February, April, May, June, October, and December 1905): 1–4, 43–46, 143–146, 179–190, 220–223, 449–452, 519–521; "Personal," *JEPG*, 15 (January 1905): 53; Hughes, *Networks of Power*, 145, 158, 379; "Ryan Laboratory at Stanford Dedicated as Two Million Volt Tests are Conducted," *JEPG*, 57, October 1, 1926): 256–257; J.T. Lusignan, Jr., "Ryan Laboratory Opens Untouched Field of Electrical Investigation," *JEPG*, 57 (December 21, 1926): 403–407; Walter Dreyer, "Mechanical Features of Modern High Voltage Transmission Lines," *California Engineer*, 5 (December 1926): 110–111.

44. Perrine participated in AIEE meetings regularly during the late 1890s and was designated west coast "honorary secretary" of the AIEE before 1900. See "Personal," *JEPG*, 9 (January 1900): 17; "Engineering Associations in the West," 69; Coleman, *PG&E*, 150–151, 157; Hughes, *Networks of*

Power, 380–384. Baum's publications included many journal articles, two books on AC regulation, and a book proposing a national super-power system. Hughes, "Science-Technology Interaction," 650–658, discusses the careers of Ryan and Peek.

45. R.W. Sorenson, H.H. Cox, and G.E. Armstrong, "California 220,000-Volt, 1100-Mile, 1,500,000-KW. *Transmission Bus," Transactions of the American Institute of Electrical Engineers,* 38, pt. 2 (1919): 1237–1268; Royal W. Sorenson, "California Institute of Technology's Million-Volt Laboratory," *JEPG,* 53 (October 1, 1924): 242. For examples of this continued cooperative work, see H.A. Barre, "New Transmission Line Construction in 1920," *JEPG,* 45 (December 15, 1920): 566–567; John A. Koontz, "220,000-Volt Transmission Progress," *JEPG,* 46 (May 15, 1921): 477–479; G.H. Stockbridge, "Avoiding Flashovers on 220-Kv. Transmission Lines," *EW,* 85 (March 21, 1925): 611–612; Boone, "Future Electrical Transmission," 167–169; "Hydro-Electric Energy as a Conserver of Oil," *JEPG,* 34 (October 1, 1917): 290–291; "Electrical Interconnection to Conserve Fuel," *EW,* 71 (January 5, 1918): 12–14; "Fuel Problems on the Pacific Coast," *Mechanical Engineering,* 41 (February 1919): 264–269; Frank R. Devlin, "The Importance of Development Our Hydro-Electric Resources," *Pacific Municipalities,* 35 (July 1921): 255–256; "The Basic Role of Electricity in Western Growth," *JEPG,* 46 (June 15, 1921): 597–611.

46. In the early 1970s, of the ninety-five major dams in California higher than 190 feet or with a reservoir capacity of at least 100,000 acre-feet, most were not directly the result of hydroelectric power projects. Thirteen were strictly for hydroelectricity, twenty-eight were for hydroelectric power as well as irrigation or municipal water storage, and the remaining fifty-four were parts of irrigation and/or municipal water systems. Sixty-five were constructed after 1940. See California, Statistical Abstract, 1970 175–177, Table G–10.

47. Donald C. Jackson, "Considering the Multiple Arch Dam: Theory, Practice and the Ethics of Safety in a Case of Innovative Hydraulic Engineering," *Natural Resources Journal,* 32 (Winter 1992): 77–100; Williams, *Evaluation of the Historic Resources of the Lee Vining Creek . . . and Rush Creek,* A–64–A70.

Chapter X/Electrifying the City and the Home

1. David E. Nye, *Electrifying America: Social Meaning of a New Technology, 1880–1940* (Cambridge, MA: The M.I.T. Press, 1990), 32, 57. Vaclav Smil, *Energy in World History* (Boulder, CO: Westview Press, 1994), 211–212, summarizes the general experience in the diffusion of electricity.

2. R.H. Thurston, "The Border-Land of Science," *The North American Review,* 150 (January 1890): 67; S.E. Doane, "Electric Light—A Factor in Civilization," *Journal of the Western Society of Engineers,* 20 (January 1915): 17; Frank Baum, "The Effect of Hydro-Electric Power Transmission upon Economic and Social Conditions, with Special Reference to the U.S. of America," *Transactions of the International Engineering Congress, 1915* (San Francisco, September 20–25, 1915): 247.

On the centrality of electricity to modernity, see James W. Carey and John J. Quirk, "The Mythos of the Electronic Revolution," *The American Scholar,* 39 (Spring 1970): 219–241, and (Summer 1970): 395–424; David Nasaw, "Cities of Light, Landscapes of Pleasure," *The Landscape of Modernity: Essays on New York City, 1900–1940,* ed. David Ward and Oliver Zunz (New York: Russell Sage Foundation, 1992), 273–286. On modernity, see David Harvey, *The Condition of Postmodernity: An Enquiry into the Origins of Cultural Change* (Oxford: Basil Blackwell, 1989), ch. 2, especially 31–32; and Marshall Berman, *All That Is Solid Melts Into Air: The Experience of Modernity* (New York: Penguin Books, 1988), 16–36.

3. E.H. Mullin, "The City of the Future," *Cassier's Magazine,* 13 (November 1897): 28; "Street Lighting in San Francisco," *JEPG,* 21 (December 5, 1908): 383; "Street Lighting in Alameda, Cal.," *EW,* 60 (November 2, 1912): 946–947; "Ornamental Street Lighting in San Francisco," *EW,* 61 (May 17, 1913): 1043; "The Electric City," *Energy,* 1 (August 1906): n.p.

4. "The Portola Festival at San Francisco," *EW,* 54 (December 9, 1909): 1422–1423.

5. "Lighting of the Panama-Electric Exposition," *EW,* 61 (May 17, 1913): 1043–1044; Hamilton Wright, "Electrical Features Panama-Pacific International Exposition," *JEPG,* 27 (September 23, 1911): 261–264; Nye, *Electrifying America,* 33–47, 63–65; William F. Raber, "Architecture of the Night," *California Journal of Development,* 20 (June 1930): 28.

6. Quoted in "San Francisco's Path of Gold," *JEPG*, 37, (October 14, 1916): 297. "A New Note in Street Lighting" and Walter D'Arcy Ryan, "Downtown Lighting System for San Francisco," *EW*, 68 (September 2, 1916): 451, 457–459. "San Francisco's Path of Gold," *EW*, 68 (October 14, 1916): 751. Nye, *Electrifying America*, 57.

7. James C. Williams, "The Trolley: Technology and Values in Retrospect," *San José Studies*, 3 (November 1977): 84–85.

8. On the general development of electric power in manufacturing, see Louis C. Hunter and Lynwood Bryant, *A History of Industrial Power in the United States, 1780–1930, Volume Three: The Transmission of Power* (Cambridge, MA: The M.I.T. Press, 1991), 224–237; Richard B. Du Boff, "The Introduction of Electric Power in American Manuafacturing," *Economic History Review*, 20 (December 1967): 509–518; and Duboff, *Electric Power in American Manufacturing, 1889–1958* (New York: Arno Press, 1979).

9. Examples of mining use of electricity are found in A.M. Hunt and Wynn Meredith, "Electric Power: Its Generation and Utilization for Mining Work on the Pacific Coast," *California Mines and Minerals* (San Francisco: L. Roesch Co., 1899), 73–87; Rudolph W. Van Norden, "Mountain King Mining Company Power Plant," *JEPG*, 36 (February 12, 1916): 125–132; E.B. Criddle, "Hydro-Electricity and Mining," *California Journal of Development*, 15 (July 1925): 6–7, 28. Copper and iron smelting in Northern California Power's territory is discussed in Terry S. Reynolds and Charles Scott, *The Battle Creek Hydroelectric System: An Historical Study* (Washington, D.C., [1980]), 21–158 *passim;* Ralph L. Phelps, "The Electrical Smelting of Iron Ore," *JEPG*, 19 (July 27, 1907): 69–71; C.F. Elwell, "The Reduction of Iron Ores by Electricity," *JEPG*, 21 (July 18, 1908): 33–36; Rudolph W. Van Norden, "Electric Iron Smelter at Heroult on the Pitt," *JEPG*, 29 (November 23, 1912): 453–459. On electric power in the iron and steel fabrication industry, see W.W. Hanscom, "Electric Equipment of Columbia Steel Plant," *JEPG*, 24 (April 8, 1911): 301–304; Rudolph W. Van Norden, "Pacific Coast Steel Co.'s San Francisco Plant," *JEPG*, 29 (October 5, 1912): 295–297; R.H. Fenkhausen, "Electricity in the Union Iron Works," *JEPG*, 31 (November 15, 1913): 435–442; W.M. McKnight, "Stassano Electric Furnace at Redondo," *JEPG*, 35 (July 17, 1915): 37–28.

10. "Electricity in the Cement Industry [editorial]" and L.D. Gilbert, "Electricity in the Cement Industry," *JEPG*, 30 (April 5, 1913): 318, and 309–310; Lloyd W. Chapman, "Electrified Cement Industry Aids Western Construction," *JEPG*, 47 (July 1, 1921): 7–9. Examples of other industrial electric power use are in J.W. Swaren, "Motor Drive in the Dow Pump Works," *JEPG*, 26 (April 22, 1911): 345–350; C.R. Owens, "Industrial Electric Heating Installations in Northern California," *JEPG* 59 (September 1, 1927): 140–143; Archie Rice, "Electricity Applied to Paper-Making," *JEPG*, 26 (January 14, 1911): 43–44. Oil field use of electricity is discussed in A.E. Wishon, "Electrical Energy in the Oil Field," *JEPG*, 30 (April 26, 1913): 375–379; H.A. Russell, "Electricity in the Oil Fields," *Western Engineering*, 2 (April 1913): 269–275; S.G. Gassaway, "Cost of Pumping Oil by Electricity in the California Oil Fields," *Western Engineering*, 3 (October 1913): 273–278; "Developing Oil Field Load," *JEPG*, 63 (September 1, 1929): 109–112. On contemporary interest in the electrochemical industry, see Romaine W. Myers, "Pacific Coast Electro-Chemical Possibilities," *JEPG*, 34 (March 27, 1915): 243–248; J.W. Beckman, "Pacific Coast Electro-Chemical Possibilities," *JEPG*, 35 (September 18, 1915): 209–213; "Great Western Electro-Chemical Company Plant, Pittsburg, California," *Western Construction News*, 4 (June 25, 1929): 319–320.

11. The 1920 industrial survey is found in "The Industrial Load in California," *EW*, 76 (August 28, 1920): 428–430. Suggested strategies to further build industrial electric power use are given in J.O. Case, "Analyzing Unelectrified Possibilities in California," *JEPG*, 44 (April 15, 1920): 364–367. On the motion picture industry, see Carl M. Heintz, "Electricity in the Making of Moving Pictures," *JEPG*, 43 (July 1, 1919): 9–11. On the development of the microwave electronics industry, see James C. Williams, "Frederick E. Terman and the Rise of Silicon Valley," *Technology in America*, 2nd ed., Carroll Pursell, ed. (Cambridge: The M.I.T. Press, 1990), 276–291.

12. Mark H. Rose, *Cities of Light and Heat: Domesticating Electricity in Urban America* (University Park: Pennsylvania State University Press, 1995), focuses on the domestic adoption of electricity in Denver, Colorado, and Kansas City, Missouri.

13. Thomas P. Hughes, *Networks of Power: Electrification in Western Society, 1880–1930* (Baltimore: Johns Hopkins University Press, 1983), 218–219.

14. S.M. Kennedy, "Sale of Electrical Appliances for Regular Lamp Circuits and Their Effect on Load and Income," *EW*, 60 (December 7, 1912): 1210.

15. *Ibid.*, 1210–1213; S.M. Kennedy, "Southern California Edison Company's Method of Disposing of Electric Appliances," *EW*, 62 (November 22, 1913): 1062–1063.

16. Kennedy, "Sale of Electrical Appliances for Regular Lamp Circuits," 1210–1211; S.M. Kennedy, "Selling Lamp-Socket Appliliances," *EW*, 65 (May 29, 1915): 1412–1414; A.W. Childs, "Building an Appliance Load," *JEPG*, 49 (December 1, 1922): 404. General Electric was so impressed by one photograph of a pile of two thousand flatirons taken in trade by Southern California Edison, that it circulated it widely within the industry (see Nye, *Electrifying America*, 264). Chicago's Commonwealth Edison successfully adopted the technique from SCE of lending flatirons in late 1908 (see Richard F. Hirsh, *Technology and Transformation in the American Electric Utility Industry* [Cambridge, MA: Cambridge University Press, 1989], 212, n. 18). See Rose, *Cities of Light and Heat*, ch 3, for an analysis of marketing in Kansas City and Denver. On lamp socket appliances, see Fred E.H. Schroeder, "More 'Small Things Not Forgotten': Domestic Electrical Plugs and Receptacles, 1881–1931," *Technology and Culture*, 27 (July 1986): 530–533.

17. E.A. Wilcox, *Electric Heating* (San Francisco: Technical Publishing Company, 1916). *JEPG* summarized Wilcox's book between November 1916 and April 1917, ensuring it a wide regional audience. In 1928, New York's McGraw-Hill Book Company brought out an updated edition, in which Wilcox addressed building and industrial contractors, not salespeople working directly with householders. This volume undoubtedly reached a national audience. The Hotpoint Electrical Company, started in Ontario, California, became so successful with its iron that General Electric acquired the firm and its line of products during the 1920s (Nye, *Electrifying America*, 262, 265). Rose, *Cities of Light and Heat*, ch. 4, focuses on the role of public schools in the diffusion of electricity.

18. *JEPG*, 48 (February 15, 1922): 173.

19. *JEPG*, 41 (October 15, 1918): 347. "Public Relations from the Woman's Standpoint," *JEPG*, 54 (June 1, 1925): 530; Ruth E. Creveling, "Telling the Story of Electric Service to Women by Women," *JEPG*, 60 (March 1, 1928): 139.

20. "Public Relations from the Woman's Standpoint," 54:530–534.

21. *Ibid.*, 532; "Space in Free Market Utilized for Display Booth," *JEPG*, 47 (October 1, 1921): 274; Garnett Young, "Placing Electrical Convenience Outlets in Every Room," *JEPG*, 50 (March 15, 1923): 208.

22. "Space in Free Market," *JEPG*, 47 (October 1, 1921): 274; "A Cooperative Vacuum Cleaner Campaign," *JEPG* (May 1, 1918), 40: 444–445; "The Southern California Edison Company's Appliance Campaign," *JEPG*, 52 (January 1, 1924): 7–8; "Selling Appliances for Use Every Day in the Year," *JEPG*, 54 (January 1, 1925): 26; "Better Kitchen and House Lighting Literally 'Brought Home' in Fresno," *JEPG*, 55 (November 1, 1925): 321–322; "If Summer Comes—Then Refrigerators," in "Better Merchandising" section, *JEPG*, 56 (April 15, 1926): 304–307; "'Free Electricity' Offer Stimulates Appliance Business in San Joaquin Valley," in "Sales of the Month" section, *JEPG*, 72 (April 1934): 18; "Building the Agricultural Load," *JEPG*, 45 (November 15, 1920): 458–461; Charles M. Coleman, *PG&E of California: The Centennial Story of Pacific Gas and Electric Company, 1852–1952* (New York: McGraw-Hill, 1952), 206. *JEPG* and *EW* carried many other articles on this topic through the 1930s.

23. Ben M. Maddox, "Data on Electric Cooking and Heating," *JEPG*, 30 (February 1, 1913): 110–111.

24. "Cooperation Between Central Station and Jobbers on the Pacific Coast," *EW*, 66 (November 27, 1915): 1224; "Co-operative Sign Selling on Pacific Coast," *EW*, 67 (February 26, 1916): 487–489; H.T. Matthews, "Electrical Prosperity Week in the Pacific Coast States," *JEPG*, 35 (October 16, 1915): 302.

25. S.M. Kennedy, "California's Electrical Co-operative Campaign," *JEPG*, 72 (July 27, 1918): 155.

26. *Ibid.*

27. "Plans for the California Campaign," *JEPG*, 42 (February 15, 1919): 157. See also Kennedy, "Co-operative Campaign," 42:157–158; S. M. Kennedy, "Co-operative Advertising in California," *EW*, 73 (January 11, 1919): 80; A.W. Childs, "Building an Appliance Load," *JEPG*, 49 (December 1, 1922): 404.

28. Kennedy, "Co-operative Advertising," 73: 80–83.

29. *EW*, 75 (February 14, 1920): 389–390; "California Electrical Co-operative Campaign on Larger Scale," *EW*, 75 (May 15, 1920): 1158–1159; "Plans for the California Campaign," 42:157–158.

30. Walter F. Price, "Origin and Development of California 'Electrical Homes,'" *JEPG*, 47 (July 1, 1921): 11–13; Clifford Edward Clark, Jr., *The American Family Home, 1800–1960* (Chapel Hill: University of North Carolina Press, 1986), 86, 162–163; Garnett Young, "Placing Electrical Convenience Outlets in Every Room," *JEPG*, 50: 207–212; "Report of Electrical Homes Exhibited During 1920," *JEPG*, 46 (January 15, 1921): 80; "Electrical Home Idea Promotes Electric Service in California," *EW*, 76 (August 28, 1920): 436–437; Fifth Annual Report, *The California Electrical Co-operative Campaign* (Oakland, CA, 1922), 13, on file in "A Record of achievement and contributions during the year 1922," v. 2, PG&E shelf, PG&E Corporate Library, San Francisco. Where the first model electric home in the United States was actually opened is not clear, but California clearly was in the forefront of the effort. A review of electric homes in *EW* revealed that over one million people visited more than sixty models across the nation in 1922 alone. See Nye, *Electrifying America*, 266.

31. Fifth Annual Report, *Co-operative Campaign*, 11.

32. Young, "Placing Electrical Convenience Outlets," 50:207–212; Fifth Annual Report, *Co-operative Campaign*, 10–11, 14–16, 27–30.

33. "Traveling Electric Home Planned by Association," *JEPG*, 51 (July 1, 1923): 27.

34. "Furthering the Electrical Idea in California," *JEPG*, 51 (September 1, 1923): 180; "Electrical Exhibit to Go to Small California Towns," *JEPG*, 81 (June 23, 1923): 1486; "Portable Window-Lighting Exhibit," *EW*, 82 (July 7, 1923): 32; "Results Tell the 1923 Story for the Cooperative Campaign," *JEPG*, 52 (March 1, 1924): 168–171; "Red Seal Electric Home Opened in Santa Cruz," *JEPG*, 55 (August 1, 1925): 101; "Plans Completed for Execution of Red Seal Plan in California," *JEPG*, 55 (December 1, 1924): 423.

35. "California Campaign Studied Nationally," JEPG, 44 (February 15, 1920): 149; "California Co-operative Campaign to be Tried in Ohio," *EW*, 76 (July 24, 1920): 203; "California Electrical Men Continue Cooperative Effort," *EW*, 77 (May 28, 1921): 1266; "California Cooperative Campaign Changes Name," *EW*, 85 (February 21, 1925): 423.

36. "Electric Service in the American Home," *EW*, 75 (May 15, 1920): 1133–1134; "Electric Power Consumption in California," *California Journal of Development*, 20 (June 1930): 40.

37. Kennedy, "Selling Lamp-Socket Appliances," 65:1412.

Chapter XI/Electrifying the Farm

1. *PRP*, 38 (November 9, 1889): 432; "Small Pelton Water-Motors," *PRP*, 38 (September 7, 1889): 214; P.J. O'Gara, "Harnessing Streams: How Otherwise Wasted Energy May Be Utilized for Ranch Heating, Lighting or Pumping," *Sunset*, 20 (March 1908): 452–453; Charles C. Moore and Co., Engineers, Inc., *Electricity in the "City of the Holy Cross"* [reprint from *JEPG*, May 1905] (San Francisco, 1905): 14.

2. P.J. O'Gara, "A Farm Hydro-Electric Plant Without Operating Attendant," *EW*, 53 (June 3, 1909): 1374–1376.

3. "Eight-Horse-Power Electric Motor," *Industry*, 3 (October 1890): 41; "Developing Plants by Electricity," *PRP*, 43 (June 4, 1892): 534; C.W. Haskins, "The Philosophy of Electrical Action on Vegetable Growth," *PRP*, 43 (February 27, 1892): 222; "Farming By Electric Power," *EW*, 8 (November 1899): 11.

4. "Electricity on the Farm," *EW*, 18 (January 26, 1907): 76–77.

5. *PRP*, 94 (July 7, 1917): 19; Western Electric Company, *Electric Light Plants: Power and Water Systems* catalog [1910], and Fairbanks, Morse and Company, *Fairbanks-Morse Home Light Plant:*

The Double Duty Unit, catalog [n.d.], "Electrical Equipment: Company progress reports, histories" box, Romaine Trade Catalog Collection, University of California, Santa Barbara; F.J. Pancratz, "Wind Power for Farm Electric Plants," *Mechanical Engineering,* 46 (November 1924): 675–682; C.D. Turnbull, "Farm Hydro-Electric Power Plants," *Farm Engineering,* 2 (May 1915): 248–249; Fredrick Irvine Anderson, *Electricity for the Farm: Light, Heat and Power by Inexpensive Methods from the Water Wheel or Farm Engine* (New York: MacMillan Co., 1916); Harry C. Ramsower, *Equipment for the Farm and Farmstead* (Boston: Ginn and Co., 1917), 124–144, 194.

6. *CREA News Letter,* No. 10 (December 15, 1930): 8, Table 5, and No. 12 (April 1, 1935): 3, Table 1; U.S. Department of Agriculture, Rural Electrification Administration, Statistical Services Section, Administrative and Loan Accounting Branch, *Number and Percentage of Farms Electrified with Central Station Service by States* (October 1964) [single sheet publication].

7. Discussion by Frank Baum in P.M. Lincoln, "Choice of Frequency for Very Long Lines," *Transactions of the American Institute of Electrical Engineers,* 22 (1903): 378. In 1910, Paul M. Downing, also of PG&E, noted that had he to do it over again, he "would seriously consider a lower frequency before adopting 60 cycle," but that 60 Hz was adopted because all the early plants were conceived originally as operating independently of one another. Downing said that, in the network which existed by 1910, "there is no question in my mind but that the lower frequency would be better." See Paul M. Downing, "The Developed High Tension Net-Work of a General Power System," *Transactions of the American Institute of Electrical Engineers,* 29, Pt. 1 (1910): 727–728. Also see Thomas P. Hughes, *Networks of Power: Electrification in Western Society, 1880–1930* (Baltimore: Johns Hopkins University Press, 1983), 128.

8. A.H. Markwart, "Power in California," *Journal of the Franklin Institute,* 204 (August 1927): 158, 163; H.A. Barre, "Water Power and Steam Power in California Utilities," *Mechanical Engineering,* 48 (June 1926): 662. The first tapping of a transmission line appears to have occurred in 1898. A farmer along the route of the 11,000-volt Folsom line to Sacramento purchased three General Electric transformers to step down power from the line to 2,200 volts to run irrigation pumps. David E. Nye, *Electrifying America: Social Meanings of a New Technology, 1880–1940* (Cambridge, MA: The M.I.T. Press, 1990), 293.

9. "Electricity in Agriculture," *EW,* 41 (October 1, 1918): 294–295. The literature on wartime fuel conservation and hydroelectric development is profuse. For example, see "Electrical Interconnection to Conserve Fuel," *EW,* 71 (January 5, 1918): 12–14; R.J.C. Wood, "Conservation of Fuel in California," *EW,* 72 (August 24, 1918): 348–350; Robert Sibley, "War Service of Electical Energy," *EW,* 40 (February 1, 1918): 112–118; "Fuel Problems of the Pacific Coast," *Mechanical Engineering,* 41 (February 1919): 264–269; "Sources of Power on the Pacific Coast," *Western Engineering,* 9 (March 1918): 103–104; Paul M. Downing, "The Water Power Situation on the Pacific Coast," *Transactions of the Commonwealth Club of California,* 15 (December 1920): 326–327; Markwart, "Power in California," 158.

10. National Electric Light Association, Hydroelectric Section, *Report of the Sub-committee on Water Power Development on the Pacific Coast* (New York, 1915), 77–79; Hughes, *Networks of Power,* 220–221.

11. U.S. Geological Survey, *Hydroelectric Power Systems of California and Their Extensions into Oregon and Nevada,* Water Supply Paper 493, by Frederick Hall Fowler (Washington, D.C., 1923), 134.

12. "Irrigation Pumping in the Coast States," *EW,* 65 (May 29, 1915): 1399–1408; Sidney Bretherton, Jr., "Electric Progress in a Model Colony," *EW,* 45 (August 15, 1920): 172–173; "Building the Agricultural Load," *EW,* 45 (November 15, 1920): 459; Al.C. Joy, "How the Power Company Aids Farm Land Settlement," *EW,* 52 (March 1, 1924): 159–161; Markwart, "Power in California," 161–163; Rudolph W. Van Norden, "San Joaquin Light and Power Corporation," *EW,* 28 (May 11, 1912): 415–449 *passim;* Van Norden, "Outdoor Substations in San Joaquin Valley," *EW,* 28 (June 29, 1912): 633, and "Editorial," 649; "California's New Industry [Cotton]," *EW,* 44 (April 1, 1920): 305–306; Putnam A. Bates, "Electricity on the Farm," *Transactions of the American Institute of Electrical Engineers,* 31, pt. 2 (1912): 1987.

13. W.E. Camp, "Electricity Makes Rice Industry Possible," *EW,* 43 (August 1, 1919): 102–104;

R.A. Balzari, "Reclamation Pumping in Sacramento Valley," *EW*, 35 (September 11, 1915): 191; A.F. Bridge and Ralph Reynolds, "Costs of Pumping Water at Isleton," *EW*, 30 (February 25, 1913): 131–132; "Selling Electrically Irrigated Farms in California," *EW*, 63 (February 14, 1914): 375; Fowler, *Hydroelectric Power Systems*, 34; "Irrigation Pumping Is Basic Rural Load," *EW*, 62 (June 1, 1929): 579.

14. R.B. Mateer, "Readiness to Serve Methods: Modernizing the House," *EW*, 30 (January 11, 1913): 44–45; "Electric Irrigation Plants," *EW*, 28 (March 2, 1912): 209. Also see G.E.P. Smith, "The Utilization of Ground Waters by Pumping for Irrigation," *Transactions of the International Engineering Congress, 1915* (San Francisco, 1915): 2:438–442; Ben D. Moses and George C. Tenney, "Rural Electrification in California," *EW*, 54 (June 15, 1925): 581–582. Charles M. Coleman, *PG&E of California: The Centennial Story of Pacific Gas and Electric Company, 1852–1952* (New York: McGraw-Hill, 1952); 205; Barre, "Water Power and Steam Power," 662.

15. E.P. Gibson, "What Electricity Can Do for the Farm Wife," The *U.C. Journal of Agriculture*, 3 (February 1916): 195. Also see "Wash Day No Longer Dreaded," *PRP*, 94 (July 14, 1917): 32–33. According to Moses and Tenney, "Rural Electrification," 582, one of the state's larger power companies reported farms accounted for 90 percent of the electric ranges on its lines.

16. University of California, Agricultural Experiment Station, *Electrical Statistics for California Farms*, Circular No. 316, by B.D. Moses (Berkeley, October 1929), 9–10, 16–17; Ben D. Moses, "Statistical Study of Uses of Electricity in California Agriculture," *EW*, 58 (June 1, 1927): 460–461; "Editorial," *EW*, 20 (February 15, 1908): 102; "Electricity in Agriculture," *EW*, 26 (March 25, 1911): 273; "Electricity for Cultivating Farm Lands," *Western Engineering*, 9 (June 1918): 248; J.H. Davidson and F.E. Boyd, "The Electric Tractor on the Farm," *EW*, 41 (October 1, 1918): 297–299.

17. "Editorial," *EW*, 20 (February 15, 1908): 102; "Dairy Farming by Electricity," *EW*, 19 (August 24, 1907): 154; "Electricity for Irrigation," *EW*, 31 (July 26, 1913): 93; University of California, Agricultural Experiment Station, *Power Requirements of Electrically Driven Dairy Manufacturing Equipment*, Bulletin No. 433, by A.W. Farrall (Berkeley, September 1927); Moses and Tenney, "Rural Electrification," 584–585.

18. "Editorial," *EW*, 20 (February 15, 1908): 102; Edgar A. Wilcox, *Electric Heating*, 93, 263–264; Also see Paul J. Denniger, "Hatching Eggs Electrically," *EW*, 44 (January 15, 1920): 73.

19. Wilcox, *Electric Heating* (1916), 191–196. Also see his *Electric Heating* (New York: McGraw-Hill, 1928), ch. 20.

20. Denniger, "Hatching Eggs Electrically," *EW*, 44 (January 15, 1920): 73; M.J. Brooks, "The World's Largest Electric Chick Hatchery," *EW*, 54 (January 15, 1925): 48; L.A. Bourk, "Commercial Poultry Raising," *California Journal of Development*, 16 (November 30, 1926): 7–8; University of California, Agricultural Experiment Station, *The Electric Brooder*, Bulletin 441, by B.D. Moses and T.A. Wood (Berkeley, November 1927), 3.

21. Moses, "Development of the Electric Brooder," 145; Wilcox, *Electric Heating* (1916), 185–186. Moses and Tenney, "Rural Electrification," 583; Page Smith and Charles Daniels, *The Chicken Book* (Boston: Little, Brown and Company, 1975), chs. 12–14 *passim*.

22. "Electric Energy for Frost Protection," *EW*, 28 (March 9, 1912): 230–231; "Frost Protection by Electrical Methods," *EW*, 28 (April 6, 1912): 318–319; University of California, Agricultural Experiment Station, *Orchard Heating in California*, Bulletin No. 398, by Warren R. Schoonover and Robert W. Hodgson (Berkeley, December 1925), 34.

23. Albert Marple, "Electricity an Aid in Farming," *Farm Engineering*, 3 (June 1916): 281; B.D. Moses, "Orchard Blowers for Frost Protection," *CREA News Letter*, No. 14 (December 19, 1938): 9–10.

24. Wilcox, *Electric Heating* (1928), 197; E.N. Cable, "Electricity in the Orchard Homes: Comforts with Electricity," *American Fruit Grower*, 39 (July 1919): 29; R.C. Griffin, "Dehydration of Fruit in California," *JEPG*, 51 (November 1, 1923): 335; University of California, Agricultural Experiment Station, *Walnut Dehydraters: Characteristics, Heat Sources, and Relative Costs*, Bulletin 531, by P.F. Nichols, B.D. Moses, and D.S. Glenn (Berkeley, June 1932), 6, 32; L.E. Holmes, "Chili Pepper Dehydrator," *CREA News Letter*, No. 14 (December 10, 1938): 20–22.

25. Bates, "Electricity on the Farm," 1986; C.D. La Moree, "The Cotton Industry in Califor-

nia," *The Electric Journal,* 11 (August 1914): 436; "Electricity in Fruit Packing," *EW,* 44 (April 1, 1920): 303; U.S. Census Office, Eleventh Census, 1890, *Report on the Manufacturing Industries of the United States, Pt. I, Totals for States and Industries* (Washington, D.C., 1895), 6:942; U.S. Census Office, Thirteenth Census, 1910, *Manufacturers, 1909: General Report and Analysis* (Washington, D.C., 1913), 8:352; U.S. Bureau of Census, Fourteenth Census, 1920, *Manufacturers, 1919: Reports for States* (Washington, D.C., 1923), 9:113; U.S. Bureau of Census, Fifteenth Census, 1930, *Manufacturers, 1929* (Washington, D.C., 1933), 16:66.

26. Lillian D. Clark, "California: Brief History of Early Home Demonstration Work," *California Annual Narrative Report, State Home Demonstration Leader, December 1, 1917–November 30, 1918* (Berkeley, College of Agriculture Extension Service, December 15, 1918), 3–8 (typewritten); *California Annual Narrative Report, Acting Extension Specialist in Home Management, December 1, 1927–November 30, 1928* (Berkeley, College of Agriculture Extension Service, July 19, 1929), 2; George Collins, "All 'Power on the Farm' Machinery to be One Big Exhibit at the State Fair," *Gilroy Advocate,* 28 August 1920, 6; "California State Fair Exhibits Latest Farm Machinery," *EW,* 47 (September 15, 1921): 241. Articles from the University's *Journal of Agriculture* and Experiment Station bulletins and circulars are cited throughout this chapter.

27. Clark, "Early Home Demonstration Work," 8, 18; "Building the Agricultural Load," *EW,* 45 (November 15, 1920): 458; Coleman, *PG&E,* 106.

28. B.D. Moses and L.J. Fletcher, "The California Project," *CREA Bulletin,* 1 (March 20, 1925): 1–3, 9.

29. *Ibid.,* 3–4, 9. L.J. Fletcher, "Electricity in California Agriculture," *CREA Bulletin,* 2 (December 28, 1952): 7–8; *CREA Newsletter,* no. 10 (December 15, 1930): 8. The members of the Executive Committee included H.M. Crawford (*PG&E*), A.M. Frost (San Joaquin Light and Power), J.W. Nelson and J.J. Deuel (California Farm Bureau Federation), N. R. Sutherland (*PG&E*), B.M. Maddox (Southern California Edison), E. A. Shreve (General Electric Company), C.A. Utley (Pelton Water Wheel Company).

30. Moses and Fletcher, "The California Project," 5–9; Moses, *Electrical Statistics;* John W. Otterson, "Snook O'Brien's Model Diversified Electrical Farm," *EW,* 54 (June 15, 1925): 591–594; J.P. Fairbank, "Taking Electricity to California Farms and Farm Homes," *EW,* 54 (January 1, 1925): 12–15; J.E. Dougherty, "'Lighting' Hens in Winter," *EW,* 62 (March 1, 1929): 144–145; *EW,* 64 (January 1, 1930): 41; University of California, Agricultural Experiment Station, *Construction and Operation of Mechanical Refrigerators for Farms,* Circular No. 329, by James R. Tavernetti (Berkeley, September 1933); University of California, Agricultural Experiment Station, *Electric Heat for Propagating and Growing Plants,* Circular No. 335, by B.D. Moses and James R. Tavernetti (Berkeley, August 1934); California Committee on Relation of Electricity to Agriculture, *Progress Report,* University of California Agricultural Experiment Station, Nos. 4–22 (Berkeley, August 1925 to February 1940) (mimeographed).

31. Data from two comparable surveys from California and Iowa as reported in *Extension Specialist Narrative Report, December 1, 1933–November 30, 1934,* 7–9.

32. L.S. Wing, "Rural Electrification from an Economic and Engineering Standpoint—I," *EW,* 57 (September 1, 1926): 164; James W. Carey and John J. Quirk, "The Mythos of the Electronic Revolution," *The American Scholar,* 39 (Spring 1970): 234–235; Carroll W. Pursell, Jr., "Government and Technology in the Great Depression," *Technology and Culture,* 20 (January 1979): 172–173; California, State Assembly, Interim Committee on Conservation, Planning, and Public Works, *Electric Power and Development of Northern California,* by Warren S. Gramm (Sacramento, May 1957), 30–31; D. Clayton Brown, *Electricity for Rural America,* 66; *Electrical West,* 104 (February 1950): 56; Melosi, *Coping With Abundance,* 134–137; Nye, *Electrifying America,* 287–335.

33. "Editorial," *EW,* 20 (February 15, 1908): 102.

Chapter XII/Resolving the Power Network

1. A number of studies have addressed these changes in American society. In particular, see the Antheneum edition preface to Samuel P. Hays, *Conservation and the Gospel of Efficiency: The Progressive Conservation Movement, 1890–1920* (New York: Antheneum, 1974). For a good intro-

duction to localism versus centralism in California politics, see Robert F. Kelley, *Battling the Inland Sea: American Political Culture, Public Policy, and the Sacramento Valley, 1850–1986* (Berkeley: University of California Press, 1989).

2. Terry S. Reynolds and Charles Scott, *The Battle Creek Hydroelectric System: An Historical Study* (Washington, D.C., [1980]), 14–15. The legal doctrine of riparian water rights, giving owners of land fronting on a stream the right to all the water in the stream and thereby preventing waters upstream from being diverted or diminished, applied in the eastern United States. California's first state legislature had simplified its early tasks, in 1850, by adopting English Common Law for the state; unbeknownst to many of them, it included the riparian doctrine. As time went by, a number of battles occurred between advocates of the conflicting doctrines, with courts often ruling in favor of riparian rights. See Donald J. Pisani, *From the Family Farm to Agribusiness: The Irrigation Crusade in California and the West, 1850–1931* (Berkeley: University of California Press, 1984), ch. 2; Laurence D. Shoup, *"The Hardest Working Water in the World": A History and Significance Evaluation of the Big Creek Hydroelectric System* (Fair Oaks, CA: Theodoratus Cultural Research, October 1988, for Southern California Edison Company), 38; State of California, Conservation Commission, *Report of the Conservation Commission of the State of California, January 1, 1913* (Sacramento, 1912), 27–31; and Norris Hundley, Jr., *The Great Thirst: Californians and Water, 1770s–1990s* (Berkeley: University of California Press, 1992), 69–73.

3. Hays, *Conservation and the Gospel of Efficiency*, 28–81, 121–146 *passim*. Also see Robert C. Catren, "A History of the Generation, Transmission, and Distribution of Electrical Energy in Southern California," (Ph.D. diss., University of Southern California, 1951), 300–309.

4. Charles H. Shinn to Gifford Pinchot, April 3, 1907, photocopy in Charles H. Shinn file, Sierra National Forest Headquarters, Fresno, California.

5. Frank G. Baum, "Water Power Development in the National Forests: A Suggested Government Policy," *Transactions of the American Institute of Electrical Engineers*, 27, Pt. 1 (1908): 483, 475–484 *passim*; Hays, *Conservation and the Gospel of Efficiency*, 75–76. Californian John Coffee Hays presented a position much like Baum's in a letter he sent as comment on another paper presented at a New York meeting of the AIEE in March 1909. See "Discussion on 'Electricity and the Conservation of Energy,'" *Transactions of the American Institute of Electrical Engineers*, 28, Pt. 1 (1909): 179–187.

6. I am indebted to Robert Kelley for shaping my understanding of cultural politics in America through his *The Cultural Pattern in American Politics: The First Century* (New York, Alfred A. Knopf, 1979). As this applies to California, particularly in terms of flood control, see his *Battling the Inland Sea*, 25–43, 119–125, 249–255, 319–338. On the hydraulic mining episode, see Kelley's *Gold vs. Grain: The Hydraulic Mining Controversy in California's Sacramento Valley* (Glendale, CA: Arthur H. Clark Co., 1959).

7. Baum, "Water Power Development," 485–502. John A Britton, a PG&E vice president, was Baum's only supporter, and he suggested a conference should be held in California to discuss the issue. Western Democratic opposition to Roosevelt's natural resource policies grew strongest among independent loggers, miners, and livestock grazers, all of whom faced conservation charges similar to hydroelectric developers. See Hays, *Conservation and the Gospel of Efficiency*, 256–260.

8. Donald J. Pisani, "Water Law Reform in California, 1900–1913," *Agricultural History*, 54 (March 1980): 295–310, traces this development. William Kerckhoff, one of Huntington's partners in Pacific Light and Power, and among the most influential electric power people in southern California, also was an active conservationist, a supporter of Pinchot's principles of scientific conservation, and president of the Water and Forest Association after 1900.

9. John Debo Galloway to Geo. C. Pardee, January 5, 1911, in the George C. Pardee Papers, Bancroft Library, Berkeley, CA. Also see Gerald D. Nash, *State Government and Economic Development: A History of Administrative Policies in California, 1849–1933* (Berkeley: Institute of Governmental Studies, University of California, 1964), 323.

10. Pisani, "Water Law Reform," 313–315; *Report of the Conservation Commission, 1913*, 8–41 *passim*; "Minority Report on the Water Commission Act," *Transactions of the Commonwealth*

Club of California, 9 (October 1914): 587–595; Nash, *State Government and Economic Development,* 324.

11. John A. Britton, "Pacific Coast Companies Meet War Problems," *EW,* 71 (January 5, 1918): 15–16; Britton, "Plea for Constructive Water Power Policy," *JEPG,* 41 (July 1, 1918): 18; P.M. Downing, "Interconnection and New Power Development," *JEPG,* 40 (March 15, 1918): 283. A number of other articles in *EW* and *JEPG* between 1916 and 1919 cite the problem of financing projects and the need for affirmative federal government action to deal with the problem. Also see Hays, *Conservation and the Gospel of Efficiency,* 259.

12. Britton, "Pacific Coast Companies," 16; [NELA convention summary], "Public Policy on the Pacific Coast," *EW,* 71 (June 1, 1918): 1151; Britton, "Plea for Constructive Water Power Policy," 20; Hays, *Conservation and the Gospel of Efficiency,* 240–241, 260. Following passage of the Water Power Act, a flood of permit applications inundated the new Federal Power Commission, testifying to the real uncertainty that had existed during the previous decade. Within five months, California companies had filed eighty applications embodying an estimated 4 million horsepower. Nationally, applications represented 16.8 million horsepower. See Catren, "Electrical Energy in Southern California," 309.

13. *Report of the Conservation Commission, 1913,* 17; A.M. Markwart, "Production of Energy," *Mechanical Engineering,* 41 (February 1919): 265; California, State Council of Defense, *Report of the Committee on Petroleum* (Sacramento, July 7, 1917), 169–172; "Hydro-Electric Energy as a Conserver of Oil," *JEPG,* 39 (October 1, 1917): 299–300; "A New Trail Blazed in California," EW, 69 (June 23, 1917): 1193–1194.

14. California, Railroad Commission, *Annual Report for 1918* (Sacramento, 1918), 54, 55–67; "Public Policy on the Pacific Coast," 1150; "Advantages of Tying in the California Systems," *EW,* 71 (January 26, 1918): 219; "Progress Toward Tying in the California Systems," *EW,* 71 (February 16, 1918): 374–375; "War Service Engineering Problems," *JEPG,* 40 (May 15, 1918): 520–524; J.P. Jollyman, "Operation of Interconnected Systems," EW, 71 (May 18, 1918): 1020–1022; "Pacific Coast Plants Fully Loaded," *EW,* 73 (May 17, 1919): 1031–1037; G.R. Kenney, "Emergency Interchange of Power," *JEPG,* 42 (April 15, 1919): 347–349. A similar interconnection effort was underway in New York; however, unlike California, where all companies except Southern California Edison transmitted power at 60 Hz, New York companies used 25 Hz, 40 Hz, and 60 to 62½ Hz, making interconnection more difficult. See "Electrical Interconnection to Conserve Fuel," *EW,* 71 (January 5, 1918): 12–14.

15. Record Group 57, Box 1, Entry 265, Folder P.R. 230, Appropriations, and Folder L C B, Department of the Interior, Records of the Geological Survey, Water Resources Division, Superpower Survey, Records concerning the Organization of the Superpower Survey, 1918–1919, National Archives, Washington, D.C.; Frank Baum, "The Effect of Hydro-Electric Power Transmission upon Economic and Social Conditions, with Special Reference to the U.S. of America," *Transactions of the International Engineering Congress, 1915* (San Francisco, September 20–25, 1915): 254; "Large California Systems United for Period of War," EW, 71 (January 26, 1918): 218; U.S. Geological Survey, *A Superpower System for the Region between Boston and Washington,* Professional Paper No. 123, by Murray, *et al.* (Washington, 1921); Thomas P. Hughes, "Technology and Public Policy: The Failure of Giant Power," *Proceedings of the IEEE,* 64 (September 1976): 1361–1371; Jean Christie, "Giant Power," *Pennsylvania Magazine of History and Biography,* 96 (October 1972): 480–507.

16. C. Edward Magnusson, "Trunk Transmission Lines," *JEPG,* 45 (December 1, 1920): 519–521; "The Pacific Coast Superpower Zone," *JEPG,* 50 (February 1, 1923): 98–99; George C. Tenney, "Superpower on the Pacific Coast," *JEPG,* 52 (February 1, 1924): 83–91; Stephen Q. Hayes, "Interconnection, Super-Power, and Electrification," *The Electric Journal,* 21 (May 1924): 247–254 with quotation on 247; W.G. Vincent, Jr., "The Interconnected Transmission System of California," *JEPG,* 54 (June 15, 1925): 571–573; A.H. Markwart, "Power in California," *Journal of the Franklin Institute,* 204 (August 1927): 163. The west coast Super-Power Committee recommendation is found in A.H. Markwart and H.A. Barre, "Regional Review of Power Resources, Distribution and Utilisation for Pacific Coast States, U.S.A.," *Transactions of the First World Power Confer-*

ence (1924): 1:596. Baum's national superpower plan is described in his *Atlas of U.S.A. Electric Power Industry* (New York: McGraw-Hill, 1923), and a synopsis of it is in "National Superpower Scheme," *EW*, 81 (June 2, 1923): 1273–1275. Also see Guy E. Tripp, *Super-Power as an Aid to Progress* (New York: G.P. Putnam's Sons, 1924).

17. H.G. Butler, "Our Undeveloped Water Resources," *Transactions of the Commonwealth Club of California*, 15 (December 1920): 340. Butler's thoughts on power industry consolidation are found in his "Does Interconnection Go Far Enough?" *JEPG*, 41 (October 15, 1918): 365–366, and Butler, "Electric Consolidations and Their Relations to Fuel Conservation," *Mechanical Engineering*, 41 (February 1919): 66–267. Also see Markwart and Barre, "Regional Review of Power Resources," 597; Martin V. Melosi, *Coping With Abundance: Energy and Environment in Industrial America* (Philadelphia: Temple University Press, 1985), 117–121; and Louis C. Hunter and Lynwood Bryant, *A History of Industrial Power in the United States, 1780–1930, Volume Three: The Transmission of Power* (Cambridge, MA: The M.I.T. Press, 1991), 368–369. See David E. Nye, *Electrifying America: Social Meanings of a New Technology, 1880–1940* (Cambridge, MA: The M.I.T. Press, 1990), ch. 4, especially 176–184, for a good discussion of peoples' varying perceptions about electricity and growing distrust of public utilities.

18. Charrles M. Coleman, *PG&E of California: The Centennial Story of Pacific Gas and Electric Company, 1852–1952* (New York: McGraw-Hill, 1952), 320–321; "Water and Power," *Transactions of the Commonwealth Club of California*, 19 (December 1924): 444; "Has Public Utility Regulation Succeeded?" *Transactions of the Commonwealth Club of California*, 27 (June 21, 1932): 167; U.S. Federal Power Commission, *Report to the Federal Power Commission on the Water Powers of California*, by Frank E. Bonner (Washington, D.C., 1928), 21–23. Frederick L. Bird and Francis M. Ryan, *Public Ownership on Trial: A Study of Municipal Light and Power in California* (New York: New Republic, Inc., 1930), a contemporary look at municipal ownership in California, published through the *New Republic Magazine*, is the only book-length work on the debate over publicly versus privately owned utilities in the region.

19. James C. Williams, "Engineering California Cities," in *Science-Technology Relationships: Relations Science-Technique*, Alexandre Herléa, ed. (San Francisco: San Francisco Press, 1993), 398. The best overall accounts of the Owens Valley aqueduct project are found in William L. Karhl, *Water and Power: The Conflict over Los Angeles' Water Supply in the Owens Valley* (Berkeley: University of California Press, 1982), and Hundley, *The Great Thirst*, 135–168.

20. Douglas R. Littlefield and Tanis C. Thorne, *The Spirit of Enterprise: The History of Pacific Enterprises from 1886 to 1889* (Los Angeles: Pacific Enterprises, 1990), 52–54; Karhl, *Water and Power*, 153, 155; Nelson Van Valen, "A Neglected Aspect of the Owens River Aqueduct Story: The Inception of the Los Angeles Municipal Electric System," *Southern California Quarterly*, 59 (Spring 1977): 85–109.

21. Karhl, *Water and Power*, 154–157; Van Valen, "A Neglected Aspect of the Owens River Aqueduct Story," 92–98.

22. Quoted in Frank G. Tyrrell, "The Los Angeles Experiment in Municipal Power," *JEPG*, 52 (May 1, 1924): 311.

23. Littlefield and Thorne, *Spirit of Enterprise*, 59–71, tell this general story. The public utilities perspective is presented in Tyrrell, "The Los Angeles Experiment," 311–314. The perspective of municipal ownership proponents is printed in the conference proceedings of the League of California Municipalities devoted to the 1920s state Water and Power Act in *Pacific Municipalities*, 35 (October 1921): 373–426. Both perspectives on Los Angeles are found in the Commonwealth Club's debates over the Water and Power Act, *Transactions of the Commonwealth Club of California*, 17 (July 1922), and v. 19 (October 1924).

24. Hundley, *The Great Thirst*, 175–176.

25. *Ibid.*, 169–183. Williams, "Engineering California Cities," 397–398. A number of historians have studied the Hetch Hetchy controversy. See Kendrick A. Clements, "Politics and the Park: San Francisco's Fight for Hetch Hetchy, 1908–1913," *Pacific Historical Review*, 47 (May 1979): 185–216, and his "Engineers and Conservationists in the Progressive Era," *California History*, 58 (Winter 1979–1980): 282–303. Elmo R. Richardson, "The Struggle for the Valley: California's

Hetch Controversy," *California Historical Society Quarterly,* 38 (September 1959): 249–258; Hays, *Conservation and the Gospel of Efficiency,* 192–198.

26. M.M. O'Shaughnessy, "Construction Progress of the Hetch Hetchy Water Supply of San Francisco," *Transactions of the American Society of Civil Engineers,* 85 (1922): 871, 903; "Hetch Hetchy Power," *Transactions of the Commonwealth Club of California,* 18 (December 1923): 288; "The Hetch Hetchy Water and Power Project for San Francisco," *EW,* 60 (December 7, 1912): 1185–1186.

27. This debate over municipal electric ownership is drawn from the 1923 Commonwealth Club debate in "Hetch Hetchy Power," *Transactions of the Commonwealth Club,* 283–372. For Phelan, see 316; Leurey, 330–331; Boyen, 334; Galloway, 372; O'Shaughnessy, 349.

28. Coleman, *PG&E,* 320–323; "Remarks by Hon. William Kent," *Transactions of the Commonwealth Club of California,* 18 (December 1923): 337. Hundley, *The Great Thirst,* 187–189. R.W. Jimerson, "Power Politics," *JEPG,* 87 (December 1941): 27–29, contains an interesting chronology of the Hetch Hetchy-public vs. private power debate in California, stretching from defeat by San Francisco voters of a city charter amendment, in May 1935, that would permit issuance of public power bonds to the defeat, in November 1941, of a major power bond issue.

29. Charles Eugene Coate, "Water, Power and Politics in the Central Valley Project, 1933–1967" (Ph.D. diss., University of California, Berkeley, 1969), 7–9. Also see Hundley, *The Great Thirst,* 239; Mary Montgomery and Marion Clawson, *History of Legislation and Policy Formation of the Central Valley Project* (Berkeley: U.S. Department of Agriculture, Bureau of Agricultural Economics, 1946); *Engineers and Irrigation: Report of the Board of Commissioners on the Irrigation of the San Joaquin, Tulare, and Sacramento Valleys of the State of California, 1873,* Engineer Historical Studies No. 5, Office of History, United States Army Corps of Engineers, annotated and with an introduction by W. Turrentine Jackson, Rand F. Herbert, and Stephen R. Wee (Fort Belvoir, VA, 1990); Kelley, *Battling the Inland Sea,* on flood control and reclamation; and Pisani, *From Family Farm to Agribusiness,* on irrigation.

30. "Proceedings of the Special Meeting of the League of California Municipalities," *Pacific Municipalities,* 35 (May, 1921): 183–187.

31. "Composition of the Committee that Prepared the Act," *Pacific Municipalities,* 35 (August 1921): 297–298.

32. "Synopsis of the Proposed Act," *Pacific Municipalities,* 35 (August 1921): 296–305; "Remarks by Mr. Horace Porter," *Pacific Municipalities,* 35 (October 1921): 419.

33. The views of the League's members are presented in "Proceedings of the Twenty-third Annual Convention of the League of California Municipalities, September 26–29, 1921," *ibid.,* 373–426. Quotations cited are found on 377, 382, 410.

34. "Address by Gifford Pinchot," *ibid.,* 399–407.

35. "Address by A.H. Redington," *ibid.,* 395.

36. *Ibid,* 425. Commonwealth Club's debates over the Water and Power Act, *Transactions of the Commonwealth Club of California,* 17 (July 1922): 181–294.

37. "Proceedings of the Twenty-fourth Annual Convention of the League of California Municipalities, September 19–23, 1922: Debate on the Water and Power Act" *Pacific Municipalities,* 36 (October 1922): 368; "Water and Power," *Transactions of the Commonwealth Club of California,* 19 (October, 1924): 335–452; "Reintroducing the Proposed California $500,000,000 Socialistic Venture, *JEPG,* 52 (May 1, 1924): 317–319; Catren, "Electrical Energy in Southern California," 168; Hundley, *The Great Thirst,* 240–241.

38. Linda J. Lear, "The Boulder Canyon Project: A Reexamination of Federal Resource Management," *Materials and Society,* 7 (1983): 329. The following broad account of Boulder Dam and the Colorado River is drawn from Frederick D. Kershner, "George Chaffey and the Irrigation Frontier," *Agricultural History,* 27 (October 1953): 115–122; Paul L. Kleinsorge, *The Boulder Canyon Project: Historical and Economic Aspects* (Stanford, CA: Stanford University Press, 1941); Norris Hundley, *Water and the West: The Colorado River Compact and the Politics of Water in the American West* (Berkeley: University of California Press, 1975); and Hundley, *The Great Thirst,* 203–232. Also see Beverly B. Moeller, Phil Swing and Boulder Dam (Berkeley: University of California

Press, 1971), and Joseph E. Stevens, *Hoover Dam: An American Adventure* (Norman: University of Oklahoma Press, 1988), the latter focusing primarily on the construction effort.

39. Similar arguments were offered elsewhere in the state. See "Remarks by Thomas H. Means on the Colorado River Problem," *Transactions of the Commonwealth of California*, 21 (April 14, 1926): 97; and John D. Galloway, "The Swing-Johnson Bill," *Transactions of the Commonwealth Club of California*, 22 (September 27, 1927): 484–487.

40. Lear, "The Boulder Canyon Project," 331.

41. Galloway, "The Swing-Johnson Bill," 481–482; R.H. Ballard, "The Power Interest in the Colorado River," *Transactions of the Commonwealth Club of California*, 21 (April 13, 1926): 81–86.

42. George H. Nash, *The Life of Herbert Hoover: The Engineer, 1874–1914* (New York: W.W. Norton, 1983); Herbert C. Hoover, "State Versus Federal Regulation," *JEPG*, 55 (July 15, 1925): 43–47; Lear, "The Boulder Canyon Project," 330–331; Catren, "Electrical Energy in Southern California," 230; The Twentieth Century Fund, *Electric Power and Government Policy: A Survey of the Relations Between the Government and the Electric Power Industry* (New York: Twentieth Century Fund, 1948), 502, n. 74. The remaining 36 percent of Boulder Dam power was reserved for Nevada and Arizona, neither of which would need it for some time. Consequently, Los Angeles and SCE agreed to purchase it until it was needed by the two nonparticipating states.

43. Lear, "The Boulder Canyon Project," 332–333. Eugene Coate, "Water, Power and Politics," 96–98, suggests that Ickes did not have as strong an interest in public power as Lear attributes to him, at least in the Central Valley Project. On the growth of planning in the federal government, efforts during the Roosevelt administration, see Otis L. Graham Jr., *Toward a Planned Society: From Roosevelt to Nixon* (New York: Oxford University Press, 1976), chs. 1–2.

44. The following account of the Central Valley Project is drawn largely from Coate, "Water, Power and Politics." Also see Hundley, *The Great Thirst*, 233–272.

45. "California Votes Water and Power Act," *JEPG*, 72 (January 1934): 36.

46. Jimerson, "Power Politics," 28–29. On the TVA concept for the project, see Twentieth Century Fund, *Electric Power and Government Policy*, 541–542.

47. Lawrence B. Lee, "California Water Politics: Opposition to the CVP, 1944–1980," *Agricultural History*, 54 (July 1980): 409.

Chapter XIII/Energy in Abundance

1. The impact of World War II on California and the west is carefully studied by Gerald D. Nash in *The American West Transformed: The Impact of the Second World War* (Bloomington: Indiana University Press, 1985), and *World War II & the West: Reshaping the Economy* (Lincoln: University of Nebraska Press, 1990). Also see Sterling L. Brubaker, "The Impact of Federal Government Activities on California's Economic Growth, 1930–1956" (Ph.D. diss., University of California, Berkeley, 1959).

2. James L. Clayton, "The Impact of the Cold War on the Economies of California and Utah, 1946–1965," Pacific Historical Review, 36 (November 1967): 449–473; James C. Williams, "The Economic Impact of Defense Spending in the San Francisco Bay Area Since World War Two," (unpublished M.A. thesis, San José State University, 1971).

3. James C. Williams, "Frederick E. Terman and the Rise of Silicon Valley," *Technology in America*, 2nd ed., Carroll Pursell, ed. (Cambridge: The M.I.T. Press, 1990), 276–291.

4. U.S. Energy Research and Development Administration, *Distributed Energy Systems in California's Future, A Preliminary Report*, by M. Christensen, *et al.*, 2 vols. (Washington, D.C., September 1977), 1:64–70; Nash, *The American West Transformed*, 228, n. 1.

5. "Western Power and Fuel Outlook—3. Petroleum," *Federal Reserve Bank of San Francisco Monthly Review* (March 1950): 40–41; Arthur M. Johnson, "California and the National Oil Industry," *Pacific Historical Review*, 34 (May 1970): 167; "California's Fuel Outlook," Transactions of the Commonwealth Club of California, *The Commonwealth*, 24 (June 21, 1948): 47–50.

6. *Ibid.* "Western Power and Fuel Outlook—3. Petroleum," 42. *Mineral Commodities of California*, Bulletin No. 156, 94–97.

7. Johnson, "California and the National Oil Industry," 167–169; White, "California's Other

Mineral," *Pacific Historical Review,* 34 (May 1970): 153–154. M. Christensen, *et al., Distributed Energy Systems in California's Future, A Preliminary Report,* 2 vols. (Washington, D.C., September 1977), 1:40–43; W.H. Hutchinson, *California: The Golden Shore by the Sundown Sea,* 2nd ed. (Belmont, CA: Star Publishing Company, 1988), 235.

8. "California's Fuel Outlook," 52–65. U.S. Bureau of Mines, *Minerals Yearbook, 1940,* 1037; California, Railroad Commission, *Annual Report, 93rd Fiscal Year* (Sacramento, 1942), 34. Wartime transportation shortages stimulated vegetable dehydration so as to reduce both tonnage and bulk for shipment. See *Minerals Yearbook, 1941,* 1122, and *1942,* 1134, *1943,* 1178. Elizabeth M. Sanders, *The Regulation of Natural Gas: Policy and Politics, 1938–1978* (Philadelphia: Temple University Press, 1981), 61, points out that residential use of natural gas in most of the United States occurred after 1945.

9. *Minerals Yearbook, 1941,* 1124, *1942,* 1135–1136, *1943,* 1180, *1944,* 1037, and *1946,* 814; California, Railroad Commission, *Annual Report, 93rd Fiscal Year* (Sacramento, 1942), 33–34, *Annual Report, 94th Fiscal Year* (Sacramento, 1943), 34–35, and *Annual Report, 95th Fiscal Year* (Sacramento, 1944), 34–35; "California's Fuel Outlook," 52–65. Douglas R. Littlefield and Tanis C. Thorne, *The Spirit of Enterprise: The History of Pacific Enterprises from 1886 to 1889* (Los Angeles: Pacific Enterprises, 1990), 129–130.

10. *Ibid.,* 130–136. *Minerals Yearbook, 1944,* 1064, *1945,* 1181, *1946,* 827, *1947,* 809, *1948,* 838, and *1951,* 879; "Western Power and Fuel Outlook—2. Natural Gas," 53, 58–59; "California's Fuel Outlook," 65–67. Charles M. Coleman, *PG&E of California: The Centennial Story of Pacific Gas and Electric Company, 1852–1952* (New York: McGraw-Hill, 1952), 306–308. To reach Blythe, a 1,020-foot long, thirty-inch diameter "Biggest Inch" pipeline suspension bridge was constructed. The first of its kind in the southwest, it was designated a civil engineering landmark by the southern California section of the American Society of Civil Engineers. See *Historic Civil Engineering Landmarks of Southern California* (Los Angeles, January 1974), 43.

11. Christensen, *Distributed Energy Systems,* 44–45; California Energy Commission, *1979 Biennial Report* (Sacramento 1979), 15, and *Energy Tomorrow, Challenges and Opportunities for California, 1981 Biennial Report* (Sacramento 1981), 117–120; California, State Assembly, Interim Committee on Conservation, Planning, and Public Works, *Electric Power and Development of Northern California,* by Warren S. Gramm (Sacramento, May 1957), 62; Littlefield and Thorne, *Spirit of Enterprise,* 142–151.

12. F.H. Fowler, "Three Major Power Possibilities in California," *JEPG,* 41 (July 1, 1918): 13; "Hydro-Electric Energy as a Conserver of Oil," *JEPG,* 34 (October 1, 1917): 299–300; W.F. Durand and C.H. Delany, "Steam-Electric Generation in the Far West," *EW,* 75 (May 15, 1920): 1123–1127. Rapid population growth in southern California between 1900 and 1920, coupled with hydroelectric uncertainties while the federal government worked out waterpower policies, prompted Pacific Light and Power to install a steam plant at Redondo in 1906. Expanded in 1908 to 65,000 horsepower, this "standby" power plant was the largest steam plant outside New York City. The same factors also prompted SCE to build an auxiliary steam plant in Long Beach in 1912. See Robert C. Catren, "A History of the Generation, Transmission, and Distribution of Electrical Energy in Southern California" (Ph.D. diss., University of Southern California, 1951), 93; and "Auxiliary Oil-Burning Station for Southern California District," *EW,* 59 (March 9, 1912): 535–536. The role of thermal electric power generation developments in the national industry before World War II is discussed by Richard F. Hirsh, *Technology and Transformation in the American Electric Utility Industry* (Cambridge, MA: Cambridge University Press, 1989), chs. 1–3; Louis C. Hunter and Lynwood Bryant, *A History of Industrial Power in the United States, 1780–1930, Volume Three: The Transmission of Power* (Cambridge, MA: The M.I.T. Press, 1991), ch. 6.

13. U.S. Federal Power Commission, *Report to the Federal Power Commission on the Water Powers of California,* by Frank E. Bonner (Washington, D.C., 1928), 25. Also see U.S. Geological Survey, *Hydroelectric Power Systems of California and Their Extensions into Oregon and Nevada,* Water Supply Paper 493, by Frederick Hall Fowler (Washington, D.C., 1923), 863; "City Asked to Cut Down Waste of Electricity" and "San Jose Puts Ban on Electric Signs," *Gilroy Advocate,* 28 August 1920, 1; "Power Curtailment," *Gilroy Advocate,* 4 September 1920, 2, citing the *Sacramento*

News. California power companies' efforts to store waters directly conflicted with downstream holders of riparian water claims, and this mitigated against development of hydroelectric storage reservoirs. See the comments of F.B. Lewis, SCE vice-president, in "Why Steam Ascendency in California," *JEPG,* 65 (August 1, 1930): 64. Precipitation data is drawn from California, Public Utilities Commission, *Report on Investigation Conducted by Public Utilities Commission, State of California, Pursuant to Senate Resolution No. 32; Northern California Generating Facilities; Special Study No. S–685* (San Francisco, 1948), 21. Precipitation varies greatly throughout the state, from highs of forty inches in the northwest to seven or eight inches in parts of the south. Also see H.D. McGlashan, "Records Show 1924 was Year of Lowest Stream Flow in California," *JEPG,* 54 (January 15, 1925): 46–47.

14. Hunter and Bryant, *Transmission of Power,* 354–355, 361–264; Hirsh, *Technology and Transformation,* 4; U.S. Federal Power Commission, *Consumption of Fuel for Production of Electrical Energy* (Washington, 1954), 5–6; "Why Steam Ascendency in California," *JEPG,* 65 (August 1, 1930): 62–65; A.H. Markwart, "Aspects of Steam Power in Relation to a Hydro Supply," *Transactions of the American Society of Mechanical Engineers,* 48 (1927): 187–208, and his "Relation of Steam to Hydro Power," *JEPG,* 59 (September 1, 1927): 131–135; H.A. Barre, "Water Power and Steam Power in California Utilities," 662; Bonner, *Water Powers of California,* 25–26; California, Railroad Commission, *Annual Report for 1925–1926* (Sacramento, 1927), 109.

15. "Steam Will Lead in 1930," *JEPG,* 64 (February 1, 1930): 68, 69; A.H. Markwart, "Power in California," *Journal of the Franklin Institute,* 204 (August 1927): 151–153; Catren, "Electrical Energy in Southern California," 205–207; Barre, "Water Power and Steam Power," 662; "1928 Steam Plants Account for 45 Per Cent of New Generating Capacity," *JEPG,* 62 (February 1, 1929): 79–82; "A New San Francisco Steam Plant," *California Engineer,* 8 (February 1930): 8–9, 32; W.C. McKay and C.E. Steinbeck, "Station 'A' Initiates Use of High-Pressure Steam in the West," *JEPG,* 67 (August 31, 1931): 68–71.

16. Markwart is quoted in "'Mingled Energy' Energy Cheapest for California," *JEPG,* 65 (July 1, 1930). Also see "Review of New Plants Shows Continued Growth," *JEPG,* 70 (February 1, 1933): 46–48; P.B. Garrett, "Automatic Electric Stations," *California Engineer,* 9 (May 1931): 5–8, 32; E.A. Crellin, "Automatic Hydro Plants," *JEPG,* 63 (April 1, 1929): 195–200; A.H. Markwart, "Engineering and Economic Aspects of the Power Surplus," *JEPG,* 72 (June 1934): 42–46; "Surplus Power: A Nightmare or a Sales Opportunity," *JEPG,* 72 (February 1934): 23–25. "Air Conditioning Symposium . . . A Market that Beckons with Profits and Loads," *JEPG,* 76 (April 1936): 39–43.

17. Annual year-end reviews in *JEPG,* 84 (January 1940): 37; v. 86 (February 1941): 48–49; v. 88 (February 1942): 42–43; and v. 90 (February 1943): 53. Also California, Railroad Commission, *Annual Report, 94th Fiscal Year* (Sacramento, 1943), 27–29; Public Utilities Commission, *Special Study No. S–685, 5–47 passim;* "Western Power and Fuel Outlook—1. Electric Power," *Federal Reserve Bank of San Francisco, Monthly Review* (November 1948): 99–100; P.M. Downing, "Multiple Service Steam-Electric Stations," *JEPG,* 87 (October 1941): 31–39; Coleman, *PG&E,* 316–317; California, Railroad Commission, *Report on Electric Power Survey of the Major Utilities Serving the California Power Market prepared for State Council of Defense* (San Francisco 1941), 38–45; Pacific Gas and Electric Company, *37th Annual Report* (1942), 21; U.S. Bureau of Reclamation, *Central Valley Project Studies; The War Program, Problems 1–5* (Washington, D.C., 1947), 71–106; Charles Eugene Coate, "Water, Power and Politics in the Central Valley Project, 1933–1967" (Ph.D. diss., University of California, Berkeley, 1969), 132–133. Work on PG&E's Pulgas and Cresta plants was slowed because of material shortages and the continuing battle over public versus private disposition of power from the Central Valley Project. Particularly see R.W. Jimerson, "Power Politics," *JPEG,* 87 (December 1941): 28–29.

18. "A Million and a Half Kw. For Sale: How the Industry Plans to Tackle Postwar Job," *JEPG,* 92 (February 1944): 77–78; Pacific Gas and Electric Company, *37th Annual Report* (1942), 24; California, Railroad Commission, *Annual Report, 96th Fiscal Year* (Sacramento, 1945), 50, *Annual Report, 98th Fiscal Year* (Sacramento, 1947), 50, and *Annual Report, 99th Fiscal Year* (Sacramento, 1948), 53–55; Public Utilities Commission, Special Study No. S-685, Gilbert Short, "Power Crisis," California Engineer, 26 (March 1948): 15, 22, 36. To keep up with postwar demand in isolated

northwest California, PG&E purchased the stern section of a Russian tanker that had broken up at sea in February 1946, beached it at Eureka, and harnessed the ships engine to generate 6,700 horsepower of electricity for its local system (Pacific Gas and Electric Company, *41st Annual Report* [1946], 23).

The "grow-and-build" strategy was a national phenonmenon, hardly unique to California's power companies. Hirsh, *Technology and Transformation*, 21, who discusses its historical development at length, explains that, "based on the presumption that new technology would become more efficient and less costly (per unit of power output), the strategy simply encouraged growth in customer usage of electricity so that utility companies would need to install new power units."

19. Richard Gordon, *U.S. Coal and the Electric Power Industry* (Baltimore: Published for Resources of the Future by the Johns Hopkins University Press, 1975), 25, n. 4; "Western Power and Fuel Outlook—1. Electric Power," 101–104; Short, "Power Crisis," 22; Harold W. Thompson, "Kilowatts, Water and Population," *California—Magazine of the Pacific*, 38 (June 1948): 14, 29; N.B. Hinson, "Pacific Southwest Electric Power Supply," *California—Magazine of the Pacific*, 40 (March 1950): 30–31; Dunlap D. Smalley, "The 'New' Power of the Pacific Southwest," *California—Magazine of the Pacific*, 42 (March 1952): 28, 52–53; "California's Fuel Outlook," 71–72; Coleman, *PG&E*, 333–334; "Who Should Develop California's Water Resources for Water and Power?," Transactions of the Commonwealth Club of California, *The Commonwealth*, 26 (March 13, 1950): 183–222 *passim;* Hirsh, *Technology and Transformation*, shows the growth of thermal efficiencies (p. 4), and discusses the postwar optimism of managers for continued improvements (ch. 4). He also offers an interesting discussion of the development of a culture within the power industry (ch. 2). With the caveat that hydropower, not steam, dominated in the Pacific Coast electric power industry, the culture which Hirsch describes for the industry at large fully applies to California's utilities.

20. Walter Dreyer, "Integrating Steam and Hydro Power in Northern and Central California," *Proceedings of the American Power Conference* (1957): 19:360. Dreyer drew heavily from his predecessor A.H. Markwart's view of "mingled power," first presented during the 1920s (p. 353). General data on thermal and hydroelectric capacity added after the war by California public utilities, municipal producers, and the Bureau of Reclamation are found under the Electric Division's "power survey" section in annual reports of the California Public Utilities Commission. Waterpower projects reported largely represent the completion of plants that had been projected years before. SCE, for example, expanded and modernized its Big Creek complex through the 1950s, and PG&E continued work on its Hat Creek, Pit River, and Kings River systems, as well as upgrading plants such as Potter Valley. See Laurence D. Shoup, "*The Hardest Working Water in the World*"*: A History and Significance Evaluation of the Big Creek Hydroelectric System* (Fair Oaks, CA: Theodoratus Cultural Research, October 1988, for Southern California Edison Company), 175–188; and James C. Williams, "Pre-1940 Hydroelectric Developments Historic Evaluation Survey," Pacific Gas and Electric Company, San Francisco, Land Department Contract, 17–83, October 27, 1983. An example of new hydroelectric development during the 1960s was the replacement of older plants such as De Sabla, which originally was constructed in 1903. See "Historic De Sabla Powerhouse to be Replaced in 1962," *PG&E Life*, 3 (April 1960): 22; "PG&E Plans $62,197,000 Power Expansion," *Public Utilities Fortnightly* (August 18, 1960): 22.

21. "PG&E Selects Pittsburg Site for Big Plant" and "Power for Defense," *California—Magazine of the Pacific*, 41 (December 1951): 27, 31, 38–43; Ronald Loveridge and Larry Yount, "The Towering Stacks of Morro Bay," *Cry California*, 2 (Fall 1967), 34; Charles Piller, *The Fail-Safe Society: Community Defiance and the End of American Optimisim* (Berkeley: University of California Press, 1991), 4–5; Steve Cohn, "The Political Economy of Nuclear Power (1945–1990): The Rise and Fall of an Official Technology," *Journal of Economic Issues*, 24 (September 1990), 783.

22. Roger W. Lotchin, *Fortress California, 1910–1961: From Warfare to Welfare* (New York: Oxford University Press, 1992), 268. Strauss is quoted in Michael Smith, "Advertising the Atom," *Government and Environmental Politics: Essays on Historical Developments Since World War Two*, Michael J. Lacey, ed. (Washington, D.C.: The Wilson Center Press, 1989), 244. Von Neumann is cited in Basalla, "Some Persistent Energy Myths," in *Energy and Transport: Historical Perspectives*

on Policy Issues, ed. George H. Daniels and Mark H. Rose (Beverly Hills, CA: Sage Publications, 1982), 34–35.

23. Raymond Wallace of Wallace, "Atomic Power and the Calutron," *California Engineer,* 24 (October 1945): 34; R.G. Spiegleman, "Harnessing the Atom to the Wheels of Industry," *California Engineer,* 25 (August 1946): 11; [L. A. Dubridge], "Atomic Energy and the Future," *California—Magazine of the Pacific,* 38 (March 1948): 41. Also see "California's Fuel Outlook," 73; and Dreyer, "Integrating Steam and Hydro Power," 360.

24. Smith, "Advertising the Atom," 236; Lotchin, *Fortress California,* 263–279; Cohn, "The Political Economy of Nuclear Power," 784, 788. The importance of rising fuel costs as a reason to exploit nuclear energy is cited by Gramm, *Electric Power and Development of Northern California,* 145, in his 1957 report to the State Assembly.

25. *Ibid.,* 785–786; Smith, "Advertising the Atom," 240. Chet Holifield, although he put the issue of nuclear energy before the JCAE, opposed private enterprise participating in nuclear energy. Still an adherent of southern California's strong support for public over private power, he had Los Angeles's municipal water and power department officials testify on behalf of confining nuclear energy development to publicly owned power agencies. See Lotchin, *Fortress California,* 276–277.

26. Gramm, *Electric Power and Development of Northern California,* 154; C.C. Whelchel, "Humboldt Bay," *Proceedings of the American Power Conference,* 20 (1958): 55–59; Whelchel, "Atomic Energy—What Does it Hold in Your Future," *California—Magazine of the Pacific,* 44 (December 1954): 3; Walton Bean and James Rawls, *California, An Interpretive History,* 6th ed. (New York: McGraw Hill, 1993), 471–473.

Isador Perlman and Israel Cornet, "Nuclear Power," *California Development Problems and Their Implications for Research, Training and Education,* Report of Exploratory Conference, February 4 and 5, 1954, University of California (Berkeley, 1955), 122–131, touch on some of the AEC-sponsored work being done in the University of California system during the early 1950s. Martin V. Melosi, *Coping With Abundance: Energy and Environment in Industrial America* (Philadelphia: Temple University Press, 1985), ch. 12, presents a good summary of national development of nuclear power. Also see George T. Mazuzan and J. Samuel Walker, "Developing Nuclear Power in an Age of Energy Abundance, 1946–1962," *Materials and Society,* 7 (1983): 307–320.

27. Smith, "Advertising the Atom," 243–248.

28. "Mechanical Energy from Volcanic Steam," *JEPG,* 39 (September 1, 1917): 210; Susan F. Hodgson, "A History of the Geysers Geothermal Field," *Geotermia, Revista Mexicana de Geoenergíe* 8 (Mayo–Agosto 1992); Reagan B. Kidd, "Using Natural Steam Power from the Geysers," *California Engineer,* 3 (September 1924): 7; Thomas T. Reed, "Power from the Earth's Heat," *Universal Engineer,* 40 (October 1924): 17–21.

29. *Ibid.,* 7–8; Edwin H. Carpenter, Jr., "Another California Centenary—The Geysers," *Pacific Historical Review,* 16 (November 1947): 391; "'Natural Steam' Plant is Proposed at the Geysers in California," *JEPG,* 48 (March 1, 1922): 193; Eugene T. Allen and Arthur L. Day, *Steam Wells and Other Thermal Activity at 'The Geysers,' California* (Washington, D.C.: Carnegie Institution, 1927), 50–59.

30. Carpenter, Jr., "Another California Centenary—The Geysers," 391; Hodgson, "A History of the Geysers Geothermal Field"; John Debo Galloway papers, MS 71–826, No. 1, Water Resources Archive, University of California, Berkeley; "In Memoriam: John D. Galloway," *California Historical Society Quarterly,* 22 (June 1943): 187–188; J.D. Galloway, "California Steam Wells," *Power,* 62 (November 24, 1925): 802–803. A number of sources useful in assessing this early period of geothermal development are found in the Robert Vogel Collection, National Museum of American History, Smithsonian Institution, Washington, D.C.

31. Development of the Geysers since 1950 is drawn largely from Hodgson, "A History of the Geysers Geothermal Field." I am most grateful for Ms. Hodgson's permission to draw extensively from her work prior to its publication. Also see ASME Landmark Nomination Form, Appendix 1, "Nomination for Designation of Mechanical Engineering Landmark," submitted by Donald R. Mullen, Lawrence Berkeley Lab, History and Heritage Chairman, SF Section of ASME, March 19,

1980, photocopy in Geo-Thermal Power folder, Robert Vogel Collection, Smithsonian Institution.

That almost no one knew about geothermal development during the early 1950s is made clear by the fact that Warren Gramm, in his otherwise solid 1957 report to the State Assembly, *Electric Power and Development of Northern California,* made no mention of it in his section on "Electric Generation from New Energy Resources," 150–156.

32. Hodgson, "A History of the Geysers Geothermal Field"; Stanley Scott and Samuel E. Wood, "California's Bright Geothermal Future," *Cry California,* 7 (Winter 1971/72): 10–23; California, Department of Finance, *Statistical Abstract, 1978,* 126, Table K–1.

33. Loveridge and Yount, "The Towering Stacks of Morro Bay," 34.

Chapter XIV/Energy and Environmentalism

1. Samuel P. Hays, *Beauty, Health, and Permanence: Environmental Politics in the United States, 1955–1985* (New York: Cambridge University Press, 1987), 4; Michael P. Cohen, *The History of the Sierra Club, 1892–1970* (San Francisco: Sierra Club Books, 1988). Leo Marx, "The Idea of 'Technology' and Postmodern Pessimism," in *Does Technology Drive History? The Dilemma of Technological Determinism,* ed. Merritt Roe Smith and Leo Marx (Cambridge: The MIT Press, 1994), 237–257, deals with the context of technological pessimism in which the environmental movement developed. This context also is suggested by Daniel Worster in his essay "Freedom and Want: The Western Paradox," in *Under Wester Skies: Nature and History in the American West* (New York: Oxford University Press, 1992), 79–92.

2. Edward Lawless, *Technology and Social Shock* (New Brunswick, NJ: Rutgers University Press, 1977), 217–225; Martin V. Melosi, *Coping With Abundance: Energy and Environment in Industrial America* (Philadelphia: Temple University Press, 1985), 270–271.

3. *Ibid.,* 271–272; Frank M. Stead, "How to Get Rid of Smog," *Cry California,* 2 (Winter 1966–67): 335–36; Gil Bailey, "Stop Driving or Stop Breathing," *Cry California,* 8 (Fall 1973): 24. Walton Bean and James Rawls, *California, An Interpretive History,* 6th ed. (New York: McGraw-Hill, 1993), 461–463, offers an overview of California's air pollution problem.

4. Stead, "How to Get Rid of Smog," 36–37. California, Department of Finance, *Statistical Abstract, 1970,* 233–234; James S. Fay and Stephanie W. Fay, eds., *California Almanac,* 5th ed. (Santa Barbara: Pacific Data Resources, 1991), 392–399.

5. August Giebelhaus, "Petroleum Refining and Transportation: Oil Companies and Economic Development," in *Energy and Transport: Historical Perspectives on Policy Issues,* George H. Daniels, and Mark H. Rose, eds. (Beverly Hills, CA: Sage Publications, 1982), 101; Jean Lindamood, "The History of Fuel Economy" and David Bryant, "The Mobil Economy Run," *Car and Driver,* 26 (February 1981): 43–44, 45–46; Melosi, *Coping with Abundance,* 271.

6. *Ibid.,* 272; Hays, *Beauty, Health, and Permanence,* 73–74; Stead, "How to Get Rid of Smog," 37; Richard A. Grant, Jr., "Air: The Blue-Sky Promise," *Cry California,* 11 (Summer 1976): 68.

7. Reagan is quoted in Bailey, "Stop Driving or Stop Breathing," 24.

8. Tom Bourne, "Is the Energy Shortage a Setback in the Smog Battle?" *California Journal,* 11 (March 1980): 124–125; Bean and Rawls, *California: An Interpretive History,* 461–462.

9. Reed Holderman, Kevin McCauley, and Gayle Olson, "Background to the Blowout: The Oil Industry and Santa Barbara," in *Environmental Hazards and Community Response: The Santa Barbara Experience,* Public History Monograph No. 2, G. Wesley Johnson and Ronald Nye, eds. (Santa Barbara: Graduate Program in Public Historical Studies, University of California, 1979), 188–191; Paul Israel, "From Rubble to Survival: The Rebuilding of a City," in *ibid.,* 53–69; Karen M. Hermann, Katherine C. Lord, and Karen L. Smith, "Resilence and Survival, 1920–1940," in *Old Town, Santa Barbara: A Narrative History of State Street from Gutierrez to Ortega, 1850–1975,* Public History Monograph No. 1, James C. Williams, ed. and comp. (Santa Barbara: Graduate Program in Public Historical Studies, University of California, 1977), 137–166; Kenny A. Franks and Paul F. Lambert, *Early California Oil: A Photographic History, 1865–1940* (College Station: Texas A&M University Press, 1985), 46–47.

10. Holderman, McCauley, and Olson, "Background to the Blowout," 191.

11. *Ibid.*, 192–194; Richard H.K. Vietor, *Energy Policy in America Since 1945: A Study of Business-Government Relations* (Cambridge: Cambridge University Press, 1984), 18–19.

12. Holderman, McCauley, and Olson, "Background to the Blowout," 194–195; Robert B. Krueger, "State Tidelands Leasing in California," *UCLA Law Review*, 5 (1958): 427–430.

13. Holderman, McCauley, and Olson, "Background to the Blowout," 195–198.

14. *Ibid.*, 198–201; Kevin McCauley, "The Politics of Power: The Federal and State Responses," *ibid.*, 206–224; Gayle Olson, "Get Oil Out!: The Local Government and Citizen Response", *ibid.*, 225–250. One of the best accounts of the Santa Barbara tragedy is Robert Easton, *Black Tide: The Santa Barbara Oil Spill and Its Consequences* (New York: Delacorte Press, 1972). For views of the spill and its consequences on the national level, see Vietor, *Energy Policy in America Since 1945*, 229–235; and Melosi, *Coping with Abundance*, 268–269.

15. On the development of California Tomorrow, I have relied particularly on John E. Hart, ed., *The New Book of California Tomorrow: Reflections and Projections from the Golden State* (Los Gatos, CA: William Kaufmann, Inc., 1984), 1–25.

16. *Ibid.*, 6–8; Samuel E. Wood and Alfred Heller, *California Going, Going . . .* (Sacramento: California Tomorrow, 1962); Richard B. Rice, William A. Bullough, and Richard J. Orsi, *The Elusive Eden: A New History of California* (New York: Alfred A. Knopf, 1988), 560–562.

17. Hart, ed., *The New Book of California Tomorrow*, 419; Hays, *Beauty, Health, and Permanence*, 389.

18. *Ibid.*, 9–12. Heller is quoted on p. 11. Other California Tomorrow booklets include Samuel E. Wood and Alfred Heller, *The Phantom Cities of California*, (San Francisco: California Tomorrow, 1963), and Samuel E. Wood and Daryl Lembke *The Federal Threats to the California Landscape* (San Francisco: California Tomorrow, 1967). *Cry California* was published quarterly from 1965 to 1982 and carried the title *California Tomorrow* from Spring 1982 through its final number issued Spring 1983.

19. Hart, ed., *The New Book of California Tomorrow*, 13.

20. *Ibid.*, 14–16.

21. Alfred Heller, ed. *The California Tomorrow Plan* (Los Gatos, CA: William Kaufmann, Inc., 1972); Hays, *Beauty, Health, and Permanence*, 44 and 389. Otis L. Graham, Jr.'s *Toward a Planned Society: From Roosevelt to Nixon* (New York: Oxford University Press, 1976), is especially useful on the importance of the planning ethos in recent American history.

22. "California and the Energy Problem," *California Journal*, 4 (June 1973): 184–185, 188; Lester Lees, "Why the Energy Crunch Came in the '70s," *Cry California*, 7 (Fall 1972): 29–34. People also were made aware of the inevitability of resource exhaustion by Donella H. Meadows, *et al.*, *The Limits to Growth: A Report for the Club of Rome's Project on the Predicament of Mankind* (New York: Universe Books, 1972).

23. The California per gallon retail price for gasoline grew very slowly over the years. Between 1950 and 1970, it increased from $.25 to $.35, a 1.26 percent average annual rate of change. By 1975, however, the price climbed to $.59, an 11.27 percent average annual rate of change. Similarly, the price of crude oil, between between 1970 and 1975, climbed at an 18.88 percent average annual rate of change, which was reflected in utility bills. California Energy Commission, *California Historical Petroleum Prices* (Sacramento, April 1983), 4, 18.

Melosi, *Coping with Abundance*, ch. 15, provides a good summary of the 1970s energy crises. Dan Mazmanian, "Toward a New Energy Paradigm," *California Policy Choices*, 8 (1992): 195–215, provides a sound historical framework for California's energy experience from 1973 through 1990.

24. Langdon Winner, "Energy Regimes and the Ideology of Efficiency," in *Energy and Transport: Historical Perspectives on Policy Issues*, ed. George H. Daniels and Mark H. Rose (Beverly Hills, CA: Sage Publications, 1982), 272.

25. Melosi, *Coping with Abundance*, ch. 16, addresses the confluence of energy and environmental issues from a national perspective. Ian Barbour, *et. al*, *Energy and American Values* (New York: Praeger, 1982), chs. 3–10, deals with the issue of conflicting values.

26. Mazmanian, "Toward a New Energy Paradigm," 198–200; Hays, *Beauty, Health, and Permanence*, 60–62; Cohen, *The History of the Sierra Club*, 435–443; Thomas Wellock, "The Battle for

Bodega Bay: The Sierra Club and Nuclear Power, 1958–1964," *California History,* 71 (Summer 1992): 193–194, 211. The Nixon and Ford administrations's "Project Independence" for national energy self-sufficiency reflected the supply-side approach to the energy crisis at the national level. See Melosi, *Coping with Abundance,* 285. Vietor, *Energy Policy in America Since 1945,* 317–318; Gil Bailey, "Project Independence," *Cry California,* 10 (Winter 1974/75): 22–30.

27. California State Assembly, Interim Committee on Conservation, Planning, and Public Works, *Electric Power and Development of Northern California,* by Warren S. Graham (Sacramento, May 1957), 154.

28. J. Samuel Walker, "Reactor at the Fault: The Bodega Bay Nuclear Plant Controversy, 1958–1964—A Case Study in the Politics of Technology," *Pacific Historical Review,* 59 (August 1990): 329, 325–328; Wellock, "The Battle for Bodega Bay," 193–211. Walker places the Bodega Bay case in context of the larger nuclear power issue in his *Containing the Atom: Nuclear Regulation in a Changing Environment, 1963–1971* (Berkeley: University of California Press, 1992).

29. Richard L. Meehan, *The Atom and the Fault: Experts, Earthquakes, and Nuclear Power* (Cambridge: The MIT Press, 1984), 19; Walker, "Reactor at the Fault," 329–344. On the general problem of uncertainty that has been inherent historically in earthquake engineering, see James C. Williams, "Earthquake Engineering: Designing Unseen Technology Against Invisible Forces," *Icon: Journal* 1 (1995): 172–194.

30. California, State Assembly, Science and Technology Advisory Council, *Nuclear Power Safety in California* (Sacramento, May 1973), IV–25; Tom Harris, "Old Nuclear Plant to Close," *San Jose Mercury News,* 29 June 1983, 1F; Philip Sporn, "Power: Nuclear vs. Conventional," *Mechanical Engineering* (June 1966): 18–21 (photocopy in Power/Energy—General folder, Miscellaneous Energy Sources, Robert Vogel Collection, National Museum of American History, Smithsonian Institution, Washington, D.C.); David Pesonen, testimony before the California Senate Committee on Public Utilities, Transit and Energy, *Hearing on Provisions of the Nuclear Power Plants Initiative, January 27, 1976, Sacramento, California* (Sacramento 1976), 6; Modesto Peace/Life Center, Stanislaus Safe Energy Committee, newsletter (September 1, 1977): 4, in Sierra Club California Legislative Office Records, 1970–1989, Series 1, Papers of the State Director, "Stanislaus Nuclear Project, 1977," Folder 2:9, Carton 2, MSS 91/1c, Bancroft Library, University of California, Berkeley (hereafter referred to as *Sierra Club Archives*).

31. *Ibid.,* 1–17; Stanley H. Mendes, Structural Engineer, testimony before the California Senate Committee on Public Utilities, Transit and Energy, *Hearing on Provisions of the Nuclear Power Plants Initiative, March 23, 1976, Sacramento, California* (Sacramento 1976), 103–108 and Appendix A; Bruce Keppel, "Nuclear Power: California's next big initiative battle," *California Journal,* 6 (June 1975): 202–203; "[Proposition] 15: Nuclear Power Plants," *California Journal,* 7 (May 1976): 6–8; Douglas Foster, "The Energy Commission: The Board You Love to Hate," *Cry California,* 13 (September 1978): 50.

32. George Barios, "The Wasteland Playground," *Cry California,* 3 (Fall 1968): 16–19; "Argonaut's Notebook," *Cry California,* 4 (Spring 1969): 3, 38; J. George Thon, "Power Generation: Its Status and Outlook," *Civil Engineering—ASCE* (September 1967): 36–39 (photocopy in Power/Energy—General folder, Miscellaneous Energy Sources, Robert Vogel Collection, National Museum of American History, Smithsonian Institution, Washington, D.C.).

33. Quoted in Maureen Fitzgerald, "Has State Government Done Its Part?" *California Journal,* 4 (June 1973): 188.

34. *Ibid.,* 189; California State Archives, California Energy Commission, Summary of CEC Development, Master Finding Aids, D–F; "California Energy Resources Conservation and Development Act (AB 1575—Warren), Abstract, Background, and Analysis," typescript, n.d., in Folder 1:18, Carton 1, MSS 91/1c, *Sierra Club Archives.*

35. John Zierold, "Environmental Lobbyist in California's Capital, 1965–1984," Berkeley: University of California, Bancroft Library, Regional Oral History Office, Sierra Club Oral History Series, 1988, p. 93. Fitzgerald, "Has State Government Done Its Part?" 191.

36. *Ibid.,* 191–192; California Assembly, Science and Technology Advisory Council, *Meeting the Electrical Energy Requirements for California* (Sacramento 1971).

37. Fitzgerald, "Has State Government Done Its Part?" 191–194, discusses all the reports. The most influential of them was W.E. Mooz, *et al., California's Electricity Quandary,* 3 vols. (Santa Monica, CA: Rand Corporation, 1972).

38. California Resources Agency, Division of Oil and Gas, *Energy in California: Its Supply, Demand, Problems* (Sacramento 1973), quoted in Fitzgerald, "Has State Government Done Its Part?" 193.

39. Foster, "The Energy Commission," 48; William H. Press, "Power vs. Environment is the Issue," *California Journal,* 4 (June 1973): 200–202; Maureen Fitzgerald, "Who Does What in the Energy Crisis: Politicians Dramatize, Commissioners Supervise, and the Utilities Improvise," *California Journal,* 4 (December 1973): 408–409; "California Energy Resources Conservation and Development Act," Folder 1:18, Carton 1, MSS 91/1c, *Sierra Club Archives.*

40. *Ibid.;* "Concession or Compromise? How the New Energy Act Should Work," *California Journal,* 5 (July 1974): 239; Melosi, *Coping With Abundance,* 278–280.

41. Charles Warren, "An Assemblyman's View: A Legislative Response to the Energy Crisis," *California Journal,* 4 (June 1973): 196. Also "Interview: A Legislator, an Industry Spokesman, and a Conservationist Discuss the Energy Problem," *California Journal,* 4 (June 1973): 203.

42. David J. Fogarty, "A Utilities Spokesman: Keeping Pace With New Energy Demands," *California Journal,* 4 (June 1973), 198–199. Also Warren, "An Assemblyman's View," 195.

43. Foster, "The Energy Commission," 49; Zierold, "Environmental Lobbyist in California's Capital," 101–102. CEC Chairperson, Richard Maullin, had been a Rand Corporation researcher and top campaign strategist for Brown. The other appointees included former Assembly Speaker Bob Moretti, whom Brown had defeated in the Democratic gubernatorial primary; Alan Pasternak, a pronuclear University of California chemical engineer; Richard Tuttle, former PUC chief counsel in the administration of Edmund G. Brown, Sr.; and Ronald D. Doctor, who authored the 1972 Rand report that recommended slowing the electricity demand growth rate (*Slowing the growth rate,* R–1116–NSF/CSA, v. 3 of *California's Electricity Quandary*), and the Rand study, *Energy Alternatives for California: Paths to the Future* (Santa Monica, CA: Rand Corporation, 1975), that established a basis for alternative energy resource development. See "The Potential Power of the New Energy Commission," *California Journal,* 6 (April 1975): 128; "California Energy Resources Conservation Commissioner Appointment, 1976–1977," Folder 1:13, Carton 1, MSS 91/1c, *Sierra Club Archives.*

44. Quoted in Foster, "The Energy Commission," 49. Also see Jennifer Jennings, "Another Energy Crisis—A Malfunctioning Commission," *California Journal,* 7 (March 1976): 91–92.

45. Foster, "The Energy Commission," 50; "Nuclear Energy—Law and Legislation, 1976," Folder 1:34, Carton 1, MSS 91/1c, *Sierra Club Archives.* "Nuclear Energy—Law and Legislation, Lawsuits, 1978," Folder 1:36, Carton 1, MSS 91/1c, *Sierra Club Archives.*

46. Ron Roach, "Senator Frankenstein," *California Journal,* 9 (June 1978): 190–191; "California Energy Resources Conservation Commissioner Appointment, 1976–1977," Folder 1:13, Carton 1, MSS 91/1c, *Sierra Club Archives;* "Varanini, Gene, appointment to Energy Commissioner, 1977," Folder 2:12, Carton 2, MSS 91/1c, *Sierra Club Archives;* Zierold, "Environmental Lobbyist in California's Capital, 108. Foster, "The Energy Commission," 50.

47. "Stanislaus Nuclear Project, 1977," Folder 2:9, Carton 2, MSS 91/1c, *Sierra Club Archives;* David Roe, *Dynamos and Virgins* (New York: Random House, 1984), chs. 2–5 *passim.* On the momentum and resistance to change of the electric power industry, see Thomas P. Hughes, *Networks of Power: Electrification in Western Society, 1880–1930* (Baltimore: Johns Hopkins University Press, 1983), and Richard F. Hirsh and Adam H. Serchuk, "Momentum Shifts in the American Electric Utility Industry: Catastrophic Change or No Change at All?" (paper presented at the annual meeting of the Society for the History of Technology, Washington, D.C., October 1993), 2–4, 11.

48. Hayden to Pasternak, 28 February 1977, Folder 1:5, Carton 1, MSS 91/1c, *Sierra Club Archives.*

49. Foster, "The Energy Commission," 51; Roach, "Senator Frankenstein," 190.

50. Quoted in *ibid.* Also see Foster, "The Energy Commission," 51–52; "Nuclear Energy—Law

and Legislation, 1976," Folder 1:34, Carton 1, MSS 91/1c, *Sierra Club Archives*; Hal Rubin, "Guide to Nuclear Power in California," *California Journal,* 10 (June 1979): 206.

51. "Californians for Nuclear Safeguards, 1979–1980," Folder 1:15, Carton 1, MSS 91/1c, *Sierra Club Archives;* Rubin, "Guide to Nuclear Power in California," 200–202; Hal Rubin, "Decommissioning: Another Joker in the Nuclear Deck," *Cry California,* 14 (September 1979): 17; "SMUD Follows Vote, Closes Rancho Seco," *Electric Light and Power,* 67 (July 1989): 1, 34. The coincidence that Michael Douglas's thrill-packed film about a near-meltdown, *The China Syndrome,* opened just weeks before the Three Mile Island meltdown did not go unnoticed by California environmentalists. See Roe, *Dynamos and Virgins,* 123.

52. Susan R. Schrepfer, "The Nuclear Crucible: Diablo Canyon and the Transformation of the Sierra Club, 1965–1985," *California History,* 71 (Summer 1992): 213–237; Mary Barnett, "Devil of a Battle Over Diablo Canyon," *California Journal,* 10 (December 1979): 417–420; Winner, "Energy Regimes and the Ideology of Efficiency," 274.

53. Nancy Barrand, "The New Energy Conflict Over the Ages-old Fuel: Coal," *California Journal,* 10 (March 1979): 91–95; James S. Canon, "Coal-Fired Power Plants: Bet You Can't Eat Just One," *Cry California,* 14 (September 1979): 22–26; "Allen-Warner Valley Energy System, 1979–1980," Folder 1:2, Carton 1, MSS 91/1c, *Sierra Club Archives.* Various articles in the *California Journal,* such as Wesley Marx's "Offshore Oil: A California Battleground," 14 (September 1983): 323–327, outline the continuing offshore oil struggle.

54. Mazmanian, "Toward a New Energy Paradigm," 200–204; Vietor, *Energy Policy in America Since 1945,* 234–235; Roe, *Dynamos and Virgins,* 133–195, *passim;* Hirsh and Serchuk, "Momentum Shifts in the American Electric Utility Industry," 11.

Chapter XV/The Soft Energy Path

1. Christopher Flavin and Nicholas Lenssen, *Power Surge: Guide to the Coming Energy Revolution* (New York: W.W. Norton and Company, 1994), 73–76. Amory Lovins first proposed the soft energy path strategy in "Energy Strategy: The Road Not Taken," *Foreign Affairs* (October 1976), 65–96, and elaborated on it in *Soft Energy Paths: Toward a Durable Peace* (San Francisco: Ballinger Publishing Co., 1977).

2. Schumacher, *Small is Beautiful: A Study of Economics as if People Mattered* (London: Vintage, 1973); *The Last Whole Earth Catalog: Access to Tools* (Menlo Park, CA: Portola Institute; distributed by Random House, New York, 1971); *Energy Primer: Solar, Water, Wind, and Biofuels* (Menlo Park, CA: Portola Institute, 1974); Sandy Eccli, ed. *Alternative Sources of Energy: Practical Technology and Philosophy for a Decentralized Society* (New York, 1975); Carroll Pursell, "Presidential Address: The Rise and Fall of the Appropriate Technology Movement in the United States, 1965–1985," *Technology and Culture,* 34 (July 1993): 631–634. Carroll Pursell, "The American Ideal of Democratic Technology," in *The Technological Imagination: Theories and Fictions,* ed. Teresa de Lauretis, Adreas Huyssen, and Kathleen Woodward (Madison, WI: Coda Press, 1980), 11–25.

3. *Ibid.;* Joann Edin, "Sim Van der Ryn—The Small-Is-Beautiful Salesman," *California Journal,* 8 (January 1977): 25–27; Hal Rubin, "Woodchips and Windmills," *California Journal,* 9 (June 1978): 187; "What is Appropriate Technology?" *Cry California,* 12 (September 1977): 30–32; Phil Smith, "If Small Is Beautiful, Why Does OAT Keep Growing," *California Journal,* 12 (April 1981): 149–150; Santa Clara County, Office of Appropriate Technology, *A Sourcebook: Resources for Appropriate Technology in Santa Clara County* (San Jose, CA: Santa Clara Environmental Management Agency, June 1978). Relevant publications of the California Office of Appropriate Technology include *Present Value: Constructing a Sustainable Future* (Sacramento 1979), *Catalogue of California Model Solar Projects* (Sacramento 1981), *Local Energy Initiatives: A Second Look, A Survey of Cities and Counties, California 1981* (Sacramento 1981), and *Cogeneration Blueprint for State Facilities* (Sacramento, March 1981).

4. Flavin and Lenssen, *Power Surge,* 74–76; Lovins, Soft Energy Paths, xiv–xv. Lovins's critics are cited in Hugh Nash, ed., *The Energy Controversy: Soft Path Questions and Answers by Amory Lovins and His Critics* (San Francisco: Friends of the Earth, 1979). Also see Albert A. Bartlett, "Forgotten Fundamentals of the Energy Crisis," *Journal of Geological Education,* 28 (1980): 4–35.

5. Richard E. Hirsh, *Technology and Transformation in the American Utility Industry* (Cambridge, MA: Cambridge University Press, 1989), 115.

6. Laura Nader, "Barriers to Thinking New About Energy," *Physics Today* (February 1981): 9. I participated in one of Chevron Corporation's gatherings in April 1976, at a very nice resort hotel in Carmel. Approximately ten Chevron officials and geologists, including at least one vice-president, directed our group of some fifteen community college philosophy, history, political science, economics, and English professors through a two-day discussion of the real character of the energy crisis. Plentifully supplied with food and drink in a most congenial atmosphere, participants were steered deftly toward a conclusion that neighborly, caring oil firms certainly were not to blame for the energy crisis and that these firms humbly hoped humanists and social scientists would not encourage misinterpretations that energy companies were simply in the business for the money.

7. California Energy Commission, *1979 Biennial Report,* (Sacramento 1979), *passim;* Hirsh, *Technology and Transformation,* Pt. 2, particularly 155–158; Richard F. Hirsh, "Conserving Kilowatts: The Electric Power Industry in Transition," *Materials and Society,* 7 (1983): 295–305. On demand-side management, see Richard F. Hirsh and Adam H. Serchuk, "Momentum Shifts in the American Electric Utility Industry: Catastrophic Change or No Change at All?" (paper presented at the annual meeting of the Society for the History of Technology, Washington, D.C., October 1993), 11–19.

8. California Energy Commission, *1979 Biennial Report,* 11–37. The CEC impetus toward conservation and alternative technologies was furthered by a study undertaken by a research team from the Berkeley and Davis campuses of the University of California and funded by the U.S. Energy Research and Development Administration. See M. Christensen, *et al., Distributed Energy Systems in California's Future: A Preliminary Report,* 2 vols. (Washington, D.C.: National Technical Information Service, September, 1977). Paul P. Craig and Mark D. Levine, "Energy Sources: California's Non-Nuclear Future," *Cry California,* 13 (September 1978): 54–59, summarizes and highlights portions of the study.

9. *Ibid.,* 55–63; California Energy Commission, *Energy Tomorrow, Challenges and Opportunities for California, 1981 Biennial Report* (Sacramento 1981), 131–155; Patricia Washburn, "We've Still Got a Long Way to Go," *Cry California,* 14 (September 1979): 8–12; David Roe, *Dynamos and Virgins* (New York: Random House, 1984), 160–161; Mary Ellen Leary, "Energy Conservation," *California Journal,* 11 (June 1980): 219–223; Robert Feinbaum, "The Slow-but-Sure Arrival of Those New Home Energy Standards," *California Journal,* 14 (March 1983): 123–125.

10. California Energy Commission, *1979 Biennial Report,* 8–9 and 11–26. Also see Martin V. Melosi, *Coping With Abundance: Energy and Environment in Industrial America* (Philadelphia: Temple University Press, 1985), 288–293.

11. California Energy Commission, *1979 Biennial Report, passim* and *1981 Biennial Report,* 85–89 and 103–107; California Energy Commission, Summary of 1980 Energy Legislation, 1 October 1980, in Folder 1:18, Carton 1, MSS 91/1c, *Sierra Club Archives*; Hirsh, *Technology and Transformation in the American Electric Utility Industry,* 168–171; Melosi, *Coping with Abundance,* 326. The California Energy Commission published a variety of reports during the early 1980s, which encouraged entrepreneurs to undertake nontraditional energy projects.

12. Dan Mazmanian, "Toward a New Energy Paradigm," *California Policy Choices,* 8 (1992): 202–203; James Cook, "Cogeneration Gap," *Forbes,* 138 (November 3, 1996): 70–73; Roe, *Dynamos and Virgins,* 132. Christopher Flavin, "Electricity's Future: The Shift to Efficiency and Small-Scale Power," *Bulletin of Science, Technology & Society,* 5:1 (1985): 55–103, provides a solid overview of the rise of PURPA projects in California.

13. Site inspections by the author of the Worman Mill and Pyramid Creek hydro systems in August 1982; interview with Dan Worman, 5186 Worman Road, Ahwahnee, California, August 1982; Michael K. Crist, "A Cultural Resource Reconnaisance of the Pyramid Creek Hydroelectric Project, El Dorado County, California," prepared for Keating & Associates, Placerville, CA, by Buada Associates, Fresno, CA, November 1981; U.S. Forest Service, Archaeological Site Survey Record, FS No. 05–11–51–316, Graeagle Hydroelectric Project, Plumas National Forest, Quincy,

CA, July 30, 1980, and letters related to project between 1935 and 1964 in Licenses and Water Rights files, Headquarters, Plumas National Forest; Matt Herron, "The Rush is on to Find New Gold in Falling Water," *Smithsonian* (December 1982): 94.

Many other minihydro systems, such as one built in the 1920s beside Fallen Leaf Lake adjacent to Lake Tahoe, had disappeared by the time PURPA was enacted. Letters and a site map related to the Fallen Leaf Lake plant for the period 1928–1950 are in Water Power file, Headquarters, Eldorado National Forest, Placerville, CA. For early small hydroelectric development in California, see for example, Chas. H. Tallant, "Water Power Development for Small Plants," *JEPG,* 44 (April 1, 1920): 318–320.

14. Herron, "The Rush is on to Find New Gold," 89–90; Tom Harris, "Small Hydroelectric Plants Turn Water into Gold," *San Jose Mercury News,* 18 October 1983, 3E. For projects in New England, see, for example, John McPhee, "Minihydro," *The New Yorker* (February 23, 1981): 44–87; Carolyn Jabs, "The Little Power Plant that Could," *Historic Preservation* (March/April 1982): 46–51.

15. Herron, "The Rush is on to Find New Gold," 92–96; California Department of Water Resources, Bulletin 211, *Small Hydroelectric Potential at Existing Hydraulic Structures in California* (Sacramento, April 1981); Delia M. Rios, "California's Love Affair with Small Hydropower Plants," *California Journal,* 12 (February 1981): 81–82.

16. Herron, "The Rush is on to Find New Gold," 92.

17. Jim Schuler of the California Department of Fish and Game, quoted in Harris, "Small Hydroelectric Plants Turn Water into Gold," 3E.

18. *Ibid.* Herron, "The Rush is on to Find New Gold," 96. California and other states developed environmental mitigation standards for small hydro projects, sometimes coming into conflict with FERC. California and FERC came into conflict, in 1989, when the state ordered the small Rock Creek plant, a PURPA project, to maintain a water flow of thirty-cubic-feet per second, while FERC required only eleven-cubic-feet per second. FERC argued, successfully, in the U.S. 9th Circuit Court of Appeals that the state could not set water flow requirements on the project. See "High Court to Decide Who Controls Power Plant Water," *San Jose Mercury News,* 5 December 1989, 14A. On the continued struggle between the hydroelectric industry and environmentalists, see, for example, George Lagassa, "The New River Conservationists," *Independent Energy,* 21 (November 1991): 63–67.

19. Robert E. Taylor, "Municipalities Have Preference to Run Hydroelectric Plants, Court Panel Says," *The Wall Street Journal,* 23 October 1985, 6; Mike Cassidy, "PG&E Gets Dam License: Santa Clara Loses Battle," *San Jose Mercury News,* 15 April 1987, Extra section, 1–2; Brad Kava, "Santa Clara, P.G.&E. Settle; City to be Paid $1 Million," *San Jose Mercury News,* 15 March 1990, B1 and B4. The settlement between PG&E and the City of Santa Clara was worked out by FERC as ordered by Congress in its 1986 action.

20. Heiman is quoted in Dale Rodebaugh, "Morgan Hill Eyes Anderson Dam as Electricity Source," *San Jose Mercury News,* 19 February 1982, 8F. Also see Pat McKenna, "Power Plants Proposed for 2 Dams," *Valley World [Gilroy],* 10 January 1979, 1; "Hydroelectricity Plans Proceed," *San Jose Mercury News,* 22 February 1982, 1B; California, Department of Water Resources, Energy Division, "List of Potential Small Hydroelectric Projects at Existing Hydraulic Facilities in California" [1981], n.p; Herron, "The Rush is on to Find New Gold," 96; Christine W. Townsend, "Municipal Water Powers Small Hydro in California," *Modern Power Systems,* 5 (July 1985): 52–53.

21. "Windmills for Electric Lighting," *Electrical World* 23 (February 3, 1894): 157–158; James J. Parsons, "Solar Radiation, Tides, and Winds," *California Development Problems and Their Implications for Research, Training and Education,* Report of Exploratory Conference, February 4 and 5, 1954, University of California (Berkeley, 1955), 119–120; Terry J. Lodge, "The 'Windmill Case:' Facing Up to Appropriate Technology," *Environmental Affairs,* 6 (1978): 491–510; Nicholas Wade, "Windmills: The Resurrection of an Ancient Technology," *Science,* 184 (June 7, 1974): 1055–1058; Matthias Heymann, "Why Were the Danes Best? Wind Turbines in Denmark, West Germany and the USA, 1945–1985," (paper presented at the annual meeting of the Society for the History of Technology, Cleveland, October 1990), 5; Robert W. Righter, "Wind Energy in California: A New

Bonanza," *California History,* 78 (Summer 1994): 144–146. Righter describes an effort during the 1920s, by Californian Dew Oliver, to build a windpower farm at San Gorgonio Pass near Palm Springs. Whether or not it was technologically feasible, it failed financially. Also see Palmer C. Putnam, *Power from the Wind* (New York, 1948).

22. Heymann, "Why Were the Danes Best?" 4–8.

23. *Ibid.,* 8–9, 11–12; U.S. Energy Research and Development Agency, "Federal Wind Energy Program, Summary," 1 May 1975, mimeographed, in Windmills General folder, Miscellaneous Energy Sources, Robert Vogel Collection, National Museum of American History, Smithsonian Institution, Washington, D.C.; "Excellent Forecast for Wind," *EPRI Journal* (June 1990): 16; "Harnessing the Wind," *PG&E Life,* 25 (August 1982): 6; "PG&E Ranks at the Top for Power Production from the Wind," *PG&E Progress,* 61 (April 1984): 1; David Ansley, "For Windmills, Power is a Breeze," *San Jose Mercury News,* 27 December 1988, 4C. Westinghouse Corporation and General Electric Company Space Systems Organization were the only electric power industry firms that received NASA windpower contracts, and both were involved heavily in the space program.

24. California Energy Commission, *1981 Biennial Report,* 104–105; "Excellent Forecast for Wind," 15–16; Tegan M. McLane, "Winds of Change Are Shifting as Wind Power Hits Comeback Trail," *680 Corridor* (October 14, 1991): 17/SR-1; John Flinn, "Gusty Weather Brings Bonanza for Wind Farms," *San Francisco Examiner,* 29 April 1984, B1 and B5.

25. Righter, "Wind Energy in California," 152–154. Wind energy project draft grant proposal, ca. 1977, in Folder 2:14, Carton 2, MSS 91/1c, *Sierra Club Archives;* "Study Citing Wind Turbine Bird Kills," *Independent Power Report* (May 22, 1992): 6; "U.S. Windpower to Begin Tests of Bird Mitigation Techniques," *Utility Environment Report* (December 25, 1992): 14; Kerry Drager, "Windpower," *California Journal,* 12 (April 1981): 151–153; "Energy: 'Whoosh' Go the Giant Windmills of Altamont Pass, But Does Anyone Hear?" *Audubon,* 85 (July 1983): 126–127; Paul Gipe, "New Attitudes Toward Wind," *Independent Energy,* 20 (March 1990): 62, 64–65. The Benicia incident was reported on Live-on-4 News (San Francisco), October 12, 1981.

26. Sylvia White, "Taking Wind Out of the Sails of an Alternative Energy Trend," *San Jose Mercury News,* 9 December 1984, 4P. "Windmill Builders Race the Taxman," *San Jose Mercury News,* 31 December 1982, 12D, reports a rush during the end of 1982 to beat a January 1st deadline which lowered tax benefits. This may have convinced some observers that the complete end of federal credits in 1985 and state credits in 1986 would end any more development. Also see Ellen Paris, "The Great Windmill Tax Dodge," *Forbes,* 133 (March 12, 1984): 39–40.

27. Heymann, "Why Were the Danes Best?" 9–14 and figure 5; "Excellent Forecast for Wind," 16–23; Donald Marier, "Windfarm Update: Double and Double Again," *Alternative Sources of Energy,* No. 72 (March/April 1985): 12–19; Paul Gipe, "Storm Masters Rise Again," *Alternative Sources of Energy,* No. 79 (March 1986): 22–25; "Analyzing 1985 Wind Project Performance," *Alternative Sources of Energy,* No. 83 (August/September 1986): 14–16.

28. McLane, "Winds of Change Are Shifting," 17/SR1; "Excellent Forecast for Wind," 15, 17, 23–25 (Osborn is quoted on 23); Leslie H. Everett, "Altamont Wind Power—The Utility Experience," *Modern Power Systems,* 6 (September 1986): 65–69; Peter Asmus, "Winds of Change," *Comstock's* (October 1992): 18–21; Flavin and Lenssen, *Power Surge,* ch. 6.

29. Clipping of Lawrence Sandek, "Solar Furnaces: Problems and Prospects," *Research and Engineering* (February 1957): 9–11; clipping of Gaston Burridge, "The Sun's Big Muscles: Scientists to Meet in Phoenix for First World Solar Energy Symposium," *Arizona Highways* (November 1955): 6–13. Miscellaneous clippings concerning C.G. Abbot in Solar Energy folder, Miscellaneous Energy Sources, Robert Vogel Collection, National Museum of American History, Smithsonian Institution, Washington, D.C. Association for Applied Solar Energy, *Proceedings of the World Symposium on Applied Solar Energy,* Phoenix, Arizona, November 1–5, 1955, sponsored by the Association for Applied Solar Energy, Stanford Research Institute, and the University of Arizona [Menlo Park, CA: Stanford Research Institute, c. 1956]. E.J. Burda, ed., *Applied Solar Energy Research: A Directory of World Activity and Bibliography of Significant Literature* (Menlo Park, CA: Stanford Research Institute for the Association for Applied Solar Energy, Phoenix [1955]).

30. Ethan B. Kapstein, "The Transition to Solar Energy: An Historical Approach," in *Energy*

Transitions: Long-Term Perspectives, ed. Lewis J. Perelman, August W. Giebelhaus, and Michael D. York (Boulder, CO: Westview Press, Inc., 1981), 115–120; Ethan Barnaby Kapstein, "The Solar Cooker," *Technology and Culture,* 22 (January 1981): 112–121; *Supra,* Chapter 5. Parsons, "Solar Radiation, Tides, and Winds;" Kerry Drager, "Solar California," *California Journal,* 10 (November 1979): 394, quotes Kathryn Ramsay.

31. Quoted in Richard A. Grant Jr., "Solar Energy for Homes: The Slow Dawn of a New Era," *Cry California,* 12 (Summer 1977): 26–29. Also see Claudia Ricci, "Solar-Energy Conflict Ahead: Utilities versus Do-It-Yourselfers," *California Journal,* 8 (March 1977): 91–93.

32. Mirviss is quoted in Drager, "Solar California," 395. Allan Temko, "Turning Hamilton Field into a Solar Village," *San Francisco Chronicle,* 17 July 1979, 4; Sim Van der Ryn, "The Solar Village," *Cry California,* 14 (Fall 1979): 13–16.

33. California Energy Commission, *1981 Biennial Report,* 25–29 and 155–158; James Ridgeway, "Energy City," *Science Digest,* 91 (January 1983): 32–34; Charlotte Higgins, "The Appropriate Metropolis: Is Davis the City of the Future?" *Cry California,* 13 (September 1978): 13–17; Melinda Welsh, "Wide Support Saves Solar; Tax Credits Survive Duke's Axe," *California Journal,* 15 (January 1984): 37–38. Edward Vine reported in "Solar California: Data Update," *California Data Brief,* 6 (April 1982), an occasional publication of the Institute of Governmental Studies at U.C. Berkeley, that pool covers and heaters accounted for 74 percent of 78,323 residential solar installations under the state solar tax credit program between 1976–1979. Domestic water heaters accounted for 15 percent, space conditioning systems for 7 percent, and photovoltaic, solar mechanical, process heat, or wind power for the remaining 3 percent. Utility inspections of solar water heaters, however, revealed more than twice as many installed domestic solar water heaters than shown by tax credit records.

34. Grant, "Solar Energy for Homes," 28–29; Ricci, "Solar-Energy Conflict Ahead," 91–93; G.W. Braun, J.F. Doyle, and J.J. Iannucci, "Solar Power Receives New Impetus in USA," *Modern Power Systems,* 6 (November 1986): 63–69. U.S. Department of Energy, San Francisco Operations Office *Information News Summary,* No. 4 (April 1982): 1; No. 10 (November 1982): 1–2; No. 5 (May 1983): 1–2; and No. 6 (June 1983): 5. Also see Tom Harris, "Glow of 300 Suns Drives Plant," San Jose Mercury News, 16 November 1982, 1E and 5E; "'Power Tower' Would Generate Electricity from Solar Energy," *PG&E Progress,* 60 (January 1983): 2.

35. "Solar Power Plant Planned in State," *San Jose Mercury News,* 4 March 1983, 8B; "PG&E Will Tap the Sun to Generate Electricity," *PG&E Progress,* 60 (May 1983): 1–2; "Cybernetic Mushrooms Change Sunlight into Electricity," *PG&E Progress,* 61 (January 1984): 4–5. For a similar photovoltaic project sponsored by the Sacramento Municipal Utility District, see Mark W. Anderson, David Thorpe, and David Collier, "Budget Overruns Threaten First Solar 100 MWe Photovoltaic Project," *Modern Power Systems,* 5 (July 1985): 19–23.

36. Elliston is quoted in "It's Sunset for Giant Solar Energy Plant," *San Jose Mercury News,* 11 March 1991, E1. *Bringing Solar Electricity to Earth,* EPRI pamphlet No. GS.3006.11.90; Pacific Gas and Electric Company, *Annual Report 1990,* 11; Glennda Chui, "Solar Energy is Making a Comeback," *San Jose Mercury News,* 7 June 1988, 1C and 3C; Michael Dorgan, "Solar Industry May Soon Have Its Day in the Sun," *San Jose Mercury News,* 9 April 1991, 1F and 6F.

37. Gerald T. White, *Formative Years in the Far West: A History of the Standard Oil Company of California and Predecessors through 1919* (New York: Appleton-Century Crofts, 1962), 240; U.S. Geological Survey, *Hydroelectric Power Systems of California and their Extensions into Oregon and Nevada,* Water Supply Paper 493, by Frederick Hall Fowler (Washington, D.C., 1923), 375–377, 414, 874; Mark Requa, "The Fuel Resources of California," *Transactions of the Commonwealth Club of California,* 7 (June 1912): 178, 199, 202–203.

38. "New Plant May Dry Garlic, Light Homes," *Gilroy Dispatch,* 23 October 1984, 1A and 10A. "Still Cogenerates Power," *Energy News,* 6 (February 1982): 3. "Breakthrough in Cutting Power Costs," *PG&E Progress,* 59 (October 1982): 5. Bill Richards, "Cogenerated Power Irritates Utilities," *The Wall Street Journal,* 23 October 1985, 6. James Cook, "Cogeneration Gap," *Forbes,* 138 (November 3, 1986): 70–73. Paul Gipe, "Kern County Cogeneration: A Goldmine in the Oilpatch," *Alternative Sources of Energy,* No. 84 (October 1986): 8–12.

39. U.S. Bureau of Mines, *Minerals Yearbook, 1922,* 5–6; Information News Summary, No. 10 (November 1982): 3, and No. 3 (March 1983): 4; California Energy Commission, *1981 Biennial Report,* 159; William Frazer, Daniel Mahr, and Peter R. Goldbrunner, "Biomass Plant Relies on Variety of Fuels," *Power,* 136 (April 1992): 101–106.

40. *CWMB: California Waste Management Bulletin* (Sacramento: March 1986): 4–5, in "Waste-to-Energy Project, 1984–1986," Folder 3:27, Carton 3, MSS 91/1c, *Sierra Club Archives;* Richard Reinhardt, "Up the Solid Waste Stream," *Cry California,* 13 (September 1978): 45–47; Kathleen Grubb, "Trash into Power? Not in My Backyard," *California Journal,* 20 (January 1989): 39–40.

41. Miller quoted in Grubb, "Trash into Power?" 39; Letter, Dr. John Holtzclaw, Sierra Club, to Robert Mussetter, Chair, Grant Review Committee, California Energy Commission, 21 January 1988, in "Correspondence, 1988," Folder 2:16, Carton 2, MSS 91/1c, *Sierra Club Archives;* Michael T. Burr, "Fluidized Bed in the Bay Area," *Independent Energy,* 20 (September 1990): 40–44.

42. Hodgson, "A History of the Geysers Geothermal Field," *Geotermia, Revista Mexicana de Geoenergíe* 8 (Mayo–Agosto 1992): 6; *Information News Summary,* No. 11 (December 1982): 1, and No. 6 (June 1983), 3; "Santa Clara, Partners to Build Geothermal Plant in Geysers," *San Jose Mercury,* 19 July 1977, 24; "Salton Sea Unit 2 on Line," *The Geothermal Hot Line,* 20 (January 1991): 20; Keith L. Sipes and Tiffany T. Nelson, "Binary Geothermal Plant Cycle Breaks New Ground in Southern California," *Modern Power Systems,* 6 (August 1986); 73–75.

43. *Ibid.;* "Trans-Pacific Plans New Power Plant at Lake City Field," *The Geothermal Hot Line,* 20 (January 1991): 17–18; "Honey Lake Power Facility, Lassen County," 19; "San Bernardino Geothermal District-Heating System: An Update," 22–24; and "California Wells," 87–88. *Information News Summary,* No. 5 (May 1983): 4.

44. Hodgson, "A History of the Geysers Geothermal Field," 6–9, 19; "The Geysers Geothermal Plants are Losing Power-generating Steam," *San Jose Mercury News,* 11 September 1989, 5B; Brad Kava, "Geothermal Energy Idea Sputters to Close," *San Jose Mercury News,* 7 December 1990, C1 and C3; "Geothermal Tragedy of the Commons," *Science,* 253 (July 12, 1991): 134–135.

45. Pursell, "Rise and Fall of the Appropriate Technology Movement," 633.

46. Mazmanian, "Toward a New Energy Paradigm," 201, 204–205; Claudia Buck, "Whatever Happened to the Energy Crisis," *California Journal,* 20 (December 1989): 513–516.

47. Mazmanian, "Toward a New Energy Paradigm," 205–206; "$500,000 Saved by Ride-Sharing," *Energy News,* 6 (April 1982): 6, a publication of the Energy Center, Santa Cruz County. California Energy Commission, *1981 Biennial Report,* ch. 4.

48. Joanne Jacobs, "A Techno-fix for Smog May Be in the Pipeline," *San Jose Mercury News,* 3 December 1990, 7B.

49. Mazmanian, "Toward a New Energy Paradigm," 206–207; Marla Cone, "L.A. to E.P.A.: Don't Hold Your Breath," *Sierra,* 72 (November/December 1987): 29–32.

50. Mazmanian, "Toward a New Energy Paradigm," 207–211; Peter Asmus, "Diversifying California's Energy Map," *California Journal,* 22 (July 1991): 309.

Conclusion

1. "California's New Energy Technology Export Program," *Alternative Sources of Energy,* 81 (May 1986): 20; Peter Asmus, "Diversifying California's Energy Map," *California Journal,* 22 (July 1991): 310–311; Dana Gardner, "One State's Drive for Clean Energy," *Design News,* 47 (September 9, 1991): 115–117; Christopher Flavin and Nicholas Lenssen, *Power Surge: Guide to the Coming Energy Revolution* (New York: W.W. Norton and Company, 1994), 50–70.

Index

ABOUT THE AUTHOR

James C. Williams is Professor of History and Director Emeritus and Historical Advisor of the California History Center and Foundation at De Anza College. He received his B.A. at the University of Oregon, his M.A. at San Jose State University, and his Ph.D. at the University of California at Santa Barbara. Editor of several volumes of regional and state history, he has also contributed to many books and periodicals, including *ICON* and *California Historian.*